Bioinformatics

A Swiss Perspective

Bioinformatics

A Swiss Perspective

editors

Ron D Appel

Ernest Feytmans

Swiss Institute of Bioinformatics, Switzerland

World Scientific

NEW JERSEY • LONDON • SINGAPORE • BEIJING • SHANGHAI • HONG KONG • TAIPEI • CHENNAI

Published by

World Scientific Publishing Co. Pte. Ltd.
5 Toh Tuck Link, Singapore 596224
USA office: 27 Warren Street, Suite 401-402, Hackensack, NJ 07601
UK office: 57 Shelton Street, Covent Garden, London WC2H 9HE

British Library Cataloguing-in-Publication Data
A catalogue record for this book is available from the British Library.

BIOINFORMATICS
A Swiss Perspective

ISBN-13 978-981-283-877-3
ISBN-10 981-283-877-5

Typeset by Stallion Press
Email: enquiries@stallionpress.com

Printed in Singapore by Mainland Press Pte Ltd

Foreword

Computational sciences have their roots in the development of increasingly powerful computers over the last few decades. Rather rapidly, the instrumentation and the newly developed methodology with the underlying algorithms became widely appreciated and used as novel research strategies serving in many different fields of academic investigation, particularly in natural sciences and engineering sciences, but also in social sciences and the humanities. Computational sciences have been recognized for their invaluable contributions to data collection, data storage, data handling, and data analysis, thus leading to efficient strategies of modeling, prediction, and design of molecular structures and of their functional properties that are often of immediate relevance for the medical sciences. Computational comparisons of DNA sequences from different organisms provide invaluable insights into past evolutionary developments, and this has become a powerful new tool in the systematics of living organisms.

The present book on computational biology testifies to the impact that this field of investigation exerts on many pending questions in the life sciences. Particular reference is thereby made to pioneering contributions that have their roots in Switzerland. In the early 1990s, I had the privilege, in my position as a member of the Swiss Science Council, to visit a group of scientists working at the University of Geneva on computer-assisted handling and storage of data related to protein sequences. The prospective value of their work became immediately obvious. A very positive recommendation that was then formulated by the Swiss Science Council may have influenced the support provided by the political leaders, and this may have also facilitated the creation of the

Swiss Institute of Bioinformatics (SIB) 10 years ago. The SIB is an active, virtual network of scientists working in different Swiss institutions in various fields of bioinformatics. While pivotal contributions of these researchers found wide recognition, this development also favored the access of the Swiss pioneers in bioinformatics to related, often complementary studies conducted in other countries and other continents.

May this overview of important aspects of bioinformatics further contribute to strengthen international contacts and serve as a testament to such a fruitful development for the basic as well as for the applied sciences.

Werner Arber
Professor Emeritus for Molecular Microbiology
Biozentrum, University of Basel, Switzerland
Nobel Laureate in Medicine 1978

Preface

Biological research and recent technological advances have resulted in an enormous increase in research data that requires large storage capacities, powerful computing resources, and accurate data analysis algorithms. Bioinformatics is the field that provides these resources to life science researchers. Originally the prerogative of a few biologists who were computer buffs and a few computer scientists who had a passion for life science, bioinformatics has developed into a fully-fledged science with its own specialists, pregraduate and postgraduate studies, as well as international conferences and journals.

In Switzerland, bioinformatics started in the 1980s, when a handful of enthusiastic scientists started developing databases and computational tools that rapidly became accessed and used by researchers from all over the world. While essential for thousands of scientists both in Switzerland and abroad, these computational biology resources were developed for many years without any substantial means. It was only in 1998 that the need for a Swiss bioinformatics infrastructure was recognized and that the Swiss Institute of Bioinformatics (SIB) was created. Even then, the SIB founders were ahead of their time, building an organization of exceptional quality and dynamism. Starting with 20 bioinformaticians in five groups at universities and research institutes in the Lake Geneva area, they were rapidly joined by new professors in major Swiss universities, thus growing into an institution of national importance, recognized worldwide for its state-of-the-art work.

After more than 10 years of existence, the SIB is regarded as one of the leading bioinformatics institutions in the world. Organized as a federation of bioinformatics research groups from Swiss universities and research institutes, the SIB provides services to the life science community that are highly appreciated worldwide, and coordinates research and education in bioinformatics nationwide. The SIB plays a central role in life science research both in Switzerland and abroad by developing extensive and high-quality bioinformatics resources that are essential for all life scientists. It contributes to the economy and quality of life through the global distribution of its products, by providing state-of-the-art tools to the industry, and by its involvement in pregraguate and postgraduate teaching programs. Knowledge developed by SIB members in areas such as genomics, proteomics, and systems biology is directly transformed by academia and industry into innovative solutions to improve global health.

This astounding concentration of talent in a given field is unusual and unique in Switzerland. This book gives an insight into some of the key areas of activity in bioinformatics in our country, covering both research work and major infrastructural efforts in genome and gene expression analysis, investigations on proteins and proteomes, evolutionary bioinformatics, and modeling of biological systems.

We are grateful to the authors of all chapters for their efforts, patience, and goodwill, without which it would not have been possible to publish this book. We are particularly indebted towards our colleague Dr. Patricia Palagi for her careful proofreading of the whole book; and are grateful to Drs. Lydie Bougueleret, Janet James, Tania Lima, and Ms. Nicole Zaghia, who went over the text of our authors whose English is not always as outstanding as their science. We also acknowledge support from the Swiss State Secretariat for Education and Research, the Swiss universities, federal institutes of technology and research institutes, the Swiss National Science Foundation, as well as several international funding bodies such as the European Research and Development Programmes and the US National Institutes of Health. Finally, we are thankful to all members of the Swiss Bioinformatics Institute, whose relentless work constitutes the core of bioinformatic excellence in Switzerland.

We hope that this book will help readers understand some of the many facets of bioinformatics today, and encourage new scientists to get on board with such a fascinating and challenging field.

Ron D. Appel and Ernest Feytmans
July 2008

Contents

List of Contributors

Prof. Amos Bairoch
Department of Structural Biology and Bioinformatics
Faculty of Medicine
University of Geneva

Founder Member and Group Leader, Swiss-Prot group
Swiss Institute of Bioinformatics
CMU – 1, rue Michel Servet
1211 Geneva 4
Switzerland

Prof. Sven Bergmann
Department of Medical Genetics
Faculty of Biology and Medicine
University of Lausanne

Group Leader, Computational Biology Group
Swiss Institute of Bioinformatics
Quartier UNIL-Bugnon, Rue du Bugnon 27
1005 Lausanne
Switzerland

Dr. Gabriel Jîvasattha Bittar
Fondation Jîvasattha and Jîvarakkhî
Buddhâyatana, Warrawee Road, American River
Kangaroo Island
SA 5221
Australia

Dr. Philipp Bucher
Swiss Institute for Experimental Cancer Research
Ecole Polytechnique Fédérale de Lausanne

Group Leader, Computational Cancer Genomics
Swiss Institute of Bioinformatics
EPFL SV — AAB Station 15
1015 Lausanne
Switzerland

Dr. Michel Cuendet
Molecular Modeling Group
Swiss Institute of Bioinformatics
University of Lausanne
Quartier Sorge, Bâtiment Génopode
1015 Lausanne
Switzerland

Dr. Mauro Delorenzi
NCCR Molecular Oncology
Group Leader, Bioinformatics Core Facility
Swiss Institute of Bioinformatics
Quartier Sorge, Bâtiment Génopode
1015 Lausanne
Switzerland

Manuel Gil
Centre for Computational Biology and Bioinformatics
Department of Computer Science
Swiss Federal Institute of Technology, Zürich

Computational Biochemistry Research Group
Swiss Institute of Bioinformatics
CAB F 61.2
Universitätstrasse 6
8092 Zürich
Switzerland

Dr. Darlene R. Goldstein
Institut de mathématiques
Ecole Polytechnique Fédérale de Lausanne

Swiss Institute of Bioinformatics
Bâtiment MA, Station 8
1015 Lausanne
Switzerland

Prof. Gaston H. Gonnet
Center for Computational Biology and Bioinformatics
Department of Computer Science
Swiss Federal Institute of Technology, Zürich

Group Leader, Computational Biochemistry Research Group
Swiss Institute of Bioinformatics
CAB H 66, Universitätstrasse 6
8092 Zürich
Switzerland

Dr. Aurélien Grosdidier
Molecular Modeling Group
Swiss Institute of Bioinformatics
University of Lausanne
Quartier Sorge, Bâtiment Génopode
1015 Lausanne
Switzerland

Dr. Lukasz Jaskiewicz
Computational and Systems Biology
Biozentrum
University of Basel

RNA Regulatory Networks
Swiss Institute of Bioinformatics
University of Basel, Klingelbergstrasse 50/70
4056 Basel
Switzerland

Dr. Evgenia V. Kriventseva
Department of Structural Biology and Bioinformatics
Faculty of Medicine, University of Geneva
CMU – 1, rue Michel Servet
1211 Geneva 4
Switzerland

Dr. Lydie Lane
Co-director CALIPHO team, Swiss-Prot group
Swiss Institute of Bioinformatics
CMU – 1, rue Michel Servet
1211 Geneva 4
Switzerland

Dr. Frédérique Lisacek
Group Leader, Proteome Informatics Group
Swiss Institute of Bioinformatics
CMU – 1, rue Michel Servet
1211 Geneva 4
Switzerland

Prof. Olivier Michielin
Multidisciplinary Oncology Center
University Hospital, Lausanne

Ludwig Institute for Cancer Research
Lausanne Branch

Group Leader, Molecular Modelling Group
Swiss Institute of Bioinformatics
Quartier Sorge, Bâtiment Génopode
1015 Lausanne
Switzerland

Dr. Patricia M. Palagi
Proteome Informatics Group
Swiss Institute of Bioinformatics
CMU – 1, rue Michel Servet
1211 Geneva 4
Switzerland

Prof. Manuel C. Peitsch
Director, Computational Sciences and Bioinformatics
Philip Morris R&D
Neuchatel
Switzerland

Chairman of the Executive Board
Swiss Institute of Bioinformatics
Quartier Sorge, Bâtiment Génopode
1015 Lausanne
Switzerland

Prof. Marc Robinson-Rechavi
Department of Ecology and Evolution
University of Lausanne

Group Leader, Evolutionary Bioinformatics Group
Swiss Institute of Bioinformatics
Quartier Sorge, Bâtiment Biophore
1015 Lausanne
Switzerland

Dr. Ute F. Röhrig
Multidisciplinary Oncology Center,
Lausanne University Hospital

Ludwig Institute for Cancer Research
Lausanne Branch

Molecular Modelling Group
Swiss Institute of Bioinformatics
Quartier Sorge, Bâtiment Génopode
1015 Lausanne
Switzerland

Prof. Ivo F. Sbalzarini
Institute of Theoretical Computer Science
Swiss Federal Institute of Technology, Zurich

Group leader, Computational Biophysics Laboratory
Swiss Institute of Bioinformatics
CAB G 34
Universitätstrasse 6
8092 Zürich
Switzerland

Prof. Torsten Schwede
Biozentrum, University of Basel

Group Leader, Computational Structural Biology
Swiss Institute of Bioinformatics
Klingelbergstrasse 50/70
4056 Basel
Switzerland

Dr. Bernhard Pascal Sonderegger
Swiss Institute of Experimental Cancer Research
Ecole Polytechnique Fédérale de Lausanne

Computational Systems Biology Group
Swiss Institute of Bioinformatics
AAB 0 18, Station 15
1015 Lausanne
Switzerland

Prof. Christian von Mering
Group Leader, Institute of Molecular Biology
University of Zurich

Group Leader, Bioinformatics/Systems Biology Group
Swiss Institute of Bioinformatics
Room Y55-L76
Winterthurerstrasse 190
8057 Zurich
Switzerland

Robert M. Waterhouse
Division of Cell and Molecular Biology
Faculty of Natural Sciences
Imperial College London
London SW7 2AZ
United Kingdom

Pratyaksha Wirapati
NCCR Molecular Oncology
Bioinformatics Core Facility
Swiss Institute of Bioinformatics
Quartier Sorge, Bâtiment Génopode
1015 Lausanne
Switzerland

Prof. Mihaela Zavolan
Biozentrum
University of Basel

Group Leader, RNA Regulatory Networks
Swiss Institute of Bioinformatics
Klingelbergstrasse 50/70
4056 Basel
Switzerland

Prof. Evgeny M. Zbodnov
Department of Genetic Medicine and Development
University of Geneva

Division of Cell and Molecular Biology
Faculty of Natural Sciences
Imperial College London, UK

Group Leader, Computational Evolutionary Genomics Group
Swiss Institute of Bioinformatics
CMU – 1, rue Michel Servet
1211 Geneva 4
Switzerland

Dr. Vincent Zoete
Molecular Modelling Group
Swiss Institute of Bioinformatics
Quartier Sorge, Bâtiment Génopode
1015 Lausanne
Switzerland

Section I

GENES AND GENOMES

Chapter 1

Methods for Discovery and Characterization of DNA Sequence Motifs

Philipp Bucher

1. Introduction

Motif discovery is considered to be an important problem in bioinformatics, as documented by a large number of papers. It is also believed to be a hard and still partly unsolved problem, despite considerable efforts by many distinguished researchers. Finally, it is an old problem with a long tradition in bioinformatics. An early example is the discovery of the Pribnow box in *E. coli* promoters.[1] Although this motif was found by visual inspection of DNA sequences, it was probably instrumental in defining the paradigm that subsequently led to the formalization of the motif discovery problem in its modern form and to the development of algorithms to solve it.

A DNA motif, such as the Pribnow box shown in Fig. 1, is defined by a set of short subsequences from longer sequences with high similarity. The subsequences share some common features, which typically are described by a consensus sequence or weight matrix. A motif must be overrepresented in a biologically defined collection of genome sequences, i.e. it must occur more frequently than

A

```
Seq1   G  T  A  C  G  A  TT G     A
Seq2   C  T  A  T  A  A  TG A     A
Seq3   C  T  T  A  T  A  AG T  G
Seq4   G  T  G  A  T  A  C  T  G  A
Seq5   A  T  A  T  G  A  TC G     G
Seq6   C  G  T  A  T  G  T  TT G
```

B

```
T  A  T  R  A  T  G
```

C

```
Seq1   G [T  A  C  G  A  TT G]    A
Seq2   C [T  A  T  A  A  TG A]    A
Seq3   C  T [T  A  T  A  AG T  G]
Seq4   G  T [G  A  T  A  C  T  G] A
Seq5   A [T  A  T  G  A  TC G]    G
Seq6   C  G [T  A  T  G  T  TT G]
```

D

	1	2	3	4	5	6	7
A	0	6	0	3	4	0	1
C	0	0	1	0	1	0	0
G	1	0	0	3	0	0	5
T	5	0	5	0	1	6	0

E

	1	2	3	4	5	6	7
A	10	70	10	40	50	10	20
C	10	10	20	10	20	10	10
G	20	10	10	40	10	10	60
T	60	10	60	10	20	70	10

F

	1	2	3	4	5	6	7
A	−9	10	−9	5	7	−9	−2
C	−9	−9	−2	−9	−2	−9	−9
G	−2	−9	−9	5	−9	−9	9
T	9	−9	9	−9	−2	10	−9

Fig. 1. The Pribnow box, an early discovered DNA motif. The Pribnow box is a promoter element of *E. coli* promoters, originally discovered by visual inspection of six experimentally characterized promoter sequences.[1] (a) Input sequence set. (b) Consensus sequence proposed in the original paper (R is the IUPAC (International Union of Pure and Applied Chemistry) code for A or G). (c) Input sequence set with highlighted motif instances. The set of motif instances is also referred to as "motif annotation" in this chapter. (d) Base count frequency matrix. (e) Base probability matrix estimated by adding one pseudocount to each element of the base count frequency matrix (probabilities are given as percentages). (f) Weight matrix. The position-specific weights of corresponding bases are summed up to compute a score for a DNA sequence of the same length as the motif. The weights of the matrix were computed as a natural log-likelihood ratio from the base probabilities, multiplied by 10, and rounded to the nearest integer (see Chapter 2).

one would expect by chance. The DNA motif has become a central concept of molecular biology, a research field which has its roots in biology as well as in physics. In order to understand why motifs are of interest, a brief look at the leading paradigms of both disciplines will be useful.

1.1. *Motif Discovery from a Biological Perspective*

The basis of modern biology is the theory of evolution by natural selection introduced by Darwin and Wallace. One of the tenets of this theory is that any genetically encoded biological structure is subject to the randomizing forces of mutation and eventually will disappear if not conserved by natural selection. According to Williams,[2] constancy and complexity are biological proof of function, even in the absence of a conceivable mechanism by which a conserved structure might contribute to the organism's fitness. The lateral organ of fishes is cited as an example. The high complexity of this organ and its high degree of conservation across species prompted biologists to carry out experiments, which eventually led to the identification of its function as a sensory organ. This is exactly the biological motivation behind motif discovery. Sequence conservation is evidence of natural selection and thus justifies an investment of experimental work to elucidate the function of a motif. In fact, this approach has been very successful in the study of protein function. There also, the discovery of a new conserved domain has often preceded the characterization of its molecular function. Even though this chapter is focused on DNA motifs, many of the concepts and methods introduced extend readily to RNA and protein sequence motifs.

A minimal degree of complexity is an essential property of a motif, as motifs of low complexity may frequently occur by chance and thus cannot meet the condition of overrepresentation. While the complexity of a morphological structure can be judged by human visual intuition, the complexity of DNA sequence motifs is typically evaluated by a conditional entropy-based index borrowed from information theory.[3] Unlike the search for new protein motifs, DNA motif discovery is often targeted to a particular function, which may, however, be broadly defined. For instance, by searching eukaryotic promoter sequences for conserved DNA motifs, one typically expects to find the target binding sites for a variety of *a priori* unknown transcription factors.

Sequence motifs can also be viewed as taxonomic entities. Mastering a bewildering diversity of phenomena through classification is a typical biological approach that can be traced back to Carl von Linné's *Systema Naturae*. Since classification has contributed so much to our understanding of the living world, the discovery of a new species or the definition of a new medical syndrome is rightly considered as a scientific achievement in its own right. A relatively recent article in *Nature* on the discovery of a new mammalian species[4] documents this view.

1.2. *DNA Motifs from a Physical Perspective*

To a physicist, the definition of a taxonomic entity hardly represents the endpoint of a research project. The physical approach aims at causal relationships between observable events, and at quantitative models that can predict the outcome of experiments. Surprisingly, DNA motif discovery has found important applications in such a research setting too. A classical example is the characterization of transcription factor binding sites, where the DNA motif becomes a quantitative model to predict the binding energy of a protein–DNA complex (Fig. 2). In fact, the Pribnow box mentioned before is also a part of a DNA–protein binding site, the one recognized by bacterial RNA polymerase. The Berg and von Hippel[5] statistical mechanical theory of protein–DNA interactions provides a connection between motif complexity (conservation) and binding energy. Interestingly, the standard descriptor used for representing a protein binding site, the energy matrix, is mathematically equivalent to the weight matrix used in *de novo* discovery of evolutionarily conserved motifs. However, the logic of the scientific inference process that leads to the definition of the matrix is reversed; here, the starting point is a known molecular function and the endpoint is an initially unknown motif which can be considered as the "genetic code" for the function.

From a computational chemistry viewpoint, an energy matrix for a transcription factor binding site is a special case of quantitative structure–activity relationship (QSAR) model.[6] There is a wealth of literature about QSAR models that is only sparsely cited in DNA motif discovery papers. Machine learning methods exploiting quantitative activity data have been widely used in the QSAR field. Interestingly, the first weight matrix-like structure used for representation of a nucleic acid

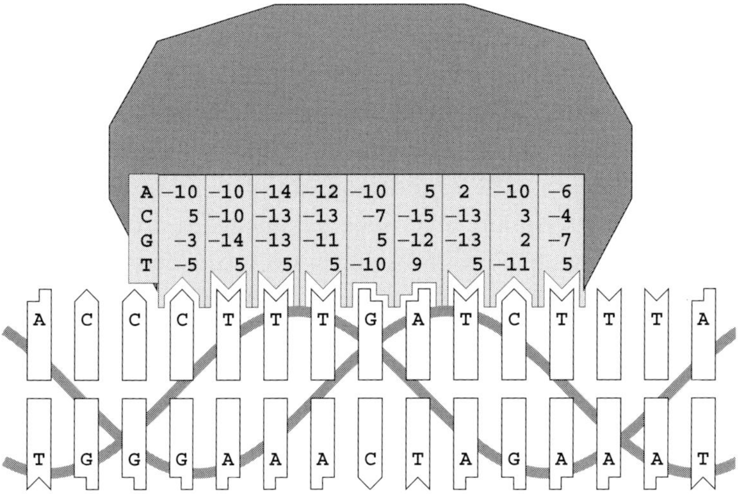

Fig. 2. Energy matrix for a transcription factor binding site. An energy matrix represents one possible physical interpretation of a weight matrix. Each element of the matrix quantitatively defines the binding energy (in arbitrary units) between a DNA base pair and a corresponding compartment of the DNA-binding surface. Energy matrices can be used to compute binding constants for DNA protein complexes, and therefore represent a special case of QSAR (quantitative structure–activity relationship) models. Note that the sign of the energy units is reversed; a high weight matrix score signifies low energy value, and thus high binding strength.

sequence motif (ribosome binding sites) was inspired by a machine learning method called "perceptron".[7]

In summary, DNA motif discovery is not an isolated, specialized topic for a closed circle of bioinformaticians. DNA motifs have many facets and have different meanings to different researchers. Mathematically equivalent descriptors have been used in many more fields, even outside life sciences. Hidden Markov models,[8] for instance, were developed in the speech recognition field.

In the following sections, a personal view of motif discovery will be presented inspired partly by the author's own work. The focus will be on essential concepts and open questions. Methods will be presented in their most basic version; a comprehensive review of current state-of-the-art motif discovery algorithms is beyond the scope of this chapter. Further references on methods can be found in recent reviews.[9,10]

2. Motif Discovery in a Nutshell

Knowing the inputs and outputs is central to the understanding of a computational problem. The data structures involved in motif discovery are shown in Fig. 1. The input consists of a set of sequences, not necessarily of fixed length. The output consists of a list of motif instances and/or a motif description: a consensus sequence, a probability matrix, or a weight matrix. The motif instances are subsequences of the input sequence, and can be defined by a sequence name and a starting position. For the type of motifs considered here, they are of fixed length. The motif description and motif annotation are intertwined entities in that the motif description defines the motif annotation of the input data set, and the set of motif instances can be used to derive the motif description.

A consensus sequence is a short sequence (k-letter word) from the DNA alphabet or from an extended alphabet containing IUPAC (International Union of Pure and Applied Chemistry) codes for incompletely specified bases in nucleotide sequences.[11] A threshold number of mismatches may be permitted. The consensus sequence, together with the maximal number of allowed mismatches, defines the motif in a deterministic and qualitative manner. Specifically, it defines the subset of all k-letter words which qualify as motif instances.

The position-specific scoring matrix, introduced in its standard form by Staden,[12] is a more flexible representation of a sequence motif. Its use is motivated by the assumption that not all mismatches to consensus sequences are equally detrimental. Therefore, the relative fit of a particular base to a given motif position is expressed by a number. Matrix descriptions for sequence motifs come in two forms: base probability matrices and additive scoring matrices, henceforth called weight matrices. The former reflects the expected frequencies of each base at each position. The latter serves to compute a motif score for a particular k-letter sequence by adding up the matrix elements corresponding to all bases at each position in the sequence. The parameters of a weight matrix can have positive or negative values.

The base probabilities are often estimated by adding one pseudo-count to the observed base count of base b at position i:

$$p(i,b) = \frac{c(i,b)+1}{4+\sum_{b'=A}^{T} c(i,b')}. \tag{1}$$

The elements of a weight matrix may be computed from a probability matrix as a log-likelihood ratio:

$$w(i,b) = \ln \frac{p(i,b)}{p_0(b)}. \tag{2}$$

Here, $w(i,b)$ and $p(i,b)$ are the weight and probability of base b at position i of the motif, respectively, and $p_0(b)$ is the background probability of base b. Note, however, that log-likelihood ratios are not universally used in the field. One of the best known motif search programs, MATINSPECTOR, uses a different way of scoring transcription factor binding sites with a base probability matrix.[13]

Like a consensus sequence, a weight matrix in conjunction with a cut-off value defines a subset of k-letter words which qualify as motif instances. However, the power of the matrix representation lies in the quantitative evaluation of candidate k-letter words, which can be exploited, for instance, for transcription factor binding site affinity prediction. On the other hand, a base probability matrix defines a motif in a probabilistic manner, a property which is exploited by probabilistic motif optimization methods such as expectation maximization.

The goal of the motif search problem is to find the best motif for a given set of input sequences. The complete statement of the problem requires the specification of a quality criterion (objective function) related to overrepresentation. A specific motif discovery method is thus characterized by three components: (a) the motif descriptor, (b) the objective function, and (c) the algorithm to scan the search space of possible motifs.

The large diversity of published motif search algorithms is readily classified along these lines. The dual nature of the motif discovery output, motif annotation and motif description, has implications on the search space that needs to be scanned. If the motif annotation is considered to be the primary result, then the search space consists of all possible motif annotations and the optimization problem becomes a gap-free local multiple alignment problem. If the motif description is considered the primary result, then the search space consists of all possible consensus sequences or weight matrices.

Some variations of the standard scheme outlined above deserve to be mentioned. Figure 1 suggests that each input sequence contains exactly one sequence motif. This is not a general requirement. In fact, motifs may occur only in a subset of the input sequences or more than once in a particular sequence. The popular motif discovery program MEME[14] has established the following nomenclature for the three different motif search modes: "oops" for one occurrence per sequences, "zoops" for zero or one occurrence per sequence, and "anr" for any number of repetitions. Furthermore, a given input sequence set may contain more than one overrepresented motif type. Many algorithms and computer programs offer, in fact, the possibility to search for multiple motifs in one run. Finally, since genomic DNA is usually double-stranded, the motif search may be extended to the reverse-complementary strands of the input sequences.

3. Overview of the Methods

3.1. *Objective Functions*

As explained before, candidate motif descriptions need to be ranked by an index that reflects the degree of overrepresentation in a statistical sense. Two very different types of probabilities are used for this purpose:

(a) the probability that a given motif occurs at least a certain number of times in the input sequence set; and
(b) the probability of the input sequences, given the motif.

Probability (a) is minimized and, from a statistical viewpoint, reflects the classical frequentist approach exemplified, for instance, by the objective function introduced by van Helden *et al.*[15] Probability (b) is maximized and is inspired by Bayesian statistics; in this case, the motif is part of a probabilistic generative model such as a hidden Markov model (HMM)[8] or a mixture model.[16] The two types of probabilities will be illustrated by examples below.

Let us first turn to the question of how to compute the probability that a motif occurs at least n times in an input sequence set. Here, and in all following examples, we will assume that motif frequencies were determined in the "anr" search mode. An exact solution to this problem is hard to obtain because of the statistical nonindependence of overlapping words.[17] In fact, the probability distribution of a k-letter word to occur zero to N times in a sequence of length $N + k - 1$ depends on the internal repeat structure of the word. To bypass this difficulty, motif discovery algorithms often rely on approximations, which are debatable from a mathematical viewpoint. According to the frequently used Poisson approximation, which assumes independence between motif occurrences, the probability of finding a motif exactly n times is given by

$$\mathrm{Prob}(n, E_i) = \frac{1}{n!} E_i^n \exp(-E_i)\cdot \tag{3}$$

Here, E_i is the expected number of occurrences of a given motif i, which is the product of the search space N and the probability p_i that a random sequence of length k constitutes an instance of motif i. The search space is the number of all possible starting positions for a motif of length k in the input DNA sequence set.

If the motif description consists of a consensus sequence based on the four-letter DNA alphabet, with a maximal number of m mismatches allowed, then the probability p_i may be computed as follows:

$$p_i = \sum_{j=0}^{m} \binom{k-j}{j} 0.25^{(k-j)} 0.75^{j}\cdot \tag{4}$$

The assumption underlying this formula is that all bases occur with an equal probability of 0.25 in random sequences. This is the simplest background (null) model that can be used in this context. Markov chains, which assume unequal probabilities for different bases and dependencies between consecutive bases, are more realistic background models for genomic DNA sequences. Algorithms have been presented for computing p_i for such a model, as well as for consensus sequences including ambiguous positions represented by IUPAC codes[18] and also for weight matrices.[19]

The Bayesian approach will be illustrated with the mixture model used by the program MEME. Again, we assume the "arn" search mode. To circumvent the mathematical difficulties of overlapping words statistics, the input sequence set is usually evaluated as if it were to consist of N nonoverlapping k-letter subsequences (N is the search space defined before). In the simplest case, the mixture model consists of two components, a motif model given by a probability matrix and a background model given by a base probability distribution. The probability of the sequences given the model is then computed as

$$\text{Prob}(\text{x}, M, M_0, q) = \prod_j \left(qP(x^j \mid M) + (1-q)P(x^j \mid M_0) \right). \quad (5)$$

In this notation, **x** denotes the total set of overlapping k-letter subsequences contained in the input sequences, and x^j is an individual member of it. $P(x^j|M)$ and $P(x^j|M_0)$ are the probabilities of subsequence x^j given the motif model and the background model, respectively. q is the mixture coefficient indicating the sequence-independent probability that a given subsequence constitutes a motif. The models M and M_0 both define probability distributions over all k-letter words. The probabilities of sequence x^j under the motif and background models, respectively, are defined as follows:

$$P(x^j \mid M) = \prod_{i=1}^{k} p(i, x_i^j) \quad (6)$$

$$P(x^j \mid M_0) = \prod_{i=1}^{k} p_0(x_i^j). \tag{7}$$

Note that the mixture coefficient q is part of the model and, thus, the target of optimization by the motif discovery algorithm. It plays a similar role as the threshold value of allowed mismatches to a consensus sequence in the frequentist motif evaluation framework. This raises the interesting question of whether the two approaches are equivalent with regard to defining the optimal threshold value for a consensus sequence or weight matrix. The answer is, to my knowledge, not known.

3.2. *Scanning the Search Space*

How do we find the best motif among all possible motifs? Three strategies can be distinguished depending on the structure of the search space, which may consist of

(a) all k-letter words of a given alphabet;
(b) all probability matrices of length k; or
(c) all motif annotations (all subsets of k-letter subsequences of input sequences).

3.2.1. *Finding the best consensus sequence*

For consensus sequences based on the four-letter DNA alphabet, the size of the search space remains computationally manageable up to a word length of about 15. The optimal motif can thus be found by enumeration, i.e. by evaluating a frequentist-type objective function for each k-letter word. Algorithms to this end are reasonably fast, as the input sequences need only to be scanned once. The word index is first initialized with zeros. Then, one word frequency is incremented each time a subsequence is processed (if mismatches are allowed, multiple motif frequencies are updated for one subsequence). For longer words, heuristic algorithms have to be used instead of exact methods. An old trick, introduced in the early 1980s,[20] is to restrict the search space to those k-letter words which actually occur in the input sequence set. A popular implementation of the word search strategy is provided by the program Weeder.[21]

For consensus motifs based on the complete 15-letter IUPAC code, the search space becomes too large for enumerative approaches. Progress has recently been reported in developing efficient heuristics under certain constraints.[22] It is debatable, however, whether such algorithms are really needed. Consensus sequences with ambiguous symbols are still less flexible than position-specific scoring matrices, and efficient and effective algorithms are readily available for optimizing such matrices.

3.2.2. *Optimizing a base probability matrix*

The search space for probability matrices is continuous and, thus, potentially infinite. Rigorous algorithms guaranteed to find the optimal motif are not available or even conceivable for now. The standard heuristic approach to optimize a probability matrix is to start from an initial motif description referred to as a seed, and to use an iterative refinement algorithm to reach a local maximum for the objective function. Expectation maximization (EM) is the classical approach used in this context, introduced to computational biology by Lawrence and Reilly.[23] The iterative part is straightforward. Based on a current model M^k, one uses Bayes' formula to compute for each subsequence x^j a weight w_j^k, which is the posterior probability that the subsequence constitutes a motif instance. For the mixture model introduced above, one obtains

$$w_j^k = P(M^k \mid x^j) = \frac{q^k P(x^j \mid M^k)}{q^k P(x^j \mid M^k) + (1 - q^k) P(x^j \mid M^0)}. \tag{8}$$

These probabilities are then used as weights to compute a new probability matrix by adding up weighted base contributions from all subsequences of the input sequence set:

$$p^{k+1}(i, b) = \frac{\sum_{j=1}^{N} w_j^k \delta(x_i^j = b)}{\sum_{j=1}^{N} w_j^k}, \tag{9}$$

where $\delta(x_i^j = b)$ is 1 if $x_i^j = b$ and 0 otherwise. Note that, in practice, only a few subsequences ... the likely motif instances ... will make significant contributions to the new model. The new mixture coefficient q is obtained as the sum of posterior probabilities for all k-letter subsequences divided by the search space:

$$q^{k+1} = \left(\frac{1}{N}\right)\sum_{j=1}^{N} w_j^k. \qquad (10)$$

Expectation maximization is a deterministic algorithm that can get trapped in a local optimum. To have a chance of reaching the globally optimal motif, it has to start from a good seed, which is already relatively close to the target. Two strategies are used for this purpose. One is to trigger the algorithm from a large number of random seeds; this is time-consuming and thus relatively inefficient. A better way is to use a consensus sequence obtained from a fast exact or heuristic word search algorithm, as described above. The combination of a word search algorithm for the seeding step and EM for refinement is probably the most effective motif discovery strategy used nowadays, and is implemented in various forms in many software tools. MEME, for instance, uses seeds obtained from exhaustive pairwise comparison of k-letter subsequences, an approach which leads to good seeds but has the unfavorable time complexity $O(N^2)$ for the seeding step.

As explained before, the mixture-model-based EM approach does not account for the nonindependent occurrence of overlapping subsequences. The HMM framework offers an explicit way of computing posterior motif probabilities, under the constraint that motif instances must not overlap, via the choice of an appropriate model architecture. In all other respects, the "training" of a HMM, i.e. the iterative optimization of the model with respect to the input sequence set, is identical to the optimization of a mixture model.

3.2.3. *Optimizing the motif annotation*

Keeping the same probabilistic mixture model framework, the motif discovery problem can also be formulated in such a way that the motif positions are the target of optimization. The search space thus becomes discrete and finite, but still remains astronomically large. The classical algorithm in this category is the Gibbs sampler,[24] a stochastic variant of EM. This approach focuses on actual motif instances rather than a generalized description of the motif. In other words, one would primarily like to know the locations of the functional elements in the given sequences rather than be able to predict motif instances in new sequences. From a theoretical and computational viewpoint, the differences are relatively minor, since both EM and Gibbs sampling update a probability matrix. However, whereas EM assigns k-letter subsequences probabilistically to the two mixtures of the model, Gibbs sampling attributes them explicitly to one or the other class. The decision whether a given subsequence x^j is included in the motif set for the next iteration is made randomly based on its current posterior probability w_j^k, as defined previously. The new probability matrix is typically estimated by a maximum *a posteriori* (MAP) likelihood method, for instance, by adding one pseudocount to each base frequency:

$$p^{k+1}(i,b) = \frac{1 + \sum_{j \in \mathrm{M}^k} \delta(x_i^j = b)}{|\mathbf{M}^k| + 4}, \tag{11}$$

where $\delta(x_i^j = b)$ is 1 if $x_i^j = b$ and 0 otherwise. $\mathbf{M}^k$ denotes the current set of motif instances, and $|\mathbf{M}^k|$ the number of elements therein.

The fact that Gibbs sampling has an in-built stochastic element helps to overcome the risk of getting trapped in a local optimum far away from the global optimum. Another advantage of a nondeterministic algorithm is that it potentially returns different results when run several times from the same seed, which also increases its chances of finding the globally optimal solution. Optimization in motif annotation space can also be done in a deterministic fashion. For instance, one could define the new motif instances by accepting all k-letter subsequences with posterior probabilities higher than 0.5. This leads to an iterative, multiple alignment algorithm, similar to the PATOP algorithm described in Sec. 5.

3.2.4. *Finding multiple motifs*

In exploratory applications, such as mining promoter sequences for new transcription regulatory motifs, one often expects to find more than one motif. For instance, a landmark paper on *Drosophila* promoters[25] reported 10 *ab intio* discovered motifs returned in one program run by MEME. Fortunately, there is a simple and efficient way to extend the basic algorithms presented above to multiple motif discovery. The principle is to proceed iteratively by searching for one motif at a time, and by progressively excluding motif instances found from subsequent iterations. More formally, this means that, after each cycle, the k-letter subsequences attributed to the newly discovered motif are removed from the search space — a process that is commonly referred to as "masking" in the sequence analysis literature. A theoretically more proper approach would use multi-component mixture models for synchronous optimization of several motifs at a time by EM, Gibbs, or a progressive local multiple alignment algorithm.

3.2.5. *Estimating the significance of a newly discovered motif*

The different types of probability values used as objective functions for motif optimization do not provide an answer to the question of whether the best motif found is significant or not, as they apply to single motifs and thus are not corrected for multiple tests. With consensus sequence motifs, a Bonferroni correction is sometimes applied; see, for instance, Xie *et al.*[26] However, this approach is likely to yield overly conservative P-value estimates, as consensus word frequencies are highly dependent on each other, especially if mismatches are tolerated. The program MEME provides significance estimates for matrix-based motif models based on a maximum likelihood ratio test (LRT), which takes into account the number of free parameters of the model.[16] This approach is quite sensitive to the properties of the null model, and in practice tends to assign low E-values to questionable motifs. A good way to corroborate the significance of a newly found motif is to rerun the motif discovery program with randomized or shuffled sequences as a control, so as to get an idea of what P-values or E-values could be expected for fortuitous motifs. Shuffling methods which preserve higher-order Markov chain properties[27] of the real sequences are recommended for this purpose.

4. Bottlenecks and Limitations

DNA motif discovery is considered to be a tough problem. This is surprising in view of the apparently simple structure of DNA motifs, as compared to protein sequence motifs. The perception that the problem is difficult is partly based on the poor track record in terms of important discoveries made by this approach, which were later confirmed by experimental follow-up studies. This contrasts with the great success of similar methods in discovering new protein sequence motifs.[28] Recent evaluation studies based on representative and realistic benchmark sequence sets indeed confirmed that current state-of-the-art motif discovery programs are highly ineffective in rediscovering experimentally characterized motif instances (transcription factor binding sites) hidden in gene regulatory sequences.[29,30] However, these results have to be interpreted with caution. Let us first have a look at the benchmarking procedure.

4.1. *Benchmarking Procedures for Motif Discovery*

Claims about poor performance of motif discovery algorithms require that the community agrees on how performance is measured. In this sense, the recent benchmarking papers have made an invaluable contribution to structuring the field by better defining the problem. It is thus of paramount importance to the newcomer to understand how these tests were set up. The procedure is schematized in Fig. 3. The benchmark sets consist of DNA sequences of a few hundred base pairs in length containing annotated transcription factor binding sites, which constitute the motifs to be discovered. The experimental motif annotations of the eukaryotic benchmarking set were taken from TRANSFAC,[31] while the prokaryotic test set is based on RegulonDB.[32] Both resources are manually curated databases relying on experimental results published in journal articles.

One test per motif is carried out. An input sequence set consists of all sequences containing a particular motif. The experimental

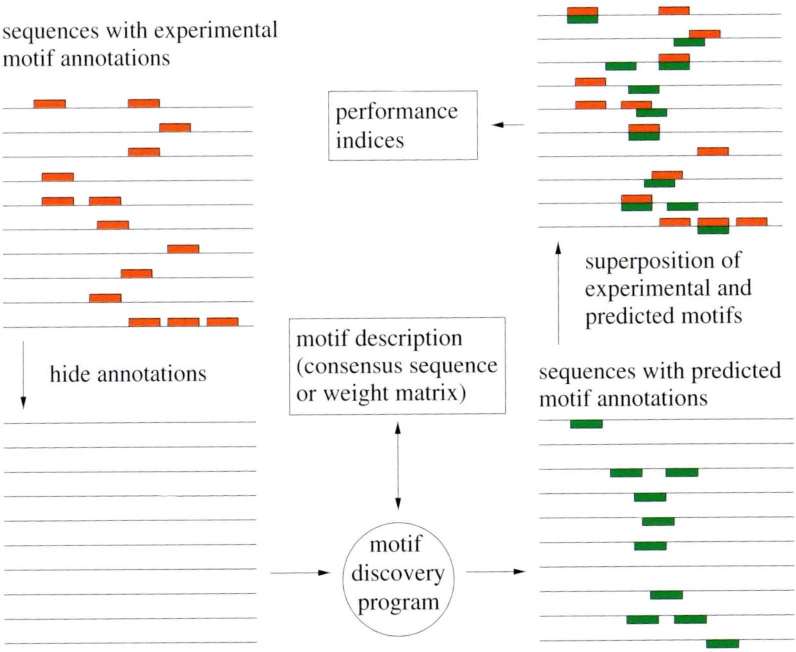

Fig. 3. Benchmarking protocol for evaluation of motif discovery algorithms. A test consists of DNA sequences which experimentally mapped binding sites to a particular transcription factor (experimental motif annotations are shown in red). The naked sequences (without motif annotations) are given as input to the motif discovery algorithm, which returns a set of predicted motif instances plus, optionally, a motif description. The experimental and predicted motif annotations are then superimposed for computation of a variety of performance indices, such as sensitivity and specificity. Partial overlaps between known and predicted motifs are usually counted as success. This protocol has been applied in the recent benchmarking studies described in Hu *et al.*[29] and Tompa *et al.*[30]

annotations, of course, are hidden to the program. The task of the motif discovery program is to rediscover the hidden motif and to return the coordinates of the corresponding motif instances. The performance is evaluated on the basis of the overlap between experimental and predicted motif annotations, and is expressed by standard measures such as sensitivity (percentage of true motif instances overlapped by predicted motif instances) and specificity (percentage of predicted motif instances overlapped by true motif instances).

As mentioned before, the results of these studies were disappointing. With the eukaryotic benchmarking set, sensitivity and specificity varied between 10% and 30%,[30] and only marginally better results were obtained with the prokaryotic test set.[29] Similar results were obtained with synthetic sequences, where the experimentally defined motif instances were hidden in computer-generated random sequences corresponding to a Markov chain model.

The primary reason for the poor performance is probably related to the characteristics of the input sets, typically consisting of only a few (in the order of 10) rather long sequences. The total number of motif instances hidden in the test sequences was also relatively low, below 20 in most cases. These are unfavorable conditions for motif discovery. A large number of short sequences, highly enriched in a given motif, would have made the task of the motif discovery program much easier.

4.2. *Why is Protein Domain Discovery Easier?*

At first glance, regulatory proteins resemble gene regulatory regions in that they have a modular architecture. The modules are motifs; in proteins, they are also called conserved domains. The key difference, however, lies in the complexity of the modules.

Regulatory DNA elements are short and based on a smaller alphabet. Complexity is expressed by the information content *IC*, which is computed from a base probability matrix as follows[3]:

$$IC = \sum_{i=1}^{k} \sum_{b=A}^{T} p(i,b) \log_2 \frac{p(i,b)}{0.25}. \tag{12}$$

The definition of information content has the form of a conditional entropy, where the null model consists of uniform base probabilities of 0.25 for each base. The information content, which is expressed in bits, indicates the random occurrence probability of a motif, which is 2^{-IC}. Typical transcription factor binding site matrices in TRANSFAC have an *IC* value of about 10 bits, which means that they are expected to occur about once in every 1000 bp. Consequently, about one match to a motif is expected to occur by chance in the input sequence sets that were used for benchmarking. Under these conditions, it is principally impossible to infer the true motif instances with high reliability. Since the probability matrices returned by motif discovery algorithms are derived from hypothetical motif instances in the input set, their quality is compromised by the contamination with false matches.

Conversely, the motifs corresponding to protein domains have a much higher complexity, often in the range of 30 bits or more. The higher information content is due to the increased length (up to 100 amino acids) and the larger size of the protein alphabet (20 instead of 4). Motifs with this degree of complexity are unlikely to occur by chance in a protein sequence of average length, and thus can be located with near certainty. The higher complexity also explains why protein domain discovery has often been initiated by a single statistically significant pairwise sequence match retrieved by database search.

4.3. *Reasons for the Limited Success of DNA Motif Discovery*

Based on the above considerations, I have doubts whether motif discovery is rightly considered to be a tough problem. The poor benchmarking results reported in recent papers are perhaps mostly due to the inadequacy of the input data sets. Shorter sequences would be needed to localize motifs with high confidence, and a larger number of motif instances would be required to obtain reliable base frequency estimates. Failure by the heuristic algorithms to find the optimal motif is unlikely to be a major reason for the poor benchmarking results. This could be tested by comparing the true versus predicted motif annotations in terms of the objective function used by the motif discovery program. My conjecture is that the true motif annotation will look less good in such a test. The fact that the consensus sequence-based motif search program Weeder[21] showed the best performance in the above-described tests supports this hypothesis.

In an overfitting situation due to sparse data, methods based on a simpler model with fewer degrees of freedom tend to perform better.

The true problem with DNA motif discovery is that biologists have published consensus sequences and weight matrices based on very few sequences for too many years, whereas computational biologists were mostly concerned with algorithmic improvements aimed at finding the globally optimal motif with higher probability and in shorter time. In the future, more efforts should be spent on analyzing the limitations of motif discovery in light of a statistical inference problem.

5. Locally Overrepresented Sequence Motifs

This last section summarizes a variant of the classical motif search problem, introduced by the author about 25 years ago for the study of promoter sequences. This method, named signal search analysis (SSA),[33] takes into account the fact that certain classes of DNA sequences such as promoters are experimentally defined by "positions" rather than by "borders". First, let us elaborate on these two concepts. A set of regulatory genomic sequence regions defined by deletion mutations, or a set of oligocleotides shown to bind a particular transcription factor *in vitro*, constitutes a DNA sequence set defined by borders. The sequences are of defined length, and the biologist has good reason to believe that a particular sequence motif is hidden anywhere within the sequences. This is exactly the experimental scenario to which the standard formulation of the motif discovery problem applies. On the other hand, promoters exemplify a sequence type defined by position; they are defined by the location of the transcription start site (TSS), which can be mapped experimentally. Promoter motifs are supposed to occur in the vicinity of a TSS, but there is no experimental protocol that would allow delineating the sequence range within which they must occur. Computational biologists have to cope with this missing information problem in some way.

5.1. *Modification of the Problem Statements*

A common way to proceed in promoter analysis is to define promoters operationally as sequences extending from arbitrarily chosen distances upstream to downstream from the TSS. For instance, Ohler *et al.*[25] used sequences between relative positions –60 and +40 for the identification of core promoter elements in *Drosophila*. A principle limitation of this approach is that it ignores a motif's specific positional distribution around the TSS, which varies widely between motifs. For instance, the eukaryotic TATA box occurs at a rather fixed distance of about 30 bp ± 5 bp upstream from the TSS; conversely, the CCAAT box occurs within a large region of about 150 bp with a maximum at –80 (Fig. 4).[34] Realistic objective functions for promoter motif discovery have to account for such differences. At equal frequency, a motif predominantly occurring within a narrow distance range should be considered more significant than one that is evenly distributed over the entire promoter region considered.

Signal search analysis (SSA) is an early method that takes positional distributions of motifs into account. It is based on the concept of a locally overrepresented sequence motif, which leads to a reformulation of the motif discovery problem, as will be explained below. The input to SSA is a set of genome sequences together with a list of so-called "functional positions", e.g. a list of TSSs. The result is a locally overrepresented motif

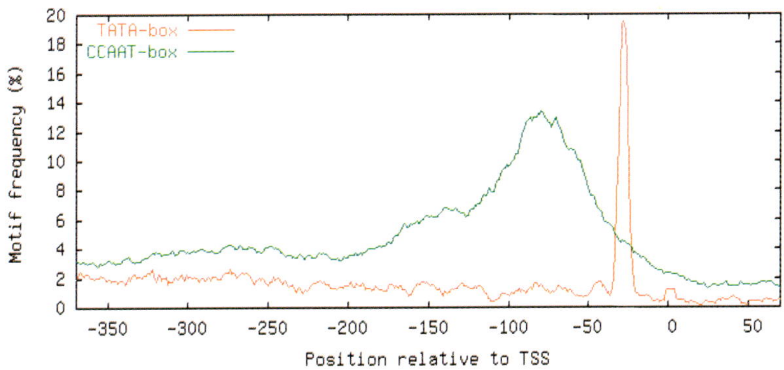

Fig. 4. Positional distributions of promoter sequence motifs around a human transcription start site (TSS). Shown are the distributions of the TATA and CCAAT boxes relative to 1867 precisely mapped TSSs from the Eukaryotic Promoter Database (EPD), release 93.[36] The plot is based on the weight matrices published in Bucher.[34] The motif frequencies were determined in overlapping windows of 20 bp for the TATA box, and 50 bp for the CCAAT box.

consisting of a motif description (consensus sequence or weight matrix + cut-off value) plus a region of preferential occurrence defined by 5′ and 3′ borders relative to the functional site. The key difference to the classical motif search problem statement is that the location of the motif relative to the reference position is transferred from input to output. As a consequence, the borders of the preferred region become targets for optimization and arguments of the objective function.

SSA uses a nonprobabilistic measure of local overrepresentation as an objective function for assessing motif quality. The computation of this measure is illustrated in Fig. 5. Briefly, the frequency of a given motif is determined in a series of adjacent, nonoverlapping windows of identical size, including the preferred region of occurrence as an individual window. The total length of the analyzed sequence region is chosen ad hoc.

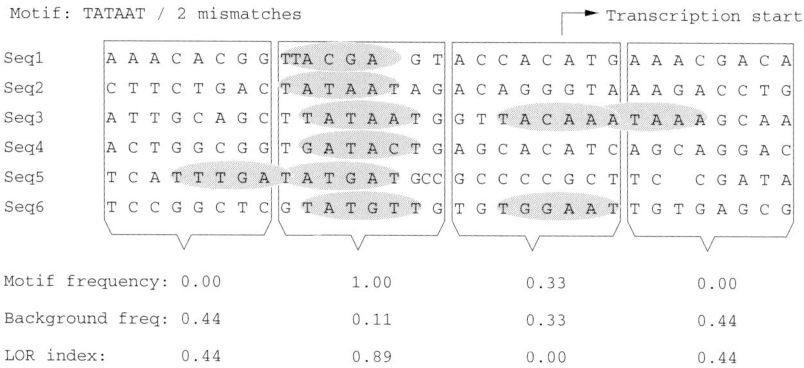

Fig. 5. Local overrepresentation — the objective function used by signal search analysis (SSA). The example sequence set consists of the six *E. coli* promoter sequences which led to the discovery of the Pribnow box motif.[1] This motif typically occurs about 10 bp upstream of the TSS. The frequency of this motif (here, defined as TATAAT with two mismatches allowed) is analyzed in a series of adjacent, nonoverlapping windows of 8 bp. The motif frequency is defined as the fraction of sequences per window that contain at least one motif instance (motifs spanning window boundaries are not counted). The background frequency for a particular window is defined as the mean of the motif frequencies in all other windows. The index of local overrepresentation (LOR) is simply the difference between the local motif frequency and the corresponding background frequency. In this example, the analyzed motif is highly overrepresented in the second window of the series, extending from relative positions −13 to −6.

The motif frequency is defined as the fraction of sequences in a window that contain at least one motif occurrence. Motifs overlapping window borders are ignored for this purpose. Local overrepresentation (LOR) is defined as the motif frequency within the window of preferential occurrence minus the average motif frequency determined in all other windows:

$$LOR_j = f_j - \frac{\sum_{i \neq j} f_i}{N - 1}. \tag{13}$$

Here, f_j is the motif frequency in window j, and N is the total number of windows. Note that the series of windows used to compute the background frequency needs to be adjusted to the specific region for which LOR is computed. The motif frequency outside the preferred regions is called background frequency, and serves the same function as the null model in the classical motif discovery framework. In fact, a major strength of SSA is its usage of a realistic null model based on natural sequences from the same genomic environment. This may explain why the weight matrices for major eukaryotic promoter elements, which were derived by this method almost 20 years ago, are still in use.

5.2. *Search Algorithms for Locally Overrepresented Sequence Motifs*

Two algorithms have been developed for the discovery of locally overrepresented sequence motifs, one for consensus sequence motifs[35] and one for weight matrices.[34] The former enumerates k-letter words, possibly containing free positions represented by a wildcard character and allowing a specified number of mismatches. The search space of the preferred region is defined by a preselected fixed window, with the 5′ and 3′ borders of the complete sequence range being taken into consideration. Using the computing power available at the time this method was conceived, the enumerative approach was possible up to a word length of about 6. To provide a heuristic search strategy for longer motifs, the original algorithm offered the possibility of restricting motif evaluation to a random subset of *k*-letter words.

The algorithm for iterative refinement of a locally overrepresented sequence motif was published under the name PATOP.[34] The three different components of the motif description — the borders of the preferred region, the weight matrix, and the cut-off value — are updated

```
Input:

T A T A A A
Cutoff: 2 mismatches

Output:

A:  -3   0 -23   0 -48   0   0   0   0  -1  -5
C:  -5  -1 -22 -41 -32 -52 -51 -31 -29 -11   0
G:  -4  -3 -26 -47 -44 -40 -54 -22 -13   0   0
T:   0  -4   0 -11   0 -23 -13 -42  -6  -7  -8
Cutoff: -74
```

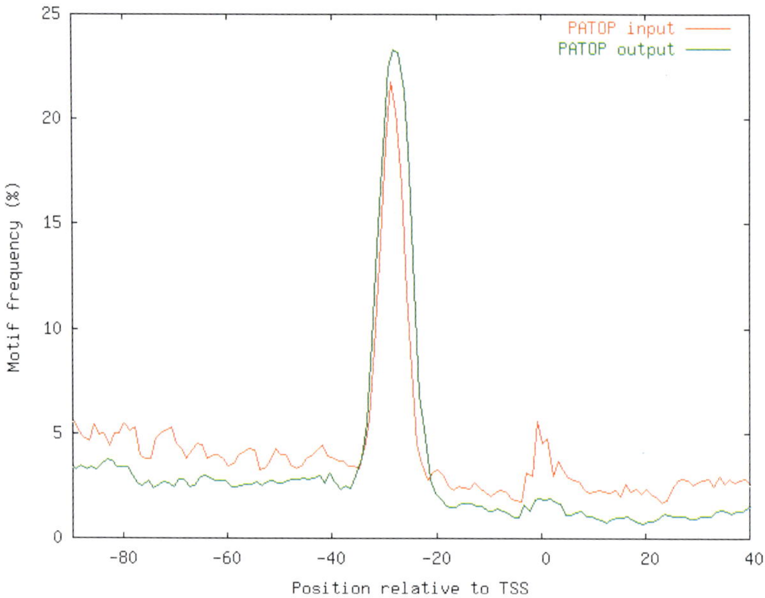

Fig. 6. Optimization of the TATA box motif by the PATOP algorithm. A new weight matrix description for the TATA box motif was optimized using local overrepresentation (LOR), as explained in Fig. 5, as an objective function. A consensus sequence-based motif description was used as the seed, and the 1867 promoter sequences from EPD (see legend for Fig. 4) served as the training set for optimization. The positional distributions of the input and output motifs are shown at the bottom. Note that the optimized weight matrix has both a higher peak frequency near position −30 and a lower background frequency elsewhere.

one at a time, in an alternating fashion. The center position and width of the preferred region, as well as the cut-off value, are optimized in an enumerative fashion based on the previously introduced objective function. User-specified upper and lower bounds and increment values define the search space of all combinations of these three parameters. The weight matrix is updated by counting the base frequencies in the current set of putative motif instances, which is unambiguously defined by the current ensemble of motif parameters. The base frequencies are converted into log-likelihood weights, as detailed in Sec. 2. A powerful additional feature of PATOP is that it can shrink or extend the length of the weight matrix on each iteration. This is achieved by including a number of additional, adjacent positions in the base frequency matrix compiled from the current motif instances. The new limits of the matrix are then defined on the basis of the observed skew of base composition in a matrix column, evaluated by a χ-squared test.

Figure 6 illustrates the effect of the refinement by PATOP with the eukaryotic TATA box as an example. The initial motif consists of the consensus sequence TATAAA with two mismatches allowed. The length of the final matrix is 11 base pairs. The DNA sequences and TSS positions used for refinement correspond to the human promoter subset of the Eukaryotic Promoter Database (EPD), release 93,[36] 1867 sequences in total. The plot shows the motif frequencies, evaluated in overlapping windows of width 8 and 20, respectively. In this example, the gain in local overrepresentation results from an increase in the peak signal frequency and from a decrease in the background frequency. In other words, the resulting optimized weight matrix has both higher sensitivity and higher specificity than the input consensus sequence.

6. Conclusions and Perspectives

The success of motif discovery depends, to a large extent, on the suitability of the input data. Ideally, the data set should consist of a large number of short sequences highly enriched in one particular motif, but otherwise random. In certain application areas, such data sets are available. For instance, the SAGE/SELEX technique[37] can produce thousands of short sequences that bind to a particular transcription factor *in vitro*.

Today, one also has access to large promoter sets defined by high-throughput methods such as expressed sequence tag (EST) sequencing of oligo-capped cDNAs[38] and CAGE.[39] However, in promoter analysis, one still faces the problem that many motifs may only be weakly over-represented and located at a highly variable distance from the TSS. An enrichment of motifs in genomic sequences can be achieved in various ways. For instance, the motif search could be targeted at specific regulatory motifs by restricting the input sequences to promoters of genes that are regulated in a particular way; see, for instance, Roth *et al.*[40] Likewise, genome-wide chromatin immunoprecipitation profiles for more than 200 transcription factors have been used to select yeast promoters that are occupied *in vivo* by a given factor in order to define its cognate binding site motifs with six different motif discovery methods.[41] A presumably very accurate binding site weight matrix has been derived from over 13 000 *in vivo* mapped sites for the insulator protein CTCF.[42] Another currently very trendy approach, exemplified in Xie *et al.*[26] is to restrict the motif search to sequence regions that are conserved across genomes. Specialized motif discovery algorithms, such as PhyloGibbs,[43] can exploit information about phylogenetic conservation contained in a multiple sequence alignment given as input to the method.

The classical weight matrix model, which assumes that motifs are of fixed length and that the contribution of individual bases at different positions are additive and independent of each other, is likely to be an oversimplification in most cases (see Benos *et al.*[44] for a critical discussion of this issue). The weight array method takes nearest-neighbor dependencies into account by scoring overlapping dinucleotides rather than individual bases.[45] Target motifs recognized by multimeric DNA-binding proteins often consist of two or more short motifs separated by spacers of slightly variable length. In fact, the Pribnow box shown in Fig. 1 is one of two conserved motifs characteristic of the major class of *E. coli* promoters. The classical EM algorithm for motif optimization is readily extensible to such bipartite, variable-length motif structures.[46] *Ab initio* discovery of composite motifs (also referred to as transcription regulatory modules), including combinations of motifs that may occur in different order and orientation, is another currently very active research direction; see Kel *et al.*[47] for an example.

A last point worth mentioning is that research on transcription factor binding sites is increasingly based on physical models of protein–DNA interactions and quantitative affinity data. Protein–binding microarrays (PBMs)[48] and related technologies[49] allow measurement of thousands of interaction energies in parallel. The TRAP model,[50] which defines the quantitative relationship between DNA sequence and PBM signal based on a weight matrix, provides a theoretical basis for inference of a binding energy matrix from quantitative affinity data. Likewise, thermodynamic modeling of the SELEX process led to the proposal of a new mathematical formula to convert base probabilities from motif discovery into base pair–protein interaction energies.[51]

There is definitely more work to be done in the field of DNA motif discovery.

References

1. Pribnow D. (1975) Nucleotide sequence of an RNA polymerase binding site at an early T7 promoter. *Proc Natl Acad Sci USA* **72**(3): 784–8.
2. Williams C. (1966) *Adaptation and Natural Selection: A Critique of Some Current Evolutionary Thought*. Princeton, NJ: Princeton University Press, pp. 10–11.
3. Schneider TD, Stormo GD, Gold L, Ehrenfeucht A. (1986) Information content of binding sites on nucleotide sequences. *J Mol Biol* **188**(3): 415–31.
4. Dung VV, Giao PM, Chinh NN *et al.* (1993) A new species of living bovid from Vietnam. *Nature* **363**: 443–5.
5. Berg OG, von Hippel PH. (1987) Selection of DNA binding sites by regulatory proteins. Statistical-mechanical theory and application to operators and promoters. *J Mol Biol* **193**(4): 723–50.
6. Selassie CD, Mekapati SB, Verma RP. (2002) QSAR: then and now. *Curr Top Med Chem* **2**(12): 1357–79.
7. Stormo GD, Schneider TD, Gold L, Ehrenfeucht A. (1982) Use of the 'Perceptron' algorithm to distinguish translational initiation sites in *E. coli*. *Nucleic Acids Res* **10**(9): 2997–3011.
8. Durbin R, Eddy S, Krogh A, Mitchison G. (1998) *Biological Sequence Analysis: Probabilistic Models of Proteins and Nucleic Acids*. Cambridge, UK: Cambridge University Press, pp. 46–79.
9. Sandve GK, Drablos F. (2006) A survey of motif discovery methods in an integrated framework. *Biol Direct* **1**: 11.
10. Das MK, Dai HK. (2007) A survey of DNA motif finding algorithms. *BMC Bioinformatics* **8**(Suppl 7): S21.

11. Cornish-Bowden A. (1985) Nomenclature for incompletely specified bases in nucleic acid sequences: recommendations 1984. *Nucleic Acids Res* **13**(9): 3021–30.
12. Staden R. (1984) Computer methods to locate signals in nucleic acid sequences. *Nucleic Acids Res* **12**(1 Pt 2): 505–19.
13. Quandt K, Frech K, Karas H. (1995) MatInd and MatInspector: new fast and versatile tools for detection of consensus matches in nucleotide sequence data. *Nucleic Acids Res* **23**(23): 4878–84.
14. Bailey TL, Williams N, Misleh C, Li WW. (2006) MEME: discovering and analyzing DNA and protein sequence motifs. *Nucleic Acids Res* **34**(Web Server issue): W369–73.
15. van Helden J, Andre B, Collado-Vides J. (1998) Extracting regulatory sites from the upstream region of yeast genes by computational analysis of oligonucleotide frequencies. *J Mol Biol* **281**(5): 827–42.
16. Bailey TL, Elkan C. (1994) Fitting a mixture model by expectation maximization to discover motifs in biopolymers. *Proc Int Conf Intell Syst Mol Biol* **2**: 28–36.
17. Pevzner PA, Borodovsky M, Mironov AA. (1989) Linguistics of nucleotide sequences. I: The significance of deviations from mean statistical characteristics and prediction of the frequencies of occurrence of words. *J Biomol Struct Dyn* **6**(5): 1013–26.
18. Atteson K. (1998) Calculating the exact probability of language-like patterns in biomolecular sequences. *Proc Int Conf Intell Syst Mol Biol* **6**: 17–24.
19. Staden R. (1989) Methods for calculating the probabilities of finding patterns in sequences. *Comput Appl Biosci* **5**(2): 89–96.
20. Queen C, Wegman MN, Korn LJ. (1982) Improvements to a program for DNA analysis: a procedure to find homologies among many sequences. *Nucleic Acids Res* **10**(1): 449–56.
21. Pavesi G, Mereghetti P, Mauri G, Pesole G. (2004) Weeder Web: discovery of transcription factor binding sites in a set of sequences from co-regulated genes. *Nucleic Acids Res* **32**(Web Server issue): W199–203.
22. Carlson JM, Chakravarty R, Khetani RS, Gross RH. (2006) Bounded search for de novo identification of degenerate cis-regulatory elements. *BMC Bioinformatics* **7**: 254.
23. Lawrence CE, Reilly AA. (1990) An expectation maximization (EM) algorithm for the identification and characterization of common sites in unaligned biopolymer sequences. *Proteins* **7**(1): 41–51.
24. Lawrence CE, Altschul SF, Boguski MS *et al.* (1993) Detecting subtle sequence signals: a Gibbs sampling strategy for multiple alignment. *Science* **262**(5131): 208–14.
25. Ohler U, Liao GC, Niemann H, Rubin GM. (2002) Computational analysis of core promoters in the Drosophila genome. *Genome Biol* **3**(12): RESEARCH0087.

26. Xie X, Mikkelsen TS, Gnirke A *et al.* (2007) Systematic discovery of regulatory motifs in conserved regions of the human genome, including thousands of CTCF insulator sites. *Proc Natl Acad Sci USA* **104**(17): 7145–50.
27. Coward E. (1999) Shufflet: shuffling sequences while conserving the *k*-let counts. *Bioinformatics* **15**(12): 1058–9.
28. Copley RR, Ponting CP, Schultz J, Bork P. (2002) Sequence analysis of multidomain proteins: past perspectives and future directions. *Adv Protein Chem* **61**: 75–98.
29. Hu J, Li B, Kihara D. (2005) Limitations and potentials of current motif discovery algorithms. *Nucleic Acids Res* **33**(15): 4899–913.
30. Tompa M, Li N, Bailey TL *et al.* (2005) Assessing computational tools for the discovery of transcription factor binding sites. *Nat Biotechnol* **23**(1): 137–44.
31. Wingender E. (2008) The TRANSFAC project as an example of framework technology that supports the analysis of genomic regulation. *Brief Bioinform* **9**(4): 326–32.
32. Gama-Castro S, Jiménez-Jacinto V, Peralta-Gil M *et al.* (2008) RegulonDB (version 6.0): gene regulation model of *Escherichia coli* K-12 beyond transcription, active (experimental) annotated promoters and Textpresso navigation. *Nucleic Acids Res* **36**(Database issue): D120–4.
33. Bucher P, Bryan B. (1984) Signal search analysis: a new method to localize and characterize functionally important DNA sequences. *Nucleic Acids Res* **12**(1 Pt 1): 287–305.
34. Bucher P. (1990) Weight matrix descriptions of four eukaryotic RNA polymerase II promoter elements derived from 502 unrelated promoter sequences. *J Mol Biol* **212**(4): 563–78.
35. Bucher P, Trifonov EN. (1986) Compilation and analysis of eukaryotic POL II promoter sequences. *Nucleic Acids Res* **14**(24): 10009–26.
36. Schmid CD, Perier R, Praz V, Bucher P. (2006) EPD in its twentieth year: towards complete promoter coverage of selected model organisms. *Nucleic Acids Res* **34**(Database issue): D82–5.
37. Roulet E, Busso S, Camargo AA *et al.* (2002) High-throughput SELEX–SAGE method for quantitative modeling of transcription-factor binding sites. *Nat Biotechnol* **20**(8): 831–5.
38. Wakaguri H, Yamashita R, Suzuki Y *et al.* (2008) DBTSS: database of transcription start sites, progress report 2008. *Nucleic Acids Res* **36**(Database issue): D97–101.
39. Kawaji H, Kasukawa T, Fukuda S *et al.* (2006) CAGE Basic/Analysis Databases: the CAGE resource for comprehensive promoter analysis. *Nucleic Acids Res* **34**(Database issue): D632–6.
40. Roth FP, Hughes JD, Estep PW, Church GM. (1998) Finding DNA regulatory motifs within unaligned noncoding sequences clustered by whole-genome mRNA quantitation. *Nat Biotechnol* **16**(10): 939–45.

41. Harbison CT, Gordon DB, Lee TI *et al.* (2004) Transcriptional regulatory code of a eukaryotic genome. *Nature* **431**(7004): 99–104.
42. Kim TH, Abdullaev ZK, Smith AD *et al.* (2007) Analysis of the vertebrate insulator protein CTCF-binding sites in the human genome. *Cell* **128**(6): 1231–45.
43. Siddharthan R, Siggia ED, van Nimwegen E. (2005) PhyloGibbs: a Gibbs sampling motif finder that incorporates phylogeny. *PLoS Comput Biol* **1**(7): e67.
44. Benos PV, Bulyk ML, Stormo GD. (2002) Additivity in protein–DNA interactions: how good an approximation is it? *Nucleic Acids Res* **30**(20): 4442–51.
45. Zhang MQ, Marr TG. (1993) A weight array method for splicing signal analysis. *Comput Appl Biosci* **9**(5): 499–509.
46. Cardon LR, Stormo GD. (1992) Expectation maximization algorithm for identifying protein-binding sites with variable lengths from unaligned DNA fragments. *J Mol Biol* **223**(1): 159–70.
47. Kel A, Konovalova T, Waleev T *et al.* (2006) Composite Module Analyst: a fitness-based tool for identification of transcription factor binding site combinations. *Bioinformatics* **22**(10): 1190–7.
48. Berger MF, Bulyk ML. (2006) Protein binding microarrays (PBMs) for rapid, high-throughput characterization of the sequence specificities of DNA binding proteins. *Methods Mol Biol* **338**: 245–60.
49. Maerkl SJ, Quake SR. (2007) A systems approach to measuring the binding energy landscapes of transcription factors. *Science* **315**(5809): 233–7.
50. Roider HG, Kanhere A, Manke T, Vingron M. (2007) Predicting transcription factor affinities to DNA from a biophysical model. *Bioinformatics* **23**(2): 134–41.
51. Djordjevic M. (2007) SELEX experiments: new prospects, applications and data analysis in inferring regulatory pathways. *Biomol Eng* **24**(2): 179–89.

Chapter 2

Comparative Genome Analysis

Robert M. Waterhouse, Evgenia V. Kriventseva
and Evgeny M. Zdobnov

1. Introduction

Comparative genomics takes advantage of whole-genome-scale sequencing projects to develop strategies to interpret patterns of natural genome sequence variations. It strives to understand the principles of molecular evolution and recognize functional elements encoded in genomes. The analyses rely on both intragenomic and cross-species comparisons to elucidate and quantify the evolutionary processes acting on genomes and how they may translate into functions and phenotypes. The large-scale nature of genomic analyses, as well as the rapidly increasing number of sequenced genomes, necessitates a computational approach to the management and interrogation of the data. Comparative genomics has revealed a great deal about the repertoire of protein-coding genes, has helped to mature our understanding of certain features such as alternative splicing of encoded proteins, regulation of gene expression, and stability of genome architecture. Nevertheless, there often appear to be even more questions raised than answered. For example, our understanding of the repertoires of non-protein-coding RNA genes or conserved noncoding sequences remains generally more limited. In conjunction with emerging functional genomics data generated through the interrogation of molecular functions at the genome scale, comparative genomics goes beyond the level of describing trends in feature similarities and differences to the level of molecular systems such as pathways, regulatory networks, and macromolecular complexes.

The availability of whole-genome-scale data provides the opportunity to identify the complete set of functional elements, and hence enables the recognition not only of what is present but also of what is absent from the genomes. This is an essential basis for many comparative techniques, for example, to decipher gene genealogies (orthology). A prerequisite, therefore, to higher-level comparative studies is the comprehensive annotation of genomes for genes and pseudogenes, both protein-coding and non-protein-coding RNAs, regulatory binding sites, conserved non-coding sequences, and repetitive elements. Conversely, comparative approaches provide evidence of evolutionary selection to guide the refinement of genome annotations. Thus, the two approaches are inextricably linked, as recently exemplified by the pilot ENCODE (ENCyclopedia Of DNA Elements) project[1] and the analysis of 12 *Drosophila* genomes.[2,3]

Although many basic biological processes are similar between human and model organisms, the detailed biology becomes less comparable between more distantly related organisms. It is therefore important to not only develop the understanding of the models as thoroughly as possible, but also establish the common basis and recognize the lineage-specific biology to make informed inferences regarding gene functions and to translate the research into new methods for maintaining human health and for diagnosing and treating diseases. As sequence data continue to be generated at an ever-increasing rate, the power of comparative analyses to illuminate evolutionary processes on a genomic scale and to facilitate the *de novo* discovery of functional elements characterized by specific evolutionary signatures is also growing. At the same time, comparative genomics methodologies continue to evolve and develop, with analytical approaches aimed at different levels of species relatedness, from deep evolutionary phylogenies to genomic variations within a population.

2. Gene Annotation

Despite the RNA origin of life, in modern organisms proteins are often considered chiefly responsible for biological function, and therefore

Table 1. The major genome annotation pipelines.

Organization	Programs	Website
Ensembl	Genscan Exonerate GeneWise Genomewise	www.ensembl.org
NCBI	Gnomon	www.ncbi.nlm.nih.gov
UCSC	BLAT BlastZ MultiZ	www.genome.ucsc.edu
Softberry Inc.	FGENESH FGENESH++ FGENESH_C	www.softberry.com
FlyBase	Genscan Genie BLASTN+Sim4	www.flybase.org

characterization of the gene content and the corresponding proteome is of particular importance. The requirements of large-scale sequence annotations have inspired the development of a handful of major computational pipelines to annotate genomic features (Table 1). This is a complex task, especially when bearing in mind that the mere definition of a gene is still being revised.[4] Although human expert interpretation generally surpasses most automated approaches to accurately predict gene/feature structures, manual curation is time-consuming and cannot be scaled up to keep pace with the ever-increasing rate of sequencing. Thus, while accurate computational feature identification from DNA sequences remains a challenging problem, substantial progress has been achieved, particularly through the exploitation of comparative genomics.

There are two main approaches in sequence analysis: (a) *ab initio* gene prediction methods that rely on statistical analysis of sequence composition to recognize features such as exons and introns; and (b) knowledge-based approaches that rely on available homology to known genes in other organisms as well as other primary data such as organism-specific expresses sequences tags (ESTs) and cDNAs. Generally, the knowledge-based approaches are more accurate when there is a sufficient amount of prior experimental data, while *ab initio* methods provide valuable unbiased information to complete the annotations. Dual or multiple genome comparisons are informative when reliable DNA-level alignments of orthologous genomic regions are available (e.g. from MULTIZ[5] or VISTA[6] pipelines), as they can be directly used to measure selective constraints which facilitate the recognition of functionally important sequences and motifs. As all gene prediction approaches have different advantages and trade-offs, new strategies have been designed to weigh up the sources of evidence and to combine the different gene predictions into a set of consensus gene models, e.g. GLEAN,[7] GeneComber[8] and JIGSAW.[9] Assessing the accuracy of the various approaches requires a gold standard annotation set against which to measure predictive performance in terms of sensitivity and specificity. Consistent benchmarking of the major gene prediction methods has recently been undertaken within the framework of the EGASP[10] (Fig. 1) and NGASP (http://www.wormbase.org/wiki/index.php/NGASP) initiatives (ENCODE and Nematode Genome Annotation Assessment Projects, respectively).

Taking advantage of the wealth of evolutionary information inherent in multi-species genomic comparisons can even challenge the gold standard of *Drosophila melanogaster* annotations,[2,3] despite the intensive human expert curation and experimental validation over many years of research. This type of comparative approach is even more critical for identification of the normally relatively elusive noncoding RNAs, the sequence conservation of which is fast diluting despite structural conservation. Although the applicability of such DNA-level comparative approaches may be limited for distantly related species, they are nonetheless becoming increasingly important as more genomic data become available and a new generation of sequence alignment and prediction tools emerges. The rapidly accumulating genome sequence data greatly facilitate the discovery and accurate annotation of genes and other functional genomic elements, while simultaneously highlighting the vital importance of computational approaches. Although comparative studies provide an expectation of the total gene count (see Fig. 3), any exact number is likely to be unreliable, as exemplified by the fact that the human gene count has been constantly and substantially revised over the

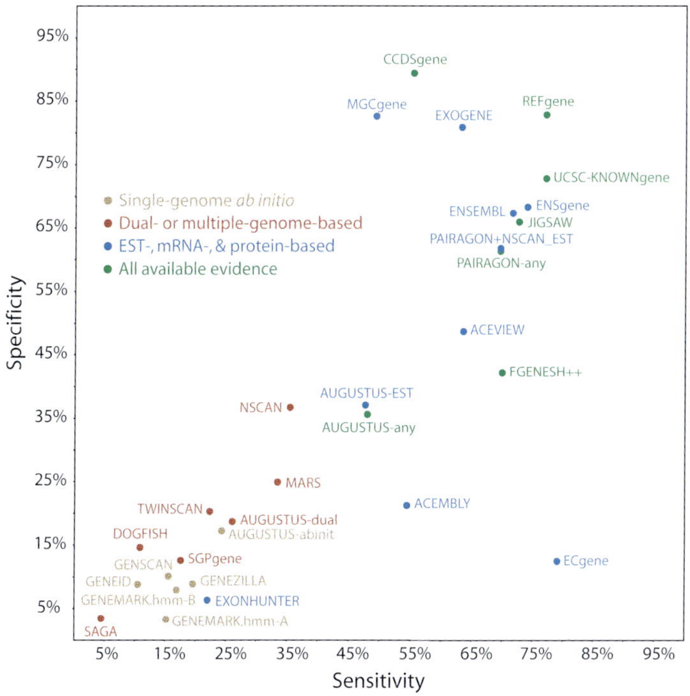

Fig. 1. The human ENCODE Genome Annotation Assessment Project (EGASP)[10] benchmarking analysis of the gene prediction accuracy (sensitivity versus specificity at the gene level) of some of the most popular gene prediction software packages. Knowledge-based approaches, which incorporate multiple lines of evidence, tend to perform better than *ab initio* methods, which can nevertheless provide complementary information for more complete genome annotations.

last several years despite the application of extensive resources. As it is not feasible to allocate comparable resources to characterize other incoming eukaryotic genomes, including some of key medical importance, the major hope remains to improve computational approaches and combine them with high-throughput experimental screens.[11]

3. Protein Families

Homology is a general concept in biology that recognizes similarities between entities as evidence of shared ancestry. This premise is particularly powerful when applied to protein sequence analysis, as the observed species sampling of sequences is usually much smaller than the total random combinations of all amino acids, thereby enabling statistically significant inference of common ancestry. With the aim of tracing evolutionary relations among genes and inferring putative functions, many techniques have been developed for sequence comparisons and assessment of the statistical significance of the homology, ranging from pairwise comparisons such as the renowned Smith–Waterman and BLAST algorithms to profiles of multiple-sequence alignments including position-specific scoring matrices and hidden Markov models. Shared functions and ancestries are confined to groups of similar genes, and homology assessment techniques can therefore be used to identify such protein families. Importantly, the approaches to define such groups are dependent on the objectives, particularly with respect to functional domain or whole gene length analyses. For example, a particular protein function could be associated with a specific stretch of amino acids, or a domain, which can be shuffled among different genes through evolution. Recognizing the sequence characteristics of this domain would fulfill an objective to group all genes with potential for this function; however, not all of the genes may share the ancestry outside of the given domain and an evolutionary objective may require approaches based on whole-length gene comparisons.

The approaches to define protein families on the basis of domains rely on comparative data, where the pattern of selection highlighting the identity of key amino acids is captured from the multiple sequence alignment. This is often achieved primarily through expert human curation, as exemplified by the PROSITE,[12] SMART,[13] Pfam,[14] SCOP,[15] and CATH[16] databases. Many such resources have joined efforts to coordinate domain annotation through the umbrella InterPro project,[17] and the unified InterProScan[18] software has been used to compare protein families for a number of genome projects. Without prior knowledge of protein domains, definition of protein families may be achieved though unsupervised clustering methods applied to all-against-all sequence comparisons. As the number of required comparisons scales dramatically with the size of the dataset, tentative cluster representatives may be used in order to reduce the number of comparisons. Homology significance scoring such as statistical error expectation values, raw similarity scores, and percentage of amino acid identities are often used to define the family cut-off thresholds that affect the granularity (tightness) of the resulting clusters. A range of cut-offs may be used to build a hierarchy of clusters so as to delegate the problem of choosing a biologically relevant cut-off to follow-up examinations. In practice, the BLASTClust utility from the BLAST package offers a straightforward solution through the implementation of single-linkage (nearest-neighbor) clustering to the BLAST scores from all-against-all sequence comparisons. More complex, derived comparison scores and different clustering techniques have been applied in projects like SYSTERS[19] and CluSTr,[20] and alternative clustering algorithms are available through specialized software like the OC program[21] or more general packages like MATLAB®. An important distinction between domain-based and unsupervised clustering techniques of gene family definition is that the former allow one gene to be classified into several groups, while the latter usually restrict gene membership to only one group.

From a comparative genomics perspective, gene family novelties and extinctions as well as significant size differences resulting from expansions or contractions of gene copy numbers can point to interesting lineage-specific biology. Indeed, comparative analyses of olfactory and other chemosensory receptors in humans and mice revealed large variations in gene family sizes that might be associated with the different lifestyles.[22] Changes in gene family sizes over evolutionary time are affected by the essentially random processes of gene duplication and loss, e.g. through pseudogenization. The development of statistical models that consider phylogenetic relationships and population genetics is essential to confidently identify any deviations from a stochastic background. Analysis of gene family dynamics in terms of duplication and pseudogenization frequencies show that functional characteristics, such as essentiality, can be predictive of evolutionary features and vice versa.[23] Gene families containing at least one essential gene are subject to stronger purifying selection than those without any essential genes; they survive longer and consequently may become more divergent in terms of sequence and upstream regulatory regions. Families without essential genes appear more dynamic, with higher rates of both fixation and pseudogenization of duplicated genes. Comparative genomics therefore provides an insight into the dynamics of gene family evolution, guiding the formation of hypotheses that can be tested experimentally to dissect the roles of different protein classes in core processes as well as in lineage-specific biology.

4. Orthologs and Paralogs

In comparative genomics, it is important to distinguish between two types of homologous genes: orthologs and paralogs.[24,25] Orthologs are genes in different species separated by the speciation event (vertical descent) from a single gene of the last common ancestor. It is therefore likely that orthologs retain the functions of the ancestral gene. Paralogs are related by gene duplication events in one lineage, and some of these additional copies may be more flexible to change their molecular functions. If duplication occurs after the speciation event, however, the arising paralogs are still also orthologous to the genes in the other species and are therefore co-orthologs (Fig. 2). In practice, such pairwise definitions are frequently expressed as one-to-one, one-to-many, or many-to-many relationships with respect to the copy number of the co-orthologs in each species. When several species are examined, all co-orthologs are considered together as one orthologous group. The explicit reference in definition relative to the last common ancestor of the considered species defines the hierarchical nature of orthologous classifications along the phylogenetic species tree. Examining closely related species produces many fine-grained orthologous groups of mostly one-to-one relations; while, inversely, considering more distantly related species results in fewer, more general (inclusive) orthologous groups containing all of the descendants of the ancestral gene. Although conservation of ortholog functions is not part of the definition, it is the most likely evolutionary scenario and provides a strong working hypothesis, particularly when the orthologs are preserved over a long period of time in single copy. Paralogs, however, are generally believed to perform distinct (though often related) functions.[26] The accurate detection of orthologous and paralogous relations among homologous genes is therefore particularly important for confident functional interpretations.

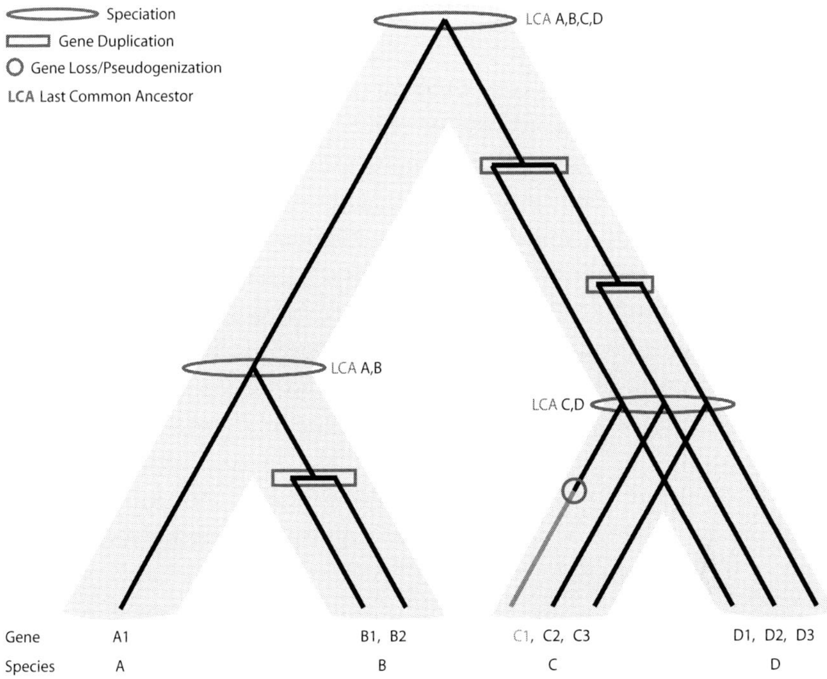

Fig. 2. Orthologous and paralogous relationships are defined by speciation and duplication events along the evolutionary path since divergence from the last common ancestor (LCA). With respect to the LCA A, B, C, D, all of the genes are co-orthologs; while with respect to the LCA C, D, there are two paralogous pairs of orthologs — C2, D2 and C3, D3 — as they were derived through a duplication event before the speciation of C and D.

In practice, delineation of orthologous relations is an intricate procedure, as it assumes knowledge of the ancestral states of the genes and requires knowledge of the complete gene repertoires. The processes of gene duplication, fusion, and shuffling, as well as pseudogenization and loss, further complicate the process, particularly with complex eukaryotic genomes. On the one hand, the ever-increasing number of completely sequenced genomes facilitates a much clearer resolution of the gene genealogies; but on the other hand, it greatly increases the computational challenges involved. There are two main approaches to delineate orthologous genes: (a) through the clustering of all-against-all pairwise gene sequence comparisons in complete genomes, and (b) using a phylogenetic framework to construct trees of all homologous genes and then reconcile these with the species phylogeny.

Identifying orthologs from gene sequence comparisons usually relies on the clustering of genes around reciprocal best hits (RBHs, also known as SymBets or best–best hits; denoting genes most similar to each other in between-genome comparisons), first introduced by the database of Clusters of Orthologous Groups (COGs).[27] Triggered by the earlier availability of much smaller and simpler bacterial genomes, the database has quickly gained wide recognition and has been extended to eukaryotic (KOGs) and archaeal (arCOGs) genomes.[28] The concept of RBHs can be interpreted in phylogenetic terms as genes from different species with the shortest connecting path over the distance-based tree. The identification of RBHs is currently widely adopted in comparative genomics for its simplicity and feasibility of application to large-scale data; however, RBH analysis in its simplest form using BLAST suffers from inaccuracies of sequence distance estimates and ignores many gene duplications after the speciation that are, in fact, co-orthologs. The inclusion of such co-orthologs can be achieved through a further step to identify genes that are more similar to the members of the RBH set in intragenome comparisons than to any other gene in the other genomes, as adopted, for example, in InParanoid/MultiParanoid,[29,30] OrthoDB,[31] and eggnog.[32] Notable alternative methodologies include a probabilistic clustering approach of OrthoMCL,[33] and the use of additional gene orthology evidence from the consideration of orthologous chromosomal regions (synteny) in BUS[34] and SYNERGY,[35] which although substantial in yeast and slowly evolving vertebrates is not very helpful, for example, in distantly related insect species.

The phylogenetic framework approach takes advantage of the well-quantified models of amino acid substitutions in the conserved cores of globular proteins to estimate evolutionary distances among genes, and to reconcile gene trees with the species phylogeny. A notable example of this tree-based approach to delineate orthologous genes is TreeFam,[36,37] adopted recently by Ensembl.[38] There are also several hybrid methods, relying on phylogenetic methods to estimate pairwise evolutionary gene distances followed by their clustering employing methods similar to those described above, such as OPTIC,[39] PhIGs,[40] and RSD.[41] Given the appropriate data, phylogenetic methods are likely to produce more accurate models of ancestral sequences and should therefore yield more accurate orthology predictions.

Appreciation of the important concept of the implicit hierarchy of orthologous relations has prompted the development of several approaches: the LOFT[42] (Levels of Orthology From Trees) package, which interactively interprets gene trees in the context of species trees; the PHOG[43] approach, which resolves orthology at each taxonomy node using explicit modeling of the ancestral genomes with sequence profiles; the eggNOG[32] database, which provides orthologous groups defined at a few major clades of organisms; and OrthoDB,[31] which delineates orthologous groups at each radiation node of the species phylogeny and provides navigation along the tree hierarchy through an interactive Web interface. Indeed, the precise delineation of such evolutionary relationships is integral to comparative genomics, where the perspective, in terms of the range of species considered, defines the types of questions that can be asked (Table 2).

Interrogating orthologous group data by phylogenetic gene copy number profiles facilitates the identification of groups with species- or lineage-specific losses or expansions as well as the identification of common single-copy genes. Such single-copy orthologs are most likely to retain the same ancestral function and therefore evolve under similar evolutionary constraints, making them an ideal data set for evolutionary studies such as the estimation of rates of molecular evolution along lineages and the dating of their radiation times (Fig. 3). Measures of genome divergence that are not obviously related, such as protein identity of orthologs, conservation of their genomic synteny, and rates of gene losses, are in fact well correlated,[44] indicating that global factors affect the fixation rate of different kinds of mutations in a population. For example, insects evolve about two to three times faster than vertebrates, resulting in both more divergent orthologs and more gene losses. Interestingly, even single-copy orthologs can be lost from some species, implying that they are not absolutely indispensable,[45] yet of course they are under substantially stronger selection than multi-copy orthologs and therefore have slower rates of losses. The accurate alignment of one-to-one

Table 2. The resources, features, and links to a selection of public databases, reflecting the principal approaches to the delineation of orthologous and paralogous relationships from large-scale genomic data.

Resource	Features	Website
InParanoid and MultiParanoid	BLAST, RBH, eukaryotes, automatic	http://inparanoid.cgb.ki.se
COG/KOG/arCOG	BLAST, RBH, semi-curated	http://www.ncbi.nlm.nih.gov/COG
OrthoDB	Smith–Waterman, RBH, hierarchical, eukaryotes, automatic	http://cegg.unige.ch/orthodb
Ensembl Compara	Smith–Waterman, ML tree reconciliation, vertebrates, automatic	http://www.ensembl.org
eggNOG	Smith–Waterman, RBH triangles, Semi-hierarchical, automatic	http://eggnog.embl.de
OrthoMCL	BLAST, Markov cluster algorithm, automatic	http://www.cbil.upenn.edu/gene-family
OPTIC	BLAST, Dn/Ds distance, automatic	http://genserv.anat.ox.ac.uk/clades
SYNERGY	FASTA, synteny-guided RBH, tree reconstruction, fungi	http://www.broad.mit.edu/regev/orthogroups

orthologs enables investigations into the evolution of gene structure in terms of the properties of their constituent exons and introns. Probabilistic modeling of intron evolution using orthologs suggests a balance of intron gain and loss across all lineages, with elevated intron gain occurring very early in the eukaryotic phylogeny and several lineages such as fungi and insects later experiencing elevated intron losses.[46]

Gene structure comparisons have also revealed the importance of alternative splicing as a major transcript diversification mechanism in animals to produce different protein isoforms from a single gene. Estimates

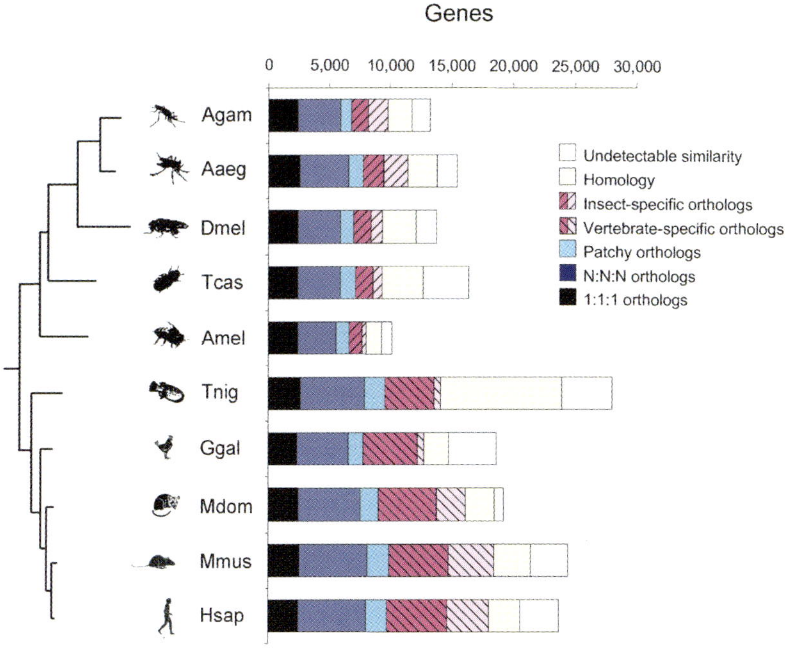

Fig. 3. Comparison of gene repertoires of 5 insect and 5 vertebrate genomes, ranging from the core of metazoan genes (black fraction on the left) to the species-unique sequences (white band on the right). The striped boxes correspond to insect- and vertebrate-specific orthologous genes, where the darker band corresponds to all insects or vertebrates (allowing one loss). "N:N:N" indicates orthologs present in multiple copies (allowing one loss), and "Patchy" indicates ancient orthologs (requiring at least one insect and one vertebrate gene) that have been differentially lost in some lineages. The species tree on the left was computed using the maximum-likelihood approach on concatenated alignments of single-copy orthologs, and it shows an accelerated rate of evolution in insects.

of somewhere between 40% and 75% of bilaterian genes are thought to undergo alternative splicing, with some genes coding for thousands of transcripts. However, human and rodent comparisons revealed at least a quarter of human splice variants that are not present in their orthologous mouse genes, and species-specific isoforms are present for about half of the alternatively spliced genes,[47] suggesting that these nonconserved isoforms may be due to "splicing noise" (or aberrant splicing).[48] Nevertheless, experimental verification has shown support for many minor splice forms with multiple ESTs that also exhibit tissue-specific expression patterns and may in fact be the major transcript in some tissues.

5. Genome Architecture

Although comparative genomics often focuses solely on gene content analysis, the study of genes in the context of their genomic order and chromosomal organization is also important. Whole-genome sequencing efforts have therefore been particularly insightful for studies of genome architectures, revealing duplications, deletions, translocations, inversions, chromosomal splitting and fusion, and even whole-genome duplications. Initially, the term "synteny" was used to denote genetic linkage of genes on the same chromosome; however, the emergence of much more detailed data on genomic gene arrangements has shifted the usage of the term. Nowadays, local conservation of gene arrangements (which are in fact orthologous genomic loci) are frequently termed regions of "microsynteny" or synteny blocks, and the term "macrosynteny" is generally used to describe longer-range orthologous regions that highlight major chromosomal rearrangements and themselves span a number of synteny blocks.

The application of comparative genomics to identify synteny blocks opens new avenues for exploring the likely chain of events of chromosomal rearrangements and for reconstructing ancestral genomes.[49] This can highlight genomic features that may contribute to genomic instability, as in the case of inversions and long-range deletions which can be driven by recombination between regions of high sequence identity, e.g. arising from transposable element activity or segmental duplications.[50] The conservation of gene arrangements by purifying selection can also reveal functional coupling between the genes as, for example, in bacteria, where genes arranged in clustered modules or operons show significant predictive power of gene functions and pathway membership.[51] Such clustering in eukaryotes, however, appears to be far less common, with only a few examples such as the HOX gene clusters, where conservation of gene order has been shown to be important for the tight regulation of gene expression.[52] Less stringent coregulation may affect larger genomic regions, which may be linked to replication properties, as in the case of so-called expression territories containing clusters of coordinately expressed genes identified in *Drosophila*. Indeed, the correlation of gene expression with gene order data indicated that eukaryotic gene arrangements are not entirely random and similar, and/or that coordinated expression patterns are often observed for physically clustered genes.[53] However, besides loosely defined gene territories, there seems to be few evolutionary constraints to preserve orthologous gene arrangements in synteny blocks. Indeed, assessing the conservation of microsynteny among several insect genomes from Diptera to Hymenoptera identified only a few hundred genes that might be linked by selection.[54] Interestingly, the size distribution of synteny blocks was found to follow the power law, which implies a nonuniform distribution of chromosomal breakpoints, i.e. exponentially clustered around breakage hot spots, rather than the commonly assumed random breakage model. It also appears that the rate of rearrangements within chromosomes (chromosomal arms in Diptera) is much higher than between chromosomes, such that orthologous relations between chromosomes can be clearly established by the excess of shared orthologous markers in comparison with a random expectation while synteny blocks gradually become almost randomly scattered along the chromosomes.[55–57]

The delineation of orthologous genomic regions at the level of multiple DNA sequence alignments provides an opportunity to identify much more illusive non-protein-coding functional sequences. For closely related species, such multiple alignments are analyzed for patterns of minor variations and population polymorphisms termed "phylogenetic footprinting" or "shadowing". For more distantly related species, the alignment of orthologous genomic blocks is more complex than that of protein sequences because nucleotides are information-poor compared to amino acids (as there are only four nucleotides and at least 20 common amino acids). For regulatory elements, the information content is generally low due to their short lengths; and for non-protein-coding RNA genes, sequence conservation is weak as structural properties of base pairing are more important. More crucially, the dynamic alignment approaches designed for protein sequence analysis cannot cope with sequence rearrangements such as duplications, inversions, and transpositions, which are common among genomic sequences. The effectiveness of DNA alignments to detect orthologous genomic regions (synteny blocks) therefore rapidly deteriorates with the rising level of divergence between the species being compared.

In practice, DNA-level whole-genome-scale alignments can be performed reasonably well when conservation is such that there are still relatively long almost-identical substrings, which are therefore unique and can be used as orthologous anchors, as implemented in CHAOS[58] or MUMmer.[59] These orthologous anchors can then be used to either guide local alignment algorithms like BLASTZ,[60] an optimized version of gapped BLAST, or define synteny blocks, as applied in Cinteny,[61] and employ global (e.g. LAGAN[62]) or semiglobal alignments (e.g. so-called "glocal" combinations of global and local methods introduced in Shuffle-LAGAN[62]). These tools, however, already begin to reach their limit at the divergence level of, for example, chicken and human or distant *Drosophila* species, when the orthologous DNA signals are mostly derived from separate exons and barely distinguishable from random noise. The delineation of synteny blocks across deeper phylogenies can therefore benefit from the use of protein translations of predicted genes to resolve complex orthologous relations. In addition, instead of conventional gap opening and extension penalties, the explicit modeling of insertions and deletions as proposed in MCALIGN[63] seems to produce more realistic alignments of noncoding regions.

While many genome arrangement variations are to some degree tolerated in populations such as segmental duplications, which lead to copy number variants, or long-scale deletions more frequently associated with genetic diseases, the most radical change of genome architecture is arguably that of whole-genome duplications (WGDs). A few WGDs have been documented in yeast and vertebrate lineages, and have contributed significantly to the current repertoire of genes and species. Current understanding of the effects of WGDs predicts a drastic loss of genes following a WGD event by pseudogenization, which subsequently slows down.[64] The recent comparative analysis of the genome of the unicellular ciliated eukaryote, *Paramecium*, revealed at least three successive WGD events, where the most recent duplication was suggested as the driving force of speciation due to incompatible differential gene losses that gave rise to a complex of 15 sibling species.[65] Functionally linked genes — genes in the same protein complex or metabolic pathway — exhibited common patterns of loss, and genes with high expression levels were less likely to be lost, implying that the retention of duplicated genes is influenced far more by important gene dosage effects than by the processes of subfunctionalization or neofunctionalization.

In addition to the dynamics of gene repertoires and their arrangements, there is a continual gain and loss of genomic DNA that appears to be mostly responsible for the extensive variation in genome sizes of even relatively closely related species. The gain of DNA is chiefly attributed to the amount of repetitive DNA arising from the activity of "selfish" or "parasitic" transposable elements (TEs), which are counterbalanced by the efficacy of the host's defense and clearance systems including the inhibition of transposition by RNAi[66] and cytosine methylation.[67] The outcome of the battle between the host genome and the mobile genetic elements is variable and depends on the history of the species population dynamics.[68] For example, recognizable TE sequences make up almost half of the genome of the dengue/yellow fever mosquito, *Aedes aegypti*, and are thought to have made a major contribution to the approximately fivefold larger genome size in comparison to the malaria mosquito, *Anopheles gambiae*. Much of the remaining species-specific junk DNA may in fact consist of ancient disabled TE copies that have diverged beyond recognition. In fact, there are usually only a handful of active TE elements and the rest are the neutrally decaying disabled copies, which are generally assumed to be unconstrained by selection, therefore providing an opportunity to calibrate an intragenomic molecular clock for comparative studies. The random insertion of proliferating TEs and their subsequent random pseudogenization lead to the effect of active elements jumping around the genome from generation to generation. This can be exploited to recognize genes originating from TEs whose function might have been recruited by the host genome, where the purifying selection of their function has kept these genes intact and immobile (thereby preserving synteny) over a long evolutionary timescale.[69] TE activity is mostly deleterious, but it provides a rich source of sequences for evolutionary innovations and may sometimes duplicate whole genes, create chimeric genes through exon shuffling, or influence gene expression by altering regulatory regions.[70] It also leads to increased frequency of recombination between nonhomologous regions of chromosomes, causing deletions, duplications, inversions, and translocations. TEs are therefore an important generator of genomic variation upon which natural selection may act.[68,71]

6. RNA Genes and Conserved Noncoding Sequences

One of the surprises of the genomics era that came with the sequencing of the human genome, followed by the mouse and rat genomes, is that only about 2% of the mammalian genomic DNA encodes proteins. The remaining regions, including TEs, without any apparent value to the host genomes were initially designated as junk DNA. Today, however, closer comparative analyses of orthologous genomic regions in multiple species have highlighted a substantial number of conserved non-protein-coding sequences, illustrating the power (and limitations) of computational comparative genomics. The recent discovery and appreciation of the biological importance of microRNA genes (miRNAs) has highlighted the need to identify and characterize non-protein-coding RNA genes (ncRNAs) beyond the well-known transfer and ribosomal RNAs, as well as the generally termed conserved noncoding sequences (CNSs) including the ultraconserved elements.[72] The pilot ENCODE project[1] now estimates that at least about 4%–7% of the human genome is under purifying selection, representing a much larger functional fraction and suggesting that the repertoire of poorly characterized functionally important elements is at least as large as the repertoire of protein-coding genes.

Knowledge of ncRNAs before whole-genome sequencing projects was limited to a handful of RNA genes including tRNAs, which transfer specific amino acids onto polypeptide chains during translation, and rRNAs, which constitute the structural basis and the peptidyl transferase activity of ribosomes. The prediction of genes encoding these well-known RNA genes, however, remains nontrivial. rRNA genes are found in multiple-copy clusters, where they experience concerted evolution keeping them highly uniform in sequence, which frequently presents a problem when assembling these genomic regions. As rRNA genes are universally present across the different domains of life, they have been widely used in phylogenetic reconstructions and taxonomic classifications; however, such studies suffer from biases of the strong structural and sequence constraints on their evolution. Studies of these classical RNA genes have contributed significantly to the development of current approaches to generalized RNA structure prediction and sequence analysis. Indeed, the analysis of multiple RNA sequence alignments for compensatory mutations (so-called "covariations" that affect sequence while preserving structure) to predict tRNA genes by scanning genomic sequences for characteristic stochastic context-free grammar (SCFG) profiles was first implemented in the tRNAscan-SE software,[73] and has now been generalized in the Infernal package[74] to scan for a variety of other ncRNA profiles. A drawback of this approach, however, is that it still relies heavily on sequence conservation, and will therefore overlook highly divergent or independently evolved ncRNAs, as exemplified in the application to miRNA gene discovery.

A great deal of work has since gone into improving methods to identify these ~22-nucleotide-long RNA molecules, which are derived from ~70-nucleotide-long genome-encoded stem-loop precursors[75] (described in detail in the previous chapter devoted to small regulatory RNAs). Methods relying heavily on sequence conservation criteria with known miRNAs appeared first, yet they were unable to capture previously unknown genes or very divergent homologs. Methods relying only on structural properties in a truly *ab initio* fashion usually identify many more putative miRNAs, but suffer from high rates of false positives. Using evolutionary conservation, however, improves the specificity of such methods, in an analogous manner to protein-coding genes. The effectiveness of such comparative approaches for miRNA identification was exemplified through the analysis of 12 *Drosophila* genomes, where the discovery power was shown to scale with the number and divergence of the species analyzed.[76] The screening of multiple alignments of orthologous DNA for patterns of compensatory mutations that preserve putative RNA secondary structure facilitates the identification of novel RNA genes.[77] This will hopefully shed some light on the possible functionality of the significant fractions of eukaryotic genomes that are transcribed into RNA, but which remain mostly unannotated and have reduced protein-coding potential.[78,79] Much of such "noisy" expression indeed appears to constitute long primary transcripts for the production of much shorter mature RNAs, e.g. a circadianly expressed 3-kb human transcript that seems to encode only one miRNA (has-mir-122).

ncRNA genes are subject to similar evolutionary processes as protein-coding genes, leading to gene duplication, diversification, pseudogenization, and loss. In a similar manner to the analysis of protein orthology, the identification of losses of ncRNA genes which are widely conserved among other organisms can be indicative of species-specific traits.

Already, early comparisons of nucleotide sequences from divergent vertebrate species identified highly conserved stretches in noncoding regions of genes with more than 70% identity over more than 100 nucleotides — an unexpected observation considering the much shorter and less specific sequence properties required for the binding of regulatory proteins.[80] The true extent of such conservation only became apparent with the sequencing and comparison of multiple mammalian genomes, where applying the same definition of conservation to human–mouse alignments identified hundreds of thousands of such CNSs constituting about 1%–2% of the genomes.[81] Interestingly, there is a higher incidence of CNSs within gene-poor regions (so-called "gene deserts"), and these sequences are not repetitive and do not share easily identifiable sequence features. The term "noncoding" was suggested to denote that there was no evidence of expression for many experimentally and computationally scrutinized CNSs on human chromosome 21. The hypothesis that CNSs merely represent regions with lower local mutation rates (mutational "cold spots") was recently rejected by the analysis of allele frequency distributions from HapMap genotype data in humans, proving that these CNSs are selectively constrained and therefore should be functional.[82] A subset of these sequences comprises the so-called ultraconserved elements that have remained mostly intact since the split of mammals and chicken and even fish.[83] Hundreds of such deeply conserved sequences have been experimentally tested within the framework of the VISTA Enhancer project, and almost half of them showed tissue-specific enhancer activity. Recent analysis of the opossum genome revealed that 20% of eutherian CNSs appear to be recent inventions after the divergence from metatheria, and that many of these eutherian-specific CNSs arose from TE sequences.[84] While the understanding is still far from comprehensive, a serious appreciation of the potentially diverse functional load of CNSs and their contribution to animal evolution is beginning to take shape.

7. Perspectives

The revolution of sequencing technologies has prompted an avalanche of molecular data, and computational comparative genomics has become instrumental for its effective interpretation. Indeed, as the latest explosion of metagenomics data from environmental sampling reveals, there is still a vast reservoir of biological diversity to be explored. Evolution appears to proceed through a succession of stochastic events that explore the available opportunities, and thus it is important to keep in mind that many observable features, including biological complexity in general, may arise by chance rather than through positive selection. Dobzhansky's famous thesis that "nothing in biology makes sense except in the light of evolution" has now been extended by Lynch to "nothing in evolution makes sense except in light of population genetics", and his book on the origins of genome architecture would make excellent further reading.[68]

References

1. Birney E, Stamatoyannopoulos JA, Dutta A *et al.* (2007) Identification and analysis of functional elements in 1% of the human genome by the ENCODE pilot project. *Nature* **447**(7146): 799–816.
2. Clark AG, Eisen MB, Smith DR *et al.* (2007) Evolution of genes and genomes on the *Drosophila* phylogeny. *Nature* **450**(7167): 203–18.
3. Stark A, Lin MF, Kheradpour P *et al.* (2007) Discovery of functional elements in 12 *Drosophila* genomes using evolutionary signatures. *Nature* **450**(7167): 219–32.
4. Snyder M, Gerstein M. (2003) Genomics. Defining genes in the genomics era. *Science* **300**(5617): 258–60.
5. Blanchette M, Kent WJ, Riemer C *et al.* (2004) Aligning multiple genomic sequences with the threaded blockset aligner. *Genome Res* **14**(4): 708–15.

6. Dubchak I. (2007) Comparative analysis and visualization of genomic sequences using VISTA browser and associated computational tools. *Methods Mol Biol* **395**: 3–16.
7. Elsik CG, Mackey AJ, Reese JT *et al.* (2007) Creating a honey bee consensus gene set. *Genome Biol* **8**(1): R13.
8. Shah SP, McVicker GP, Mackworth AK *et al.* (2003) GeneComber: combining outputs of gene prediction programs for improved results. *Bioinformatics* **19**(10): 1296–7.
9. Allen JE, Salzberg SL. (2005) JIGSAW: integration of multiple sources of evidence for gene prediction. *Bioinformatics* **21**(18): 3596–603.
10. Guigo R, Flicek P, Abril JF *et al.* (2006) EGASP: the human ENCODE Genome Annotation Assessment Project. *Genome Biol* **7** (Suppl 1): S21–31.
11. Carninci P, Kasukawa T, Katayama S *et al.* (2005) The transcriptional landscape of the mammalian genome. *Science* **309**(5740): 1559–63.
12. Hulo N, Bairoch A, Bulliard V *et al.* (2008) The 20 years of PROSITE. *Nucleic Acids Res* **36**(Database issue): D245–9.
13. Letunic I, Copley RR, Pils B *et al.* (2006) SMART 5: domains in the context of genomes and networks. *Nucleic Acids Res* **34**(Database issue): D257–60.
14. Finn RD, Tate J, Mistry J *et al.* (2008) The Pfam protein families database. *Nucleic Acids Res* **36**(Database issue): D281–8.
15. Andreeva A, Howorth D, Chandonia JM *et al.* (2008) Data growth and its impact on the SCOP database: new developments. *Nucleic Acids Res* **36**(Database issue): D419–25.
16. Greene LH, Lewis TE, Addou S *et al.* (2007) The CATH domain structure database: new protocols and classification levels give a more comprehensive resource for exploring evolution. *Nucleic Acids Res* **35**(Database issue): D291–7.
17. Mulder NJ, Apweiler R, Attwood TK *et al.* (2003) The InterPro Database, 2003 brings increased coverage and new features. *Nucleic Acids Res* **31**(1): 315–8.
18. Zdobnov EM, Apweiler R. (2001) InterProScan — an integration platform for the signature-recognition methods in InterPro. *Bioinformatics* **17**(9): 847–8.
19. Krause A, Stoye J, Vingron M. (2005) Large scale hierarchical clustering of protein sequences. *BMC Bioinformatics* **6**: 15.
20. Kriventseva EV, Fleischmann W, Zdobnov EM, Apweiler R. (2001) CluSTr: a database of clusters of SWISS-PROT+TrEMBL proteins. *Nucleic Acids Res* **29**(1): 33–6.
21. Barton G. (1993, 2002) OC — a cluster analysis program. www.compbio.dundee.ac.uk/downloads/oc/. University of Dundee, Scotland, UK.
22. Niimura Y, Nei M. (2006) Evolutionary dynamics of olfactory and other chemosensory receptor genes in vertebrates. *J Hum Genet* **51**(6): 505–17.
23. Shakhnovich BE, Koonin EV. (2006) Origins and impact of constraints in evolution of gene families. *Genome Res* **16**(12): 1529–36.

24. Fitch WM. (1970) Distinguishing homologous from analogous proteins. *Syst Zool* **19**(2): 99–113.
25. Koonin EV. (2005) Orthologs, paralogs, and evolutionary genomics. *Annu Rev Genet* **39**: 309–38.
26. Lynch M, Katju V. (2004) The altered evolutionary trajectories of gene duplicates. *Trends Genet* **20**(11): 544–9.
27. Tatusov R, Fedorova N, Jackson JD *et al.* (2003) The COG database: an updated version includes eukaryotes. *BMC Bioinformatics* **4**(1): 41.
28. Makarova KS, Sorokin AV, Novichkov PS *et al.* (2007) Clusters of orthologous genes for 41 archaeal genomes and implications for evolutionary genomics of archaea. *Biol Direct* **2**: 33.
29. Berglund AC, Sjolund E, Ostlund G, Sonnhammer EL. (2008) InParanoid 6: eukaryotic ortholog clusters with inparalogs. *Nucleic Acids Res* **36**(Database issue): D263–6.
30. Alexeyenko A, Tamas I, Liu G, Sonnhammer EL. (2006) Automatic clustering of orthologs and inparalogs shared by multiple proteomes. *Bioinformatics* **22**(14): e9–15.
31. Kriventseva EV, Rahman N, Espinosa O, Zdobnov EM. (2008) OrthoDB: the hierarchical catalog of eukaryotic orthologs. *Nucleic Acids Res* **36**(Database issue): D271–5.
32. Jensen LJ, Julien P, Kuhn M *et al.* (2008) eggNOG: automated construction and annotation of orthologous groups of genes. *Nucleic Acids Res* **36**(Database issue): D250–4.
33. Li L, Stoeckert Jr CJ, Roos DS. (2003) OrthoMCL: identification of ortholog groups for eukaryotic genomes. *Genome Res* **13**(9): 2178–89.
34. Kellis M, Birren BW, Lander ES. (2004) Proof and evolutionary analysis of ancient genome duplication in the yeast Saccharomyces cerevisiae. *Nature* **428**(6983): 617–24.
35. Wapinski I, Pfeffer A, Friedman N, Regev A. (2007) Automatic genome-wide reconstruction of phylogenetic gene trees. *Bioinformatics* **23**(13): i549–58.
36. Li H, Coghlan A, Ruan J *et al.* (2006) TreeFam: a curated database of phylogenetic trees of animal gene families. *Nucleic Acids Res* **34**(Database issue): D572–80.
37. Ruan J, Li H, Chen Z *et al.* (2008) TreeFam: 2008 Update. *Nucleic Acids Res* **36**(Database issue): D735–40.
38. Flicek P, Aken BL, Beal K *et al.* (2008) Ensembl 2008. *Nucleic Acids Res* **36**(Database issue): D707–14.
39. Heger A, Ponting CP. (2008) OPTIC: orthologous and paralogous transcripts in clades. *Nucleic Acids Res* **36**(Database issue): D267–70.
40. Dehal PS, Boore JL. (2006) A phylogenomic gene cluster resource: the Phylogenetically Inferred Groups (PhIGs) database. *BMC Bioinformatics* 7: 201.

41. Wall DP, Fraser HB, Hirsh AE. (2003) Detecting putative orthologs. *Bioinformatics* **19**(13): 1710–1.
42. van der Heijden RT, Snel B, van Noort V, Huynen MA. (2007) Orthology prediction at scalable resolution by phylogenetic tree analysis. *BMC Bioinformatics* **8**: 83.
43. Merkeev IV, Novichkov PS, Mironov AA. (2006) PHOG: a database of supergenomes built from proteome complements. *BMC Evol Biol* **6**: 52.
44. Zdobnov EM, von Mering C, Letunic I, Bork P. (2005) Consistency of genome-based methods in measuring Metazoan evolution. *FEBS Lett* **579**(15): 3355–61.
45. Wyder S, Kriventseva EV, Schröder R *et al.* (2007) Quantification of ortholog losses in insects and vertebrates. *Genome Biol* **8**(11): R242.
46. Carmel L, Wolf YI, Rogozin IG, Koonin EV. (2007) Three distinct modes of intron dynamics in the evolution of eukaryotes. *Genome Res* **17**(7): 1034–44.
47. Modrek B, Lee CJ. (2003) Alternative splicing in the human, mouse and rat genomes is associated with an increased frequency of exon creation and/or loss. *Nat Genet* **34**(2): 177–80.
48. Sorek R, Shamir R, Ast G. (2004) How prevalent is functional alternative splicing in the human genome? *Trends Genet* **20**(2): 68–71.
49. Bourque G, Zdobnov EM, Bork P *et al.* (2005) Comparative architectures of mammalian and chicken genomes reveal highly variable rates of genomic rearrangements across different lineages. *Genome Res* **15**(1): 98–110.
50. Hedges DJ, Deininger PL. (2007) Inviting instability: transposable elements, double-strand breaks, and the maintenance of genome integrity. *Mutat Res* **616**(1–2): 46–59.
51. von Mering C, Zdobnov EM, Tsoka S *et al.* (2003) Genome evolution reveals biochemical networks and functional modules. *Proc Natl Acad Sci USA* **100**(26): 15428–33.
52. Soshnikova N, Duboule D. (2008) Epigenetic regulation of Hox gene activation: the waltz of methyls. *Bioessays* **30**(3): 199–202.
53. Hurst LD, Pal C, Lercher MJ. (2004) The evolutionary dynamics of eukaryotic gene order. *Nat Rev Genet* **5**(4): 299–310.
54. Zdobnov EM, Bork P. (2007) Quantification of insect genome divergence. *Trends Genet* **23**(1): 16–20.
55. Zdobnov EM, von Mering C, Letunic I *et al.* (2002) Comparative genome and proteome analysis of *Anopheles gambiae* and *Drosophila melanogaster*. *Science* **298**(5591): 149–59.
56. Honeybee Genome Sequencing Consortium. (2006) Insights into social insects from the genome of the honeybee *Apis mellifera*. *Nature* **443**(7114): 931–49.
57. International Chicken Genome Sequencing Consortium. (2004) Sequence and comparative analysis of the chicken genome provide unique perspectives on vertebrate evolution. *Nature* **432**(7018): 695–716.

58. Brudno M, Morgenstern B. (2002) Fast and sensitive alignment of large genomic sequences. *Proc IEEE Comput Soc Bioinform Conf* **1**: 138–47.
59. Delcher AL, Kasif S, Fleischmann RD *et al.* (1999) Alignment of whole genomes. *Nucleic Acids Res* **27**(11): 2369–76.
60. Schwartz S, Kent WJ, Smit A *et al.* (2003) Human–mouse alignments with BLASTZ. *Genome Res* **13**(1): 103–7.
61. Sinha AU, Meller J. (2007) Cinteny: flexible analysis and visualization of synteny and genome rearrangements in multiple organisms. *BMC Bioinformatics* **8**: 82.
62. Brudno M, Do CB, Cooper GM *et al.* (2003) LAGAN and Multi-LAGAN: efficient tools for large-scale multiple alignment of genomic DNA. *Genome Res* **13**(4): 721–31.
63. Wang J, Keightley PD, Johnson T. (2006) MCALIGN2: faster, accurate global pairwise alignment of non-coding DNA sequences based on explicit models of indel evolution. *BMC Bioinformatics* **7**: 292.
64. Scannell DR, Byrne KP, Gordon JL *et al.* (2006) Multiple rounds of speciation associated with reciprocal gene loss in polyploid yeasts. *Nature* **440**(7082): 341–5.
65. Aury JM, Jaillon O, Duret L *et al.* (2006) Global trends of whole-genome duplications revealed by the ciliate *Paramecium tetraurelia*. *Nature* **444**(7116): 171–8.
66. Vastenhouw NL, Plasterk RH. (2004) RNAi protects the *Caenorhabditis elegans* germline against transposition. *Trends Genet* **20**(7): 314–9.
67. Bestor TH. (2003) Cytosine methylation mediates sexual conflict. *Trends Genet* **19**(4): 185–90.
68. Lynch M. (2007) *The Origins of Genome Architecture*. Sunderland, MA: Sinauer Associates.
69. Zdobnov EM, Campillos M, Harrington ED *et al.* (2005) Protein coding potential of retroviruses and other transposable elements in vertebrate genomes. *Nucleic Acids Res* **33**(3): 946–54.
70. Dooner HK, Weil CF. (2007) Give-and-take: interactions between DNA transposons and their host plant genomes. *Curr Opin Genet Dev* **17**(6): 486–92.
71. Kazazian Jr HH. (2004) Mobile elements: drivers of genome evolution. *Science* **303**(5664): 1626–32.
72. Christley S, Lobo NF, Madey G. (2008) Multiple organism algorithm for finding ultraconserved elements. *BMC Bioinformatics* **9**: 15.
73. Lowe TM, Eddy SR. (1997) tRNAscan-SE: a program for improved detection of transfer RNA genes in genomic sequence. *Nucleic Acids Res* **25**(5): 955–64.
74. Eddy SR. (2006) Computational analysis of RNAs. *Cold Spring Harb Symp Quant Biol* **71**: 117–28.
75. Ambros V. (2004) The functions of animal microRNAs. *Nature* **431**(7006): 350–5.

76. Stark A, Kheradpour P, Parts L *et al.* (2007) Systematic discovery and characterization of fly microRNAs using 12 Drosophila genomes. *Genome Res* **17**(12): 1865–79.
77. Washietl S, Hofacker IL, Lukasser M *et al.* (2005) Mapping of conserved RNA secondary structures predicts thousands of functional noncoding RNAs in the human genome. *Nat Biotechnol* **23**(11): 1383–90.
78. Maeda N, Kasukawa T, Oyama R *et al.* (2006) Transcript annotation in FANTOM3: mouse gene catalog based on physical cDNAs. *PLoS Genet* **2**(4): e62.
79. Kapranov P, Cheng J, Dike S *et al.* (2007) RNA maps reveal new RNA classes and a possible function for pervasive transcription. *Science* **316**(5830): 1484–8.
80. Duret L, Dorkeld F, Gautier C. (1993) Strong conservation of non-coding sequences during vertebrates evolution: potential involvement in post-transcriptional regulation of gene expression. *Nucleic Acids Res* **21**(10): 2315–22.
81. Dermitzakis ET, Reymond A, Antonarakis SE. (2005) Conserved non-genic sequences — an unexpected feature of mammalian genomes. *Nat Rev Genet* **6**(2): 151–7.
82. Drake JA, Bird C, Nemesh J *et al.* (2006) Conserved noncoding sequences are selectively constrained and not mutation cold spots. *Nat Genet* **38**(2): 223–7.
83. Bejerano G, Pheasant M, Makunin I *et al.* (2004) Ultraconserved elements in the human genome. *Science* **304**(5675): 1321–5.
84. Mikkelsen TS, Wakefield MJ, Aken B *et al.* (2007) Genome of the marsupial Monodelphis domestica reveals innovation in non-coding sequences. *Nature* **447**(7141): 167–77.

Chapter 3

From Modules to Models: Advanced Analysis Methods for Large-Scale Data

Sven Bergmann

1. Introduction

Microarrays have firmly established themselves as a standard tool in biological and biomedical research. Together with the rapid advancement of genome sequencing projects, microarrays and related high-throughput technologies have been key factors in the study of the more global aspects of cellular systems biology.[1] While genomic sequence provides an inventory of parts, a proper understanding of the functions and organization of those parts requires comprehensive views of the regulatory relations between them.[2] Genome-wide expression data offer such a global view by providing a simultaneous read-out of the mRNA levels of all (or many) genes in the genome.

Most microarray experiments are conducted to address specific biological issues [Fig. 1(a)]. In the simplest case, such a study may focus on the expression response to the deletion of individual genes or to specific cellular conditions. Already when extending the experimental setup to include several conditions, e.g. time points along the cell cycle[3] or several tissue samples, the sheer amount of data points necessitates computational tools to extract and organize relevant biological information. A wide range of approaches has been developed, including numerous clustering algorithms, statistical methods for detecting differential

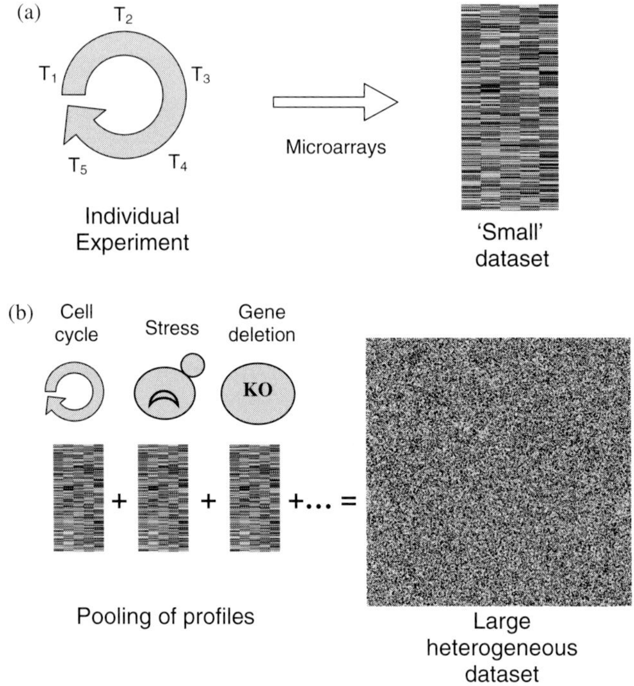

Fig. 1. Expression data (a) Individual microarray experiments addressing specific biological issues give rise to small datasets comprising only a few distinct experimental conditions (e.g. time points). (b) Large-scale expression data can be generated by pooling profiles from many such individual experiments (or by conducting dedicated comprehensive assays). Such data cover not only thousands of genes, but also many cellular states by including a heterogeneous collection of experimental conditions.

expression, and dimension-reduction techniques (reviewed by Brazma *et al.*[4] and Slonim[5]).

In addition to the specific biological questions addressed in such individual focused experiments, it is widely recognized that a wealth of supplementary information can be retrieved from a large and heterogeneous dataset describing the transcriptional response to a variety of different conditions.[2] Furthermore, the relatively high level of noise in these data sets can be dealt with most effectively by combining many arrays probing similar conditions. Comprehensive data have been used to provide functional links for unclassified genes,[3,6–9] to predict novel *cis*-regulatory elements,[7,10–12] and to elucidate the structure of the transcriptional program.[12,13]

Large-scale expression data may result from systematic efforts to characterize a range of transcription states by testing many different biological conditions.[6,13,14] In addition, large datasets can be assembled by collecting expression profiles and pooling them into one comprehensive database [Fig. 1(b)]. Until recently, these data appeared in different formats and were scattered among various internet sites (if available at all).[4] The increasing availability of microarray technology and the ensuing explosion of available expression profiles (usually obtained in different laboratories using different array technologies) have prompted the establishment of standardized annotations such as the MIAME[15] and MAGE-ML[16] standards, and a number of public repositories for chip data.[17–20]

Single microarray experiments are global only in the sense that the genes probed span the entire or most of the genome. The idea of composing large-scale expression datasets is to include a large variety of conditions in order to span also the expanse of transcriptional states of the cell. While this is a necessary step towards the elucidation of the transcription programs, such data present new and serious challenges to the mathematical and computational tools used to analyze them. In particular, the context-specific nature of regulatory relationships poses a difficult computational problem. Consequently, a sizeable variety of different approaches has been proposed in the literature (see review by Ihmels and Bergmann[21]).

1.1. *The Modular Concept*

Whenever we face a large number of individual elements that have heterogeneous properties, grouping elements with similar properties can help to obtain a better understanding of the entire ensemble. For example, we may attribute human individuals of a large cohort to different groups based on their sex, age, profession, etc. in order to obtain an overview of the cohort and its structure. Similarly, individual genes can be categorized according to their properties to obtain a global picture of their organization in the genome. Evidently, in both cases alike, the assignment of the elements to groups — or modules — depends on which of their properties are considered and on how these properties are processed in order to associate different elements with the same module.

A major advantage of studying properties of modules, rather than individual elements, relies on a basic principle of statistics. The mean variance is proportional to $1/N$, where N is the number of (statistical) variables used to compute its value, because fluctuations in these variables tend to cancel each other out. Thus, mean values over the elements of a module or between the elements of different modules are more robust measures than the measurements of each single element. This is particularly relevant for the noisy data produced by chip-based high-throughput technologies.

1.2. *Regulatory Patterns are Context-Specific*

The central challenge in the analysis of large and diverse collections of expression profiles lies in the context-dependent nature of coregulation. Usually, genes are coordinately regulated only in specific experimental contexts, corresponding to a subset of the conditions in the dataset. Such conditions could be different environments (*external* conditions) or distinct tissues or developmental stages (*internal* conditions). Most standard analysis methods classify genes based on their similarity in expression across *all* available conditions. The underlying assumption of uniform regulation is reasonable for the analysis of small datasets, but limits the utility of these tools for the analysis of heterogeneous large datasets for the following reasons. First, conditions irrelevant for the analysis of a particular regulatory context contribute noise, hampering the identification of correlated behavior over small subsets of conditions. Second, genes may participate in more than one function, resulting in one regulation pattern in one context and a different pattern in another. This is particularly relevant for splice isoforms, which are not distinguished by the probes on the array, but may differ in their physiological function or localization. Thus, combinatorial regulation necessitates the assignment of genes to several context-specific and potentially overlapping modules. In contrast, most commonly used clustering techniques yield disjoint partitions, assigning each gene to a single cluster.

Several examples of combinatorial regulation have been discussed in the literature. Yuh *et al.*[22] analyzed the combinatorial logic in the control element of a sea urchin gene. We elucidated the coregulation of the Krebs cycle in *Saccharomyces cerevisiae* and identified two subparts of the cycle that are autonomously coregulated under different sets of conditions.[7] Several examples of condition-specific regulation in yeast and the correlation with transcription factor binding sites were given by Gasch and Eisen.[23] Pilpel *et al.*[24,25] pursued a systematic approach to characterize motif combinations and their synergistic effect on expression patterns at the genomic level. Establishing computational tools that deal with context-specific coexpression is particularly relevant for higher eukaryotes,[26] since it is generally expected that the degree of combinatorial regulation is elevated in these organisms.

1.3. *Coclassification of Genes and Conditions*

To take these considerations into account, expression patterns must be analyzed with respect to specific subsets; genes and conditions should be coclassified.[7,27–32] The resulting "transcription modules" (another common term is "biclusters") consist of sets of coexpressed genes together with the conditions under which this coexpression is observed. The naive approach to evaluating the expression coherence of all possible subsets of genes over all possible subsets of conditions is computationally unfeasible, and most analysis methods for large datasets seek to limit the search space in an appropriate way. For example, Getz *et al.*[29] introduced a variant of biclustering based on the idea to perform standard clustering iteratively on genes and conditions. Their coupled two-way clustering procedure is initialized by separately clustering the genes and conditions of the full matrix. Each combination of the resulting gene and condition clusters defines a submatrix of the expression data. Instead of considering all possible combinations, two-way clustering is then applied to all such submatrices in the following iteration. Other biclustering methods, like the plaid model[30] and gene shaving,[31] aim to identify only the most dominant bicluster in the dataset, which is then masked in a subsequent run to allow for the identification of new clusters. The SAMBA (Statistical-Algorithmic Method for Bicluster Analysis) biclustering method[32] combines graph theory with statistical data modeling. While each method has its advantages and disadvantages,[33] a common drawback is their scaling properties in terms of central processing unit (CPU) time and memory usage when applied to large data.

1.4. *Complexity of the Output and Visualization*

Visualization methods like topographic maps[9] or hierarchical dendrograms have been useful for the interpretation of results from standard clustering methods. As the size and complexity of the data increase, meaningful organization and visualization of the extracted features becomes an ever more important issue. Valuable information beyond the simple enumeration of clusters can be gleaned from the data, including higher-order relationships between clusters,[7,9,33,34] analyses carried out at variable resolutions,[29,33–35] and the identification of hierarchical organization.[33–36] Visual representations can help to elucidate relationships between genes, conditions, or clusters,[7,9,33–35] and to integrate additional information about functional annotations or regulators.[7,8,24,37,38]

1.5. *From Modules to Models*

Transcription modules not only provide the basic building blocks that characterize the structure of the genome-wide transcription program under a variety of conditions, but they also supply a map for a more interpretable characterization of transcriptional changes induced by novel experiments. In particular, searching for coherent changes in expression in a larger module, one may identify patterns that are too weak to discern when considering each of its genes alone. For example, Mootha *et al.*[39] showed that the coordinate expression of a set of functionally related genes was significantly altered in human diabetic muscle, even though this effect was too subtle to be apparent at the single-gene level (see, however, comment in Damian and Gorfine[40]). Segal *et al.*[41] used expression data from almost 2000 published microarrays from human tumors to establish a compendium of modules combining genes with similar behavior across arrays. This cancer module map allowed them to characterize clinical conditions (like tumor stage and type) in terms of a profile of activated and deactivated modules. For instance, they found that a growth inhibitory module, consisting primarily of growth suppressors, was coordinately suppressed in a subset of acute leukemia arrays, suggesting a possible mechanism for the uncontrolled proliferation in these tumors.

These and other results[42–45] illustrate the value of analyzing the complex processes underlying biological conditions such as human diseases in terms of transcription modules. Yet, while a modular characterization of such processes provides a powerful tool to elucidate aspects of the normal and defective regulatory mechanisms, it is only one step towards the goal of obtaining detailed mechanistic models of processes pertaining to disease. Since cellular processes are regulated at all stages leading from DNA to functional proteins, the integration of information on regulatory sequence as well as posttranscriptional regulation is crucial in this endeavor.

1.6. *Data Integration*

For many organisms, different types of high-throughput data are rapidly accumulating, including protein–protein interaction data,[46] transcription factor binding information,[47] genomic and promoter sequence and gene ontologies,[48] and protein localization studies.[49] Yet, these data are often noisy and incomplete and cannot be used to infer coregulation directly.

Coregulation of genes is generally expected to reflect involvement in related cellular functions or pathways, and to result from shared promoter-binding sites for a common set of transcription factors or other forms of coordinated regulation. Therefore, several authors employed the above-mentioned types of additional data to provide a starting point and focus for coexpression analysis methods.[7,33,37,38,41,50] For example, target-regulator network analyses can be simplified if the set of potential regulators does not include the entire genome but can be restricted to a smaller number of candidates.[37,38] Combining gene expression and protein interaction data improves the discovery of coherent functional groups and reduces false positives.[50,51] Similarly, the GRAM (Genetic Regulatory Modules) algorithm by Bar-Joseph *et al.*[38] investigates module–regulator relationships by integrating physical regulator binding data with expression profiles. The algorithm aims to improve the reliability of binding data by allowing lower stringency for those interactions that are supported by coexpression; likewise, it restricts the analysis of coexpression to genes that are targets to a common set of transcription factors.

2. Modular Algorithms

2.1. *Signature Algorithm*

The signature algorithm[7] (see Fig. 2 for details) was designed to test whether a set of candidate genes exhibits a coherent expression over a subset of the microarray data, thus already taking context-specific regulation into account. These test sets are constructed by integration of additional biological data, including functional annotations and regulatory sequence information.

In order to provide a more global modular picture of the transcription program, this algorithm was later extended into an iterative scheme (the iterative signature algorithm, or ISA; see Fig. 3) that allows for an efficient modular decomposition of large-scale expression data (typically tens of thousands of gene probes tested over hundreds of conditions) also in the absence of any *a priori* information.[34] The ISA is one of the state-of-the-art methods for these types of data according to various performance measurements,[19,20] and has been employed in numerous biological studies (e.g. Refs. 19, 21–23). Briefly, the ISA uses a set of expression data to identify a compendium of transcription modules, consisting of coexpressed genes as well as the experimental conditions for which this coherent expression is the most pronounced.

This algorithm has the following advantages: (1) Genes and samples can be assigned to multiple modules (while standard clustering produces mutually exclusive units). This is well founded from the biological point of view, because splice variants may hybridize to the same probe and the

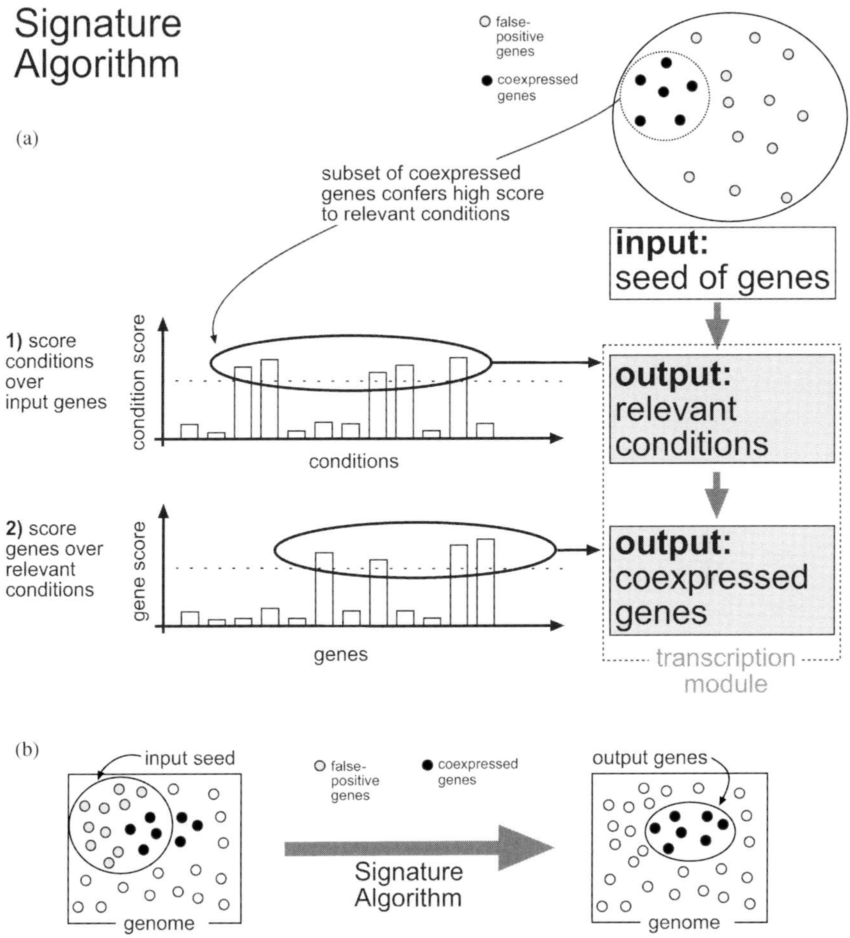

Fig. 2. The signature algorithm requires as input a set of genes, some of which are expected to be coregulated based on additional biological information such as a common promoter binding motif or functional annotation. (a) The algorithm proceeds in two steps: In the first step, this input seed is used to identify the conditions that induce the highest average expression change in the input genes. Only conditions with a score above some threshold are selected. In the second stage of the algorithm, genes that are highly and consistently expressed over these conditions are identified. The result consists of a set of coregulated genes together with the regulating conditions, and is termed as "transcription module." (b) The output contains only the coregulated part of the input seed, as well as other genes that were not part of the original input but display a similar expression profile over the relevant conditions.

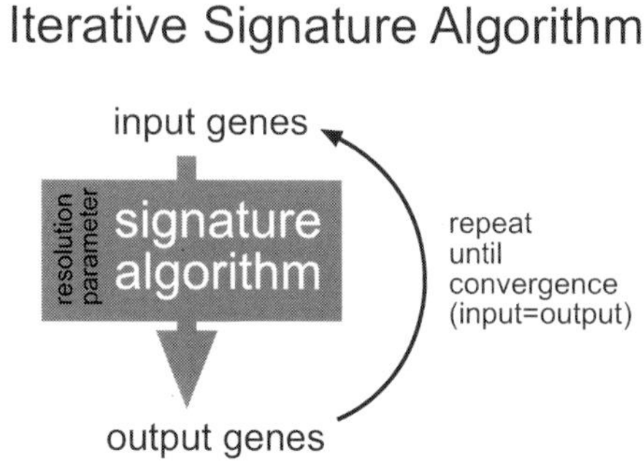

Fig. 3. The iterative signature algorithm (ISA) is an extension of the signature algorithm and is designed to reveal hierarchies of coregulatory units of varying expression coherence. This approach is applicable also in the absence of biologically motivated seeds, in which case the iterative scheme is initialized by many sets of randomly chosen input genes. The output genes determined by the signature algorithm are reused as input. This procedure is iterated until input and output converge. Each resulting transcription module is self-consistent: its genes are most coherently coexpressed over the module conditions, which in turn induce the most coherent expression of the module genes. Modules at different resolutions can be obtained by changing the coregulation threshold parameter.

same gene can function in several processes, which are induced under different experimental conditions. (2) Requiring only coherent gene expression over a subset of arrays allows for picking up subtle signals of context-specific and combinatorial coregulation. Given the experimental noise in microarray data, these signals may be too weak to be extracted from the correlations over all samples that are used by many clustering algorithms. (3) Since the ISA does not require the calculation of correlation matrices, it is highly efficient computationally and is thus applicable even to very large datasets.

2.2. *Comparative Analysis*

We used modular algorithms for a comparative analysis employing expression data from *S. cerevisiae*, *C. elegans*, *E. coli*, *A. thaliana*, *D. melanogaster*, and *H. sapiens*, and found that functionally related genes are indeed frequently coexpressed in these organisms.[35] We applied the signature algorithm

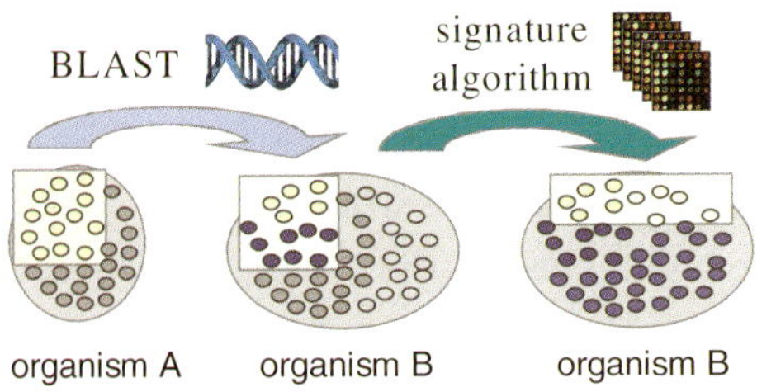

Fig. 4. Starting from a set of coexpressed genes (yellow dots in left box) associated with a particular function in organism A, we first identify the homologes in organism B using BLAST (middle box). Only some of these homologs are coexpressed, while others are not (blue dots). The signature algorithm selects this coexpressed subset, and adds further genes (light yellow) that were not identified based on sequence but share similar expression profiles (right box).

to seeds containing the orthologs of coexpressed yeast genes with known cellular functions (Fig. 4). In most cases, a coexpressed subset of these orthologs could be identified. These genes are likely to participate in a function similar to that of the original yeast genes. Moreover, this approach provides functional predictions for genes that have similar expression patterns but no sequence similarity with the original genes.

The modular structures of the expression data were characterized by first identifying the transcription modules in each dataset (using the ISA) and subsequently their organization in each transcription program (c.f. Fig. 8). The coexpression of several sets of ancient genes pertaining to fundamental cellular functions (like those coding for ribosomal proteins) has been conserved across all organisms. Yet, the relative importance of these conserved modules to the transcription program varies significantly between organisms. Moreover, a significant number of modules are composed primarily of genes that are organism-specific.[35]

Comparing the coexpression of modules (rather than of individual genes), we revealed similarities and differences in higher-order structures of the various transcription programs. Specifically, we asked whether pairs of modules that are (anti)correlated in one organism exhibit the same regulatory relationship in another organism. Studying a set of eight representative modules related to core cellular processes among the six diverse species, the available expression data indicated that relatively few of these relationships have been conserved among the six diverse species.[35]

Global properties of expression data can also be studied by constructing "expression networks," where genes with similar expression profiles are connected.[52] Despite the small proportion of conserved relationships, we found that the topological properties of such networks are conserved in all six organisms. This includes power-law connectivity distributions (with similar exponents), increased likelihood of connecting genes of similar connectivity, and a high degree of clustering. In addition, highly connected genes were significantly more likely to be essential and conserved.[35]

2.3. *Differential Clustering Algorithm*

We developed the differential clustering algorithm (DCA) to study systematically whether the coexpression of genes in one organism has been conserved fully, partially, or not at all in another organism.[53] The DCA is applied to a set of orthologous genes that are present in both organisms. Such a set can be defined according to a functional category, a common motif in the upstream sequence, or a transcription module in either dataset. As a first step, the pairwise correlations between these genes are measured in each organism separately, defining two pairwise correlation matrices (PCMs) of the same size. Next, the PCM of the primary ("reference") organism is clustered, assigning genes into subsets that are coexpressed in this organism, but not necessarily in the second ("target") organism. Finally, the genes within each coexpressed subgroup are reordered by clustering according to the PCM of the target organism. This procedure is performed twice, reciprocally, such that each PCM is used once for the primary and once for the secondary clustering, yielding two distinct orderings of the genes.

The results of the DCA are presented in terms of the rearranged PCMs (Fig. 5). Since these matrices are symmetric and refer to the same set of orthologous genes, they can be combined into a single matrix without losing information. Specifically, we join the two PCMs into one composite matrix such that the lower-left triangle depicts the pairwise correlations in the reference organism, while the upper-right triangle

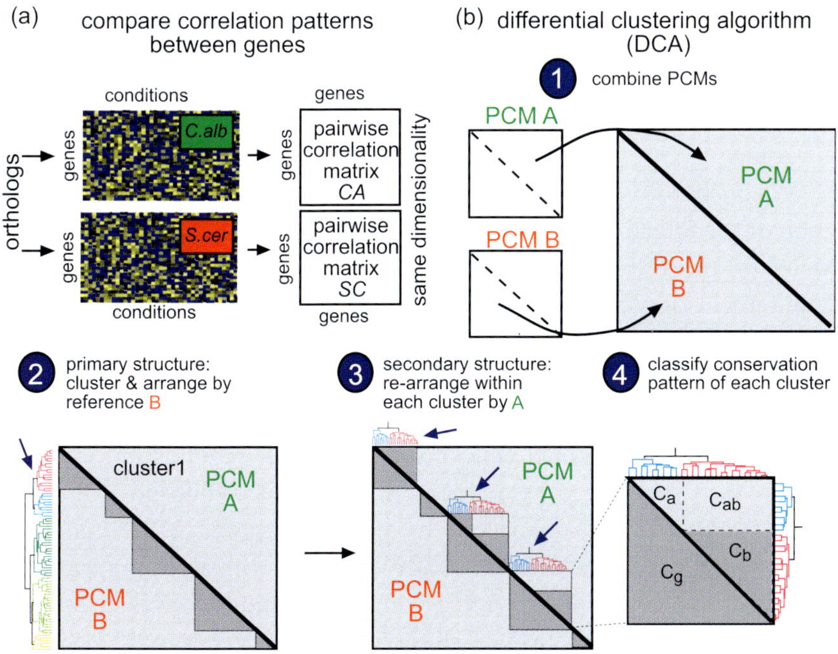

Fig. 5. (a) Pairwise correlation matrices (PCMs) are calculated from the expression data in each organism. (b) The PCMs are combined into a single matrix, where each triangle corresponds to one of the PCMs (1). The genes are then ordered in two steps: First, genes are clustered and the PCMs are rearranged according to the correlations in the reference organism ("B") (2). Second, the genes assigned to each of the resulting primary clusters are reclustered according to their correlations in the target organism "A" (secondary clustering) (3). Finally, the conservation patterns of each cluster are classified automatically into one of the four conservation classes (4).

depicts the correlations in the target organism (Fig. 5(b)). Inspection of the rearranged composite PCM allows for an intuitive extraction of the differences and similarities in the coexpression pattern of the two organisms. Specifically, the PCM is clustered and each primary cluster is subdivided into two secondary clusters *a* and *b*, and characterized by three correlation values corresponding to the average correlations of genes within (C_a, $C_b < C_a$) and between (C_{ab}) these subclusters. An automatic scoring method is then applied to classify clusters into one of four conservation categories: *full*, *partial*, *split*, or *no conservation* of coexpression. This approach can also be extended to study the higher-order correlations between groups of functionally linked genes, providing a more global view of the conserved and diverged parts of the transcription programs.[53]

Using the DCA, we systematically compared the transcription program of the fungal pathogen *Candida albicans* with that of *S. cerevisiae*.[53] Many of the differences we identified are related to the differential requirement for mitochondrial function in the two yeasts. In rich medium, *S. cerevisiae* prefers to grow anaerobically and therefore downregulates mitochondrial ribosomal protein (MRP) genes; in contrast, in *C. albicans* these genes are coexpressed with other genes required for growth (such as ribosomal genes), since it usually requires mitochondrial functions that use oxygen for its metabolism. While no overrepresented sequence was connected to the *S. cerevisiae* MRP genes, in *C. albicans* the MRP genes were clearly associated with an overrepresented upstream sequence motif (AATTTT). This sequence was previously implicated in the regulation of rRNA processing genes in *S. cerevisiae*, although its functional role was not demonstrated experimentally. Experimental work[54] showed that the AATTTT motif is indeed a *cis*-regulatory element of MRP genes in *C. albicans*. Examining the appearance of this motif in more detail, we found that its position relative to the ORF start codon is highly conserved in both *S. cerevisiae* and *C. albicans*, as well as in nine other sequenced yeast species that are considered to be intermediate in the evolution between *S. cerevisiae* and *C. albicans*. In all species examined, the AATTTT motif is significantly overrepresented in genes involved in rRNA processing. Strikingly, its over-representation in MRP promoters was found in all genomes that diverged from the *S. cerevisiae* lineage prior to the whole-genome duplication event. Thus, it appears that the emergence of anaerobic growth capacity in yeast is associated with a global rewiring of its transcriptional network involving dozens of MRP genes which lost a specific regulatory motif. Our results provide further support for the (old) idea[55] that gene duplication can facilitate the evolution of new function not only by specialization of coding sequences, but also by facilitating the evolution of gene expression.

2.4. Ping-Pong Algorithm

High-throughput technologies are now used to generate different types of data from the same biological samples. A central challenge lies in the proper integration of such data. To this end, we proposed the concept of comodules, describing coherent patterns across paired datasets, and conceived several modular methods for their identification. We proposed the ping-pong algorithm (PPA; see Fig. 6) and other modular schemes for the identification of such comodules.[56] For example, we studied the integration of gene expression and drug response data from the NCI60 project. For this study, 60 tumor cell lines were analyzed using both microarrays[57–59] and assays monitoring their growth when subjected to a large number of chemical compounds.[60,61] Thus, each cell line is described by two profiles, one for the expression of each gene and one for its resistance to each drug.

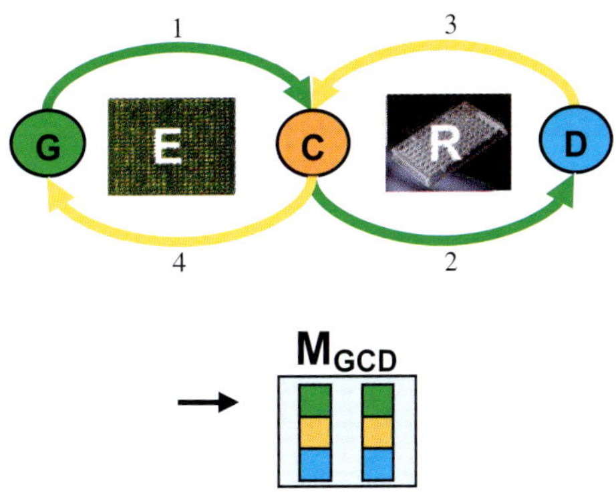

Fig. 6. The ping-pong algorithm starts with a candidate set of genes *G* and uses the available expression data *E* to identify the cell lines *C* for which these genes exhibit a coherent expression (arrow 1). In the next step, the response data *R* are employed to select drugs *D* that elicit a similar response in these cell lines (arrow 2). This set of drugs is then utilized to refine the set of cell lines by eliminating those that have an incoherent response profile and adding others that behave similarly across these drugs (arrow 3). Finally, this refined set of cell lines is used to probe for genes that are coexpressed in these lines (arrow 4). This alternating procedure is reiterated until it converges to stable sets of genes, cell lines, and drugs. We refer to these sets as comodules M_{GCD}, which generalize the concept of a module from a single dataset to multiple datasets.

For predictive purposes, it is useful to identify sets of cell lines that share similar expression *and* growth phenotype patterns. An eventual aim is to use such a compendium of sets in order to predict the growth phenotype of a new sample based on the similarity of its expression profile with a particular cell module (and suggest the adequate combination of drugs to stop it from proliferating). In order to test our predictions, we resorted to the public databases DrugBank and Connectivity Map to show that the PPA predicts drug–gene associations significantly better than other methods. Moreover, comodules not only have increased power to predict drug–gene associations, but also significantly reduce the complexity of the data and provide context in terms of the relevant experimental conditions. Our in-depth analysis of a large compendium of comodules suggests that they provide interesting new insights into possible mechanisms of action for a wide range of drugs, and suggest new targets and novel therapies.

3. Module Analysis

3.1. *Module Annotation*

The value of transcription modules or comodules depends critically on what biological insight can be gleaned from the particular combination of elements (genes, samples, conditions, drugs, etc.) of which they consist. With the growing body of information on the function of gene products,[48,62,63] it is feasible to provide an initial module annotation based on automated functional enrichment analysis. Specifically, overrepresentation of genes belonging to the same functional category in one module suggests its association with this function. Over-representation can be quantified in terms of a *p*-value, based on the total numbers of elements in the category and the module as well as their intersection. Usually, these *p*-values are computed using Fisher's exact test. Presently, several articles on online enrichment analysis (FunSpec,[64] MAPPFinder,[65] and FatiGO[66]) have been published. Functional categories for many human and mouse genes are provided, for example, by the Gene Ontology (GO) project,[67] and associations with metabolic pathways are available on the KEGG[68] database. Although functional annotations are incomplete, and sometimes even wrong, very small p-values usually indicate a functional link for the module.

3.2. Module Visualization

Standard hierarchical clustering still remains the default analysis tool for large sets of biological data, despite the limitations of this analysis method for large-scale data.[7,27–32] One reason for this is that the widely-used representation of expression data in terms of a reordered color-coded matrix with dendrograms delineating the clusters and their hierarchy has the advantage of being exceedingly simple. In particular, many biologists apparently appreciate that the original expression values (or ratios) are shown in this presentation (somewhat akin to the fact that showing the image of a gel shift experiment is still a must, although quantification software for gels has existed for some time). Accordingly, we have designed a new visualization tool, ExpressionView, which presents modules as rectangles that denote its genes and arrays on the actual expression data (see Fig. 7). Since it is in general impossible to represent more than two mutually overlapping modules in this manner, we have developed an algorithm that minimizes the fraction of genes or arrays that appear as disconnected module fragments. Thus, our tool maintains the aforementioned simplicity of the common cluster representation, while allowing for an intuitive presentation of overlapping groups of genes and arrays.

Transcription modules provide the building blocks of the transcription network. A systems biology approach aims not only at identifying (and annotating) these "blocks", but also describing the relationships between them in order to reveal the structure of the entire network. Module relationships can be defined in many ways: the extent of common genes, conditions, or functional categories describes static intermodule relations. Yet, our identification of modules in terms of gene scores and condition scores also retains information on what experiments induce or suppress the module's genes, and to what degree. Thus, correlating modules over these scores can

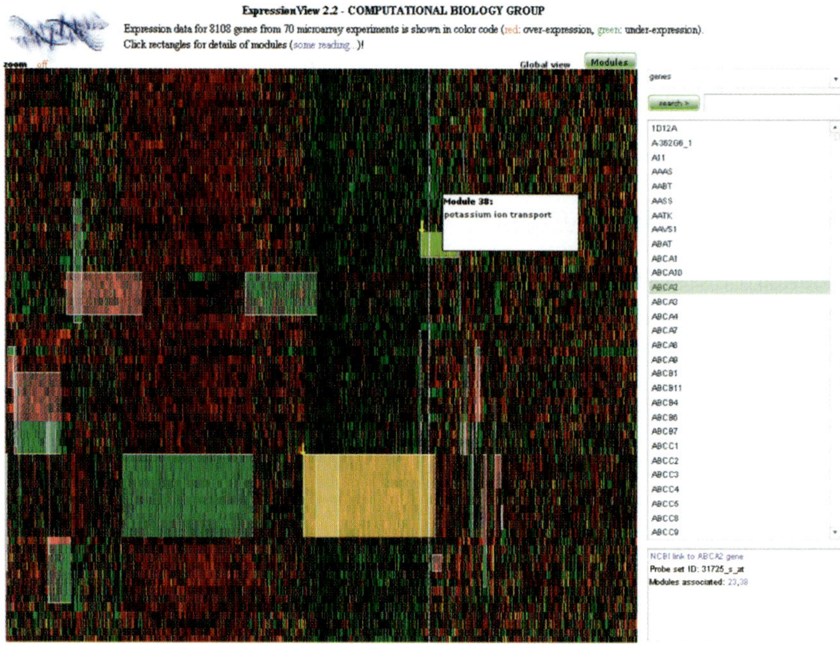

Fig. 7. Screenshot from the prototype of ExpressionView (available online at http://serverdgm.unil.ch/bergmann/ExpressionView.html). Similar to the standard biclustering, our tool visualizes the expression levels of all genes in the dataset (columns) under many experimental conditions (rows) using a color code. Yet, the order of genes and conditions has been optimized in order to highlight coherent expression patterns that are apparent only over a subset of the entire dataset (i.e. transcription modules). Genes (as well as conditions or modules) can be selected in the window on the right. Modules are clickable, providing detailed information on their genes and conditions, as well as automated annotation in terms of enriched GO categories and KEGG pathways.

provide insights into dynamic relationships. For example, studying gene expression data from baker's yeast revealed that genes involved in stress response are usually suppressed when genes involved in protein synthesis are induced, and vice versa.[7,69] This reflects a fundamental strategy of the yeast's transcription program to invest the available resources either to grow or to survive. For higher organisms, whose survival chances do not depend on the ability of its cells to outgrow others, but rather on their proper coordination, this strategy is probably not relevant. Nevertheless, comparing the relationships between the modules derived from expression data from these organisms may give us new insights into the global organization of their transcription programs.

It is useful to provide graphic visualizations of the higher-order module organization. To this end, we have proposed two complementary presentations. A module tree [Fig. 8(a)] summarizes the entire compendium of transcription modules identified by the ISA at different resolutions. Similar modules detected for different stringencies on coexpression are connected by lines and form the branches of the module tree. Branches may merge if several modules that emerge at a higher threshold converge into the same module when iterated at a lower

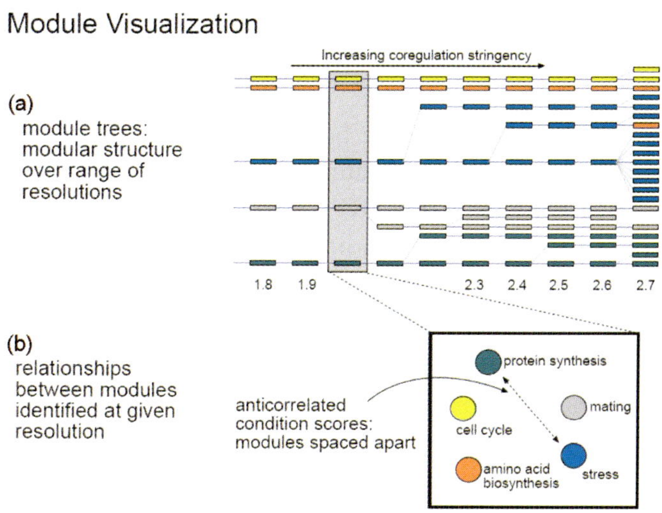

Fig. 8. Visualization of large-scale expression data. (a) Module trees summarize the transcription modules identified by the ISA at different resolutions (rectangles). Similar modules identified for different stringencies on coexpression are connected by lines and form the branches of the module tree. Branches may merge if several modules that emerge at a higher threshold converge into the same module when iterated at a lower threshold (transversal lines). (b) All modules identified at the same resolution are represented on a plane such that their distances reflect the regulatory relations. Modules induced under similar conditions are closer to each other, while large distances indicate inverse activation.

threshold (transversal lines). In the module layer presentation [Fig. 8(b)], modules identified at the same resolution are represented on a plane such that their distances reflect the regulatory relations. Modules induced under similar conditions are closer to each other, while large distances indicate inverse activation.

4. Outlook

High-throughput data acquisition technologies have created the potential for new insights into biological systems. Yet, the hope to better understand the regulation of these systems and to eventually predict their response will only materialize with adequate computational tools to process and visualize the vast amount of data that these technologies produce. Medical research is rapidly adopting high-throughput technologies to characterize clinical samples, and therefore requires appropriate computational tools to interpret the results and use them for diagnostic purposes; we believe that a modular approach to large-scale data will yield great benefits with respect to both aspects. In particular, we anticipate a great need for integrating data from multiple platforms ranging from SNP chips and CGH chips to mRNA chips and protein chips. Extending the modular approach for the analysis of multiple datasets of various types has been shown to be useful for predicting drug–gene associations from relatively inexpensive high-throughput data, as generated for the NCI60 panel.[56] Such an approach enables large-scale hypothesis generation that may allow more cost-effective targeting of research resources for direct experimental studies.

New modular approaches for integrative analysis of large-scale data are likely to be useful also for the meta-analyses of other large-scale biological data. For example, comodule analysis has great potential for the integration of expression data from two different species. In this case, the common dimension of the two sets of expression data is established by the orthologous genes (rather than the samples, as for the NCI-60 data). Here, comodule analysis could provide a sensitive means to identify not only the sets of orthologs with conserved coexpression, but also the respective experimental conditions under which this coexpression is induced. Similarly, for datasets covering different types of gene regulation (e.g. posttranscriptional modifications or protein expression), our approach allows for revealing those sets of genes that are coregulated at multiple instances. In fact, the ping-pong algorithm[56] is particularly useful for these applications because it does not require regulation data for identical sets of conditions, but automatically identifies those conditions from both datasets that give the best match for coregulation.

References

1. Kitano H. (2002) Systems biology: a brief overview. *Science* **295**: 1662–4.
2. Lander ES. (1999) Array of hope. *Nat Genet* **21**: 3–4.
3. Tavazoie S, Hughes JD, Campbell MJ *et al.* (1999) Systematic determination of genetic network architecture. *Nat Genet* **22**: 281–5.
4. Brazma A, Robinson A, Cameron G, Ashburner M. (2000) One-stop shop for microarray data. *Nature* **403**: 699–700.
5. Slonim DK. (2002) From patterns to pathways: gene expression data analysis comes of age. *Nat Genet* **32**(Suppl): 502–8.
6. Hughes TR, Marton MJ, Jones AR *et al.* (2000) Functional discovery via a compendium of expression profiles. *Cell* **102**: 109–26.
7. Ihmels J, Friedlander G, Bergmann S *et al.* (2002) Revealing modular organization in the yeast transcriptional network. *Nat Genet* **31**: 370–7.
8. Wu LF, Hughes TR, Davierwala AP *et al.* (2002) Large-scale prediction of *Saccharomyces cerevisiae* gene function using overlapping transcriptional clusters. *Nat Genet* **31**: 255–65.
9. Kim SK, Lund J, Kiraly M *et al.* (2001) A gene expression map for *Caenorhabditis elegans*. *Science* **293**: 2087–92.
10. Bussemaker HJ, Li H, Siggia ED. (2001) Regulatory element detection using correlation with expression. *Nat Genet* **27**: 167–71.
11. Hughes JD, Estep PW, Tavazoie S, Church GM. (2000) Computational identification of *cis*-regulatory elements associated with groups of functionally related genes in *Saccharomyces cerevisiae*. *J Mol Biol* **296**: 1205–14.
12. Wang W, Cherry JM, Botstein D, Li H. (2002) A systematic approach to reconstructing transcription networks in *Saccharomy cescerevisiae*. *Proc Natl Acad Sci USA* **99**: 16893–8.
13. Gasch AP, Spellman PT, Kao CM *et al.* (2000) Genomic expression programs in the response of yeast cells to environmental changes. *Mol Biol Cell* **11**: 4241–57.

14. Su AI, Wiltshire T, Batalov S *et al.* (2004) A gene atlas of the mouse and human protein-encoding transcriptomes. *Proc Natl Acad Sci USA* **101**: 6062–7.
15. Brazma A, Hingamp P, Quackenbush J *et al.* (2001) Minimum information about a microarray experiment (MIAME) — toward standards for microarray data. *Nat Genet* **29**: 365–71.
16. Spellman PT, Miller M, Stewart J *et al.* (2002) Design and implementation of microarray gene expression markup language (MAGE-ML). *Genome Biol* **3**: RESEARCH0046.
17. Ball CA, Awad IA, Demeter J *et al.* (2005) The Stanford Microarray Database accommodates additional microarray platforms and data formats. *Nucleic Acids Res* **33**: D580–2.
18. Brazma A, Parkinson H, Sarkans U *et al.* (2003) ArrayExpress — a public repository for microarray gene expression data at the EBI. *Nucleic Acids Res* **31**: 68–71.
19. Edgar R, Domrachev M, Lash AE. (2002) Gene Expression Omnibus: NCBI gene expression and hybridization array data repository. *Nucleic Acids Res* **30**: 207–10.
20. Ikeo K, Ishi-i J, Tamura T *et al.* (2003) CIBEX: Center for Information Biology gene Expression database. *C R Biol* **326**: 1079–82.
21. Ihmels JH, Bergmann S. (2004) Challenges and prospects in the analysis of large-scale gene expression data. *Brief Bioinform* **5**: 313–27.
22. Yuh CH, Bolouri H, Davidson EH. (1998) Genomic *cis*-regulatory logic: experimental and computational analysis of a sea urchin gene. *Science* **279**: 1896–902.
23. Gasch AP, Eisen MB. (2002) Exploring the conditional coregulation of yeast gene expression through fuzzy *k*-means clustering. *Genome Biol* **3**: RESEARCH0059.
24. Pilpel Y, Sudarsanam P, Church GM. (2001) Identifying regulatory networks by combinatorial analysis of promoter elements. *Nat Genet* **29**: 153–9.
25. Sudarsanam P, Pilpel Y, Church GM. (2002) Genome-wide co-occurrence of promoter elements reveals a *cis*-regulatory cassette of rRNA transcription motifs in *Saccharomyces cerevisiae*. *Genome Res* **12**: 1723–31.
26. Werner T. (2001) The promoter connection. *Nat Genet* **29**: 105–6.
27. Bittner M, Meltzer P, Trent J. (1999) Data analysis and integration: of steps and arrows. *Nat Genet* **22**: 213–5.
28. Cheng Y, Church GM. (2000) Biclustering of expression data. *Proc Int Conf Intell Syst Mol Biol* **8**: 93–103.
29. Getz G, Levine E, Domany E. (2000) Coupled two-way clustering analysis of gene microarray data. *Proc Natl Acad Sci USA* **97**: 12079–84.
30. Lazzeroni L, Owen A. (1999) Plaid models for gene expression data. Technical report, Statistics, Stanford University, USA.

31. Hastie T, Tibshirani R, Eisen MB *et al.* (2000) 'Gene shaving' as a method for identifying distinct sets of genes with similar expression patterns. *Genome Biol* **1**: RESEARCH0003.
32. Tanay A, Sharan R, Shamir R. (2002) Discovering statistically significant biclusters in gene expression data. *Bioinformatics* **18**(Suppl 1): S136–44.
33. Ihmels J, Bergmann S, Barkai N. (2004) Defining transcription modules using large-scale gene expression data. *Bioinformatics* **20**: 1993–2003.
34. Bergmann S, Ihmels J, Barkai N. (2003) Iterative signature algorithm for the analysis of large-scale gene expression data. *Phys Rev E Stat Nonlin Soft Matter Phys* **67**: 031902.
35. Bergmann S, Ihmels J, Barkai N. (2004) Similarities and differences in genome-wide expression data of six organisms. *PLoS Biol* **2**: E9.
36. Eisen MB, Spellman PT, Brown PO, Botstein D. (1998) Cluster analysis and display of genome-wide expression patterns. *Proc Natl Acad Sci USA* **95**: 14863–8.
37. Segal E, Shapira M, Regev A *et al.* (2003) Module networks: identifying regulatory modules and their condition-specific regulators from gene expression data. *Nat Genet* **34**: 166–76.
38. Bar-Joseph Z, Gerber GK, Lee TI *et al.* (2003) Computational discovery of gene modules and regulatory networks. *Nat Biotechnol* **21**: 1337–42.
39. Mootha VK, Lindgren CM, Eriksson KF *et al.* (2003) PGC-1alpha-responsive genes involved in oxidative phosphorylation are coordinately downregulated in human diabetes. *Nat Genet* **34**: 267–73.
40. Damian D, Gorfine M. (2004) Statistical concerns about the GSEA procedure. *Nat Genet* **36**: 663; author reply 663.
41. Segal E, Friedman N, Koller D, Regev A. (2004) A module map showing conditional activity of expression modules in cancer. *Nat Genet* **36**: 1090–8.
42. Lamb J, Ramaswamy S, Ford HL *et al.* (2003) A mechanism of cyclin D1 action encoded in the patterns of gene expression in human cancer. *Cell* **114**: 323–34.
43. Rhodes DR, Yu J, Shanker K *et al.* (2004) Large-scale meta-analysis of cancer microarray data identifies common transcriptional profiles of neoplastic transformation and progression. *Proc Natl Acad Sci USA* **101**: 9309–14.
44. Desai KV, Xiao N, Wang W *et al.* (2002) Initiating oncogenic event determines gene-expression patterns of human breast cancer models. *Proc Natl Acad Sci USA* **99**: 6967–72.
45. Chang CF, Wai KM, Patterton HG. (2004) Calculating the statistical significance of physical clusters of co-regulated genes in the genome: the role of chromatin in domain-wide gene regulation. *Nucleic Acids Res* **32**: 1798–807.
46. Gavin AC, Bösche M, Krause R *et al.* (2002) Functional organization of the yeast proteome by systematic analysis of protein complexes. *Nature* **415**: 141–7.
47. Lee TI, Rinaldi NJ, Robert F *et al.* (2002) Transcriptional regulatory networks in *Saccharomyces cerevisiae*. *Science* **298**: 799–804.

48. Mewes HW, Amid C, Arnold R *et al.* (2004) MIPS: analysis and annotation of proteins from whole genomes. *Nucleic Acids Res* **32** Database issue: D41–4.
49. Huh WK, Falvo JV, Gerke LC *et al.* (2003) Global analysis of protein localization in budding yeast. *Nature* **425**: 686–91.
50. Tirosh I, Barkai N. (2005) Computational verification of protein–protein interactions by orthologous co-expression. *BMC Bioinformatics* **6**: 40.
51. Segal E, Wang H, Koller D. (2003) Discovering molecular pathways from protein interaction and gene expression data. *Bioinformatics* **19**(Suppl 1): i264–71.
52. Farkas I, Jeong H, Vicsek T *et al.* (2003) The topology of the transcription regulatory network in the yeast, *Saccharomyces cerevisiae. Phys A: Stat Mech Appl* **318**: 601–12.
53. Ihmels J, Bergmann S, Berman J, Barkai N. (2005) Comparative gene expression analysis by differential clustering approach: application to the *Candida albicans* transcription program. *PLoS Genet* **1**: e39.
54. Selmecki A, Bergmann S, Berman J. (2005) Comparative genome hybridization reveals widespread aneuploidy in *Candida albicans* laboratory strains. *Mol Microbiol* **55**: 1553–65.
55. Ohno S, Wolf U, Atkin NB. (1968) Evolution from fish to mammals by gene duplication. *Hereditas* **59**: 169–87.
56. Kutalik Z, Beckmann J, Bergmann S. (2008) A modular approach for integrative analysis of large-scale gene-expression and drug-response data. *Nat Biotechnol* **26**(5): 531–9.
57. Staunton JE, Slonim DK, Coller HA *et al.* (2001) Chemosensitivity prediction by transcriptional profiling. *Proc Natl Acad Sci USA* **98**: 10787–92.
58. Ross DT, Scherf U, Eisen MB *et al.* (2000) Systematic variation in gene expression patterns in human cancer cell lines. *Nat Genet* **24**: 227–35.
59. Butte AJ, Tamayo P, Slonim D *et al.* (2000) Discovering functional relationships between RNA expression and chemotherapeutic susceptibility using relevance networks. *Proc Natl Acad Sci USA* **97**: 12182–6.
60. Scherf U, Ross DT, Waltham M *et al.* (2000) A gene expression database for the molecular pharmacology of cancer. *Nat Genet* **24**: 236–44.
61. Shi LM, Fan Y, Lee JK *et al.* (2000) Mining and visualizing large anticancer drug discovery databases. *J Chem Inf Comput Sci* **40**: 367–79.
62. Wheeler DL, Church DM, Federhen S *et al.* (2005) Database resources of the National Center for Biotechnology Information. *Nucleic Acids Res* **33**: D39–45.
63. Bairoch A, Apweiler R, Wu CH *et al.* (2005) The Universal Protein Resource (UniProt). *Nucleic Acids Res* **33**: D154–9.
64. Robinson MD, Grigull J, Mohammad N, Hughes TR. (2002) FunSpec: a web-based cluster interpreter for yeast. *BMC Bioinformatics* **3**: 35.
65. Doniger SW, Salomonis N, Dahlquist KD *et al.* (2003) MAPPFinder: using Gene Ontology and GenMAPP to create a global gene-expression profile from microarray data. *Genome Biol* **4**: R7.

66. Al-Shahrour F, Diaz-Uriarte R, Dopazo J. (2004) FatiGO: a web tool for finding significant associations of Gene Ontology terms with groups of genes. *Bioinformatics* **20**: 578–80.
67. Ashburner M, Ball CA, Blake JA *et al.* (2000) Gene ontology: tool for the unification of biology. The Gene Ontology Consortium. *Nat Genet* **25**: 25–9.
68. Kanehisa M, Goto S, Kawashima S *et al.* (2004) The KEGG resource for deciphering the genome. *Nucleic Acids Res* **32**: D277–80.
69. Ihmels J, Levy R, Barkai N. (2004) Principles of transcriptional control in the metabolic network of *Saccharomyces cerevisiae. Nat Biotechnol* **22**: 86–92.

Chapter 4

Integrated Analysis of Gene Expression Profiling Studies — Examples in Breast Cancer

Pratyaksha Wirapati, Darlene R. Goldstein
and Mauro Delorenzi

1. Introduction

The high-throughput nature of biological assays such as microarrays has contributed to their rise in importance for studying the molecular basis of fundamental biological processes and complex disease traits. The widespread use of microarrays has resulted in a large-scale, rapid expansion of data. Data from many microarray studies are deposited in databases such as Gene Expression Omnibus (GEO)[1,2] and ArrayExpress,[3] often to satisfy journal requirements to make the primary data publicly available. It is hoped that ready access to the data will facilitate the integration of information across different studies.

Information from different studies may be combined at various levels. Combining test statistics from different studies allows ready integration of diverse data types without great loss of information. We present a framework for the combination of results across possibly very heterogeneous studies. This framework is based on meta-analytic combination of z-scores derived from generalized linear models fitted to primary data for each study. Multiple hypothesis testing correction is based on the final combined Z.

2. Combining Information

For combination of separate study information to be sensible, relevant quantities emanating from the studies should be linked in some way, even if only conceptually. How to combine data appropriately depends on the (joint) alternative of interest. Below, we describe several ways in which information may be combined.

There is a recent and increasing literature on combining microarray studies via meta-analysis. The most commonly used meta-analytical methods, described, for example, by Sutton *et al.*,[4] typically combine either parameter estimates or *p*-values.

A major difficulty with synthesizing information is the occurrence of study heterogeneity. In general, studies carried out by different research groups may vary in a number of ways: scientific research goals, population of interest, design, quality of implementation, subject inclusion and exclusion criteria, baseline status of subjects (even with the same selection criteria), treatment dosage and timing, management of study subjects, outcome definition or measures, scoring of outcomes (e.g. pathologist scoring of tumors), data integrity, and statistical methods of analysis.

Additional issues more specific to the microarray context include the following: differences in the technology used for the study, heterogeneity of measured expression from the same probe occurring multiple times on the array, multiple (different) probes for the same gene, variability in the probes used by different platforms, differences in quantification of gene expression even when the same technology is used, and large "batch" effects, due, for example, to a particular laboratory and/or technician.

Genomic studies differ from traditional epidemiological or clinical trials in several respects. One obvious difference is that the number of variables measured in genomic studies is usually in the thousands per sample, rather than the perhaps tens per sample for a clinical trial. Microarray study sample sizes are also typically much smaller, putting additional impetus on effective combination. Study goals ordinarily differ as well.

In a clinical trial, the overall goal is primarily to obtain a combined estimate treatment effect. Genomic studies more often focus on combining evidence supporting the role of a gene or to rank evidence for a large number of genes. In contrast to the estimation scenario, in this case it may be advantageous rather than harmful to draw upon multiple, heterogeneous sources. Heterogeneity should tend to increase the robustness of inferences, thereby enhancing the generalizability of study conclusions. The effects of within-study bias might also be reduced, as we would expect different biases in different studies.

3. Individual Patient Data (IPD)

Data for a combined analysis can be obtained in various ways, including extraction of aggregate data (e.g. test statistics) from published reports or collection of data at the individual patient level (e.g. from publicly available databases). Systematic reviews of summaries based on individual patient data (IPD) are sometimes considered as the gold standard as compared to meta-analysis based on published summaries.[5]

Basing a meta-analysis on published summaries can be problematic for a number of reasons. Information on patient characteristics or outcomes may be lacking. In addition, the study may be ongoing after publication, so the data on which the publication is based may be out of date. When raw data are available for each study, more flexible and detailed analyses (e.g. subgroup analyses) not carried out by the original investigators can be performed. Finally, there is scope for further improvement of data quality when the original data are available for independent scrutiny.

4. SwissBrod: Swiss Breast Oncology Database

4.1. *Difficulties with Public Data Sources*

Information such as original data files from public repositories or supplementary files provided as support for a publication are largely article-oriented and tuned to support the paper's claims. Without further refinement, these are unsuitable for meta-analysis for the following reasons:

(1) *Lack of independent patient cohorts.* Multiple articles may have used some of the same arrays or the same patients using other

technologies. Resolving the dependencies by merging article-oriented datasets or eliminating duplicated patients can be tedious and unnecessarily burden downstream statistical meta-analysis.

(2) *No standard variable names or representation of values.* The same name may be used in different studies to mean different things (e.g. "survival" may mean overall time until death or recurrence-free time), or different names may be used to refer to the same entities.[6] In addition, there is a need for better documentation of the technologies used in deriving measured values. For example, several methods exist to determine estrogen receptor (ER) status, including ligand binding assay, immunohistochemistry, reverse transcription–polymerase chain reaction (RT-PCR), and microarray; it may not make sense to consider all methods as equivalent across studies. This example also highlights the need for a hierarchy of variables. In this example, ER status is at a higher level than the technology used to obtain it.

(3) *Difficulty of maintaining a consistent mapping of probes to genes.* This is essential for cross-platform matching based on genes. Since the transcriptome is still continually being updated, it must remain possible to map probes using up-to-date information sources.

(4) *Selective inclusion of information.* Some data warehouses tend to concentrate on specific platforms or repositories. Because tumor samples are nonrenewable, it is important to include all data emanating from them. This includes older samples and RT-PCR data, and not just the newest data obtained from a specific type of microarray.

(5) *Unclear or differing study design and patient selection criteria.* Most breast cancer expression data generated to date are based on samples obtained from tumor banks (i.e. population-based sampling). More recent studies may be based on patients selected for clinical trials, implying completely different inclusion/exclusion criteria. Combining studies of selected patients with population-based studies may result in biased or uninterpretable results. Furthermore, some datasets also contain multiple arrays per patient in order to yield longitudinal information on tumor progression and metastasis or chemotherapy response. This possibility implies a hierarchy of samples, analogous to

the variable hierarchy described above. Care must be taken to avoid unspecified dependencies between samples.

4.2. Aim of SwissBrod

To avoid these problems and facilitate data mining and integration while ensuring high data quality, the authors work with SwissBrod, an in-house database developed at the Swiss Institute of Bioinformatics (SIB). The aim of SwissBrod is to provide curated clinical and expression data in a form ready for detailed statistical meta-analysis.[4,7] Aiding correct application and interpretation of statistical methods is the main goal. This involves identifying the actual sampling units (patients, tissues, or arrays) and design (patient selection criteria).

In line with the precise scope of our aims, SwissBrod contains primary data on breast cancer only, and is currently limited to patient-based tumor data. Data normalization was provided by the original study authors. No other derived or summary results or statistical analyses are provided, as these may be computed at will by users.

4.3. SwissBrod Data Curation

Primary datasets are acquired from a variety of public sources: tables within journal articles, supplementary materials on a journal website, author websites, and public repositories (e.g. GEO, ArrayExpress, Stanford Microarray Database). As part of the curation process, the datasets undergo quality controls and some reconfiguration to facilitate downstream analyses.

A major aim of the curation process is to redefine and consolidate dataset variables and cohorts in a way appropriate for statistical analysis. The new cohorts must satisfy basic conditions, such as independent patients as sampling units, and may cut across articles and original dataset boundaries, as happens when a study is extended by the same investigating group. Study design is noted, along with selection criteria. In addition, stable probe identifiers are established so that mapping of probes to genes can be readily carried out.

5. Spectrum of Possible Analyses

The possibilities for combining information across studies can be viewed as occurring along a spectrum of levels of analysis, moving roughly from a combination of least to most "processed" quantities — that is, in order of decreasing information content:

(1) pooling raw data;
(2) pooling adjusted data;
(3) combining parameter estimates;
(4) combining test statistics;
(5) combining transformed p-values;
(6) combining statistic ranks; and
(7) combining decisions (e.g. via intersecting Venn diagrams).

5.1. *Pooling Raw Data*

One way of combining information across studies is to pool the raw, unadjusted data and analyze them together as a single data set. This approach is sometimes called a "mega-analysis". Here, a (fixed or random) covariate indicating study origin can be included in an overall model.

If the different datasets are sufficiently homogeneous and all measure relevant covariates for adjustment, this strategy might be viable. However, even when the raw data are available, this method has a number of drawbacks, particularly for microarray data. It is generally inappropriate to pool raw data from heterogeneous studies (e.g. Simpson's paradox[8]). If computing power is limited, pooling raw data may not even be feasible. It is difficult to imagine a microarray study for which this would be the method of choice. Even planned multi-site studies often exhibit site-specific effects, for which adjustment of some type is required.

For example, even when using the same chip in different studies, joint normalization of pooled data typically does not remove the study batch effect.[9,10] Obviously, this problem becomes even worse with the use of different arrays in different studies, and does not even make sense in the case of data from heterogeneous assays, where signals are noncommensurable.

One approach which can start at this level is hierarchical modeling. Using this approach successfully requires a good understanding of how the different data types may be jointly modeled. In many applications, this will be difficult to implement.

5.2. *Pooling Adjusted Data*

Correction before pooling (see, for instance, Benito *et al.*[9]) consists of applying an adjustment to the data separately for each study, and then combining adjusted values to analyze together as a single set. For example, rather than including a study-specific effect in addition to covariate adjustment in an overall model for pooled raw data, separate adjustments for each study could be made.

This method might not work so well for very heterogeneous microarray studies, where alternative technologies provide intrinsically different measures, e.g. single- versus dual-channel array types. There is as yet no accepted way to transform such diverse values to a single, common scale of measurement.

In addition, separate adjustment will not guarantee that the resulting adjusted observations (typically residuals) will be similarly distributed across studies. Without some further modification (e.g. scaling or other transformation), even these adjusted observations may well remain too heterogeneous for pooling.

5.3. *Combining Parameter Estimates*

Meta-analysis consists of statistical methods for combining results of independent studies addressing related questions. One aim of combining results is to obtain increased power — studies with small sample sizes are less likely to find effects even when they exist. Putting results together increases the effective sample size, thereby allowing more precise effect estimation and increasing power. The uncovering of a significant effect from a combined analysis, where individual studies do not make positive findings at the same significance level, has been referred to in the microarray meta-analysis literature as "integration-driven discovery" (IDD).[11]

Parameter estimates are typically combined using either a fixed effects or random effects model, depending on the absence or presence of heterogeneity between results of the different studies.[12] The overall estimate of the treatment effect is then a weighted average of individual effects, with weights depending on the variances within and between individual study estimates.

5.4. *Combining Test Statistics*

Taking the perspective that there is an overall concept one is interested in assessing, instead of a single parameter relevant to all studies, we can combine test statistics or other derived quantities rather than common model parameters. The combination is generally a (possibly equally) weighted average of individual study statistics. Specialized procedures have been developed for a number of common situations (see, for example, Sutton *et al.*[4] and references therein).

There are several examples in the microarray literature. Stevens and Doerge[13] provide a good exposition on standard meta-analysis, some simulations, and one application. Their work assumes commensurable expression values in Affymetrix data. In an application to pancreatic cancer, Grützmann *et al.*[14] use random effects meta-analysis for t-statistics on a consensus set of genes, with a false discovery rate (FDR) multiple testing correction.

Choi *et al.*[11] apply standard meta-analyses based on effect size to two sets of microarray studies; they also consider a Bayesian approach. Other possibilities include sample-size-weighted average of t-statistics[15] and combination of p-values on the probit scale,[16] which is equivalent to combining z-scores. Carrying out meta-analysis in the framework of permutation-based multiple testing correction is mentioned in Westfall and Young.[17] Dependence-aware false discovery rate (FDR) has also been suggested.[11,18]

A limitation of microarray meta-analysis as currently implemented is in the types of statistics that are combined. As much of meta-analysis was developed in the context of comparative studies (such as clinical trials), the combination of continuous measures has tended to focus on two-sample tests. In this case, the standardized mean difference d (see Hedges and Olkin[7]) is often used as the effect size of interest. As we demonstrate below, combining z-scores provides a flexible means of combining information from more general testing scenarios.

5.5. *Combining p-values*

Combining results using transformed p-values has a long history. It does not require knowledge of the composition of the individual datasets, which may be highly heterogeneous and with no common parameter. This level of combination is closely related to the previous level, as transformations (e.g. the probit) can yield test statistics that are straightforward to combine.

Perhaps the most widely used method for p-value combination is attributed to Fisher.[19] This method is based on the fact that under the null, the test statistic p-value is distributed as a uniform (0,1) random variable, so that the log p-value is distributed exponentially. The p-values are combined by summing the logs. Twice this sum is distributed as a χ^2 random variable with degrees of freedom two times the number of study p-values. A microarray example is given in Rhodes *et al.*,[20] who use Fisher's method to combine results for genes measured in all studies.

Although it can perform well in a variety of circumstances, a drawback of Fisher p-value combination is its treatment of conflicting studies. Given a sufficiently large χ^2 value for the first $k - 1$ studies, an additional study — regardless of its size, quality, or extent of negative findings — cannot produce an overall nonsignificant result with this method, as the smallest value that can be added to the overall χ^2 statistic is 0. In addition, conclusions may depend more heavily on the within-study choice of statistic rather than on global data characteristics. Combining inverse normal (probit)-transformed p-values (i.e. z-scores)[7,16] avoids these limitations.

5.6. *Combining Statistic Ranks*

In many microarray studies, the main result consists of an ordered gene list. One way to combine such lists across studies is by averaging percentile ranks. Although order is preserved, the magnitude of differences between genes is lost, entailing at least some, and possibly very great, loss of information. Combining ranks has been used as a strategy in meta-analysis of genetic linkage studies.[21]

Ranking has also been used in the microarray context.[22] The underlying idea is that since microarray data are noisy, the ranks may be relatively more reliable than the actual measured values. This view is somewhat undermined by the fact that the stability of the ranking depends on the rank, the differences between underlying values, and the sample size. It also ignores that the measure ranked in the original analysis will also affect the ordering, and that independent studies may well have used different statistics. The method does, however, take care of the problem of different measurement types and scales for different platforms.

We show below that platform noncommensurability can be handled by transforming to a relevant test statistic. Thus, methods combining ranks would only seem desirable if the sole information available from each study is a list of ranked genes without any numerical measurement.

5.7. *Combining Decisions*

At the crudest level, we can combine the results of statistical hypothesis tests. That is, rather than combining the p-values for the tests, we can combine the resulting decisions based on some threshold for significance. This Venn diagram method looks for overlap between lists of significant genes across studies. A combined list of genes can be ranked based on vote counting or formed based on repeated significance across studies. Tools to carry out this type of procedure have been created for microarray studies.[23,24]

This method was proposed over 50 years ago.[25] Requiring all tests to individually satisfy a significance criterion was shown to be inadmissible for testing an exponential family parameter,[26] and in general does not appear to have very good power properties.[27] Yet, despite rather poor performance, this method is intuitive and seems to be widely used.

6. Data Integration Methodology

Here, we outline the analysis strategy for combining information from heterogeneous genomic studies:

(1) beginning with raw (or preprocessed) primary data;
(2) data cleaning/quality assessment;
(3) gene matching;
(4) single-gene generalized linear model (GLM) modeling for outcome of interest;
(5) combination of *z*-statistics across studies; and
(6) *p*-value multiplicity adjustment of the final combined statistic.

6.1. *Data Acquisition*

For combining test statistics, we require access to all primary data, not just the "top genes" or *p*-values. If image analysis files are available, we could use them to preprocess the data. More typically, the available genomic data are already preprocessed (e.g. image analysis and normalization for microarrays). Where possible, it is also desirable to obtain the relevant clinical data so that covariates may be included in the data model. We are then able to model fit models within each data set, which will yield the statistics that are to be combined across studies. In our case, this acquisition step is part of SwissBrod preprocessing.

6.2. *Data Cleaning*

With the primary data in hand, we carry out data cleaning and make quality-based decisions on which studies and samples to include. Where possible, quality of the hybridizations should be assessed so that low-quality chips are removed from further analysis. However, most public data are provided as normalized expression measures, precluding rigorous assessment of hybridization quality. Other aspects of quality assessment include relevance of individual specific study questions, study design, patient inclusion/exclusion criteria, and removal of duplicated individuals. We resolve most quality issues during SwissBrod curation.

Once the within-data set preparations are complete, genes must be matched across datasets as a prelude to combination.

6.3. *Gene Matching*

Since the interpretation of study results usually makes statements about genes, matching probe/probe set expression values at the level of genes is desirable. UniGene identifiers or gene symbols can be used. Where there are multiple probes (or probe sets) of the same gene, we choose a unique one to represent the corresponding gene. We generally choose the most variable probe across samples; other common selection rules are to choose the probe with the highest or second-highest average signal across samples.

Here, we note that it is not necessary to consider only the common probes or genes common across all datasets. This unnecessarily restricts the set of genes about which inference can be made. Missing data, such as genes which are not measured in all studies, are readily accommodated in this framework for data combination.

6.4. *Outcome Modeling*

We consider within each study a generalized linear model (GLM)[28] for single genes or sets of genes, possibly including study covariates, describing the association between the measured quantity (e.g. gene expression) and an outcome of interest. The association is supposed to be captured in one (or possibly more) coefficient of the model. This class of models is sufficiently broad and flexible to handle most commonly investigated situations, including quantitative, categorical, and survival outcomes.

The models across studies are not required to contain the same or similar parameters, as we will not be combining parameter estimates. Rather, there is assumed a set of (at least conceptually) related hypotheses that can be connected through test statistics based on model coefficients. To get around heterogeneity in the underlying measurements, it is important to use relative measures within a study.

6.5. *Z-Transform for Combining Test Statistics*

Because we cannot in general expect data from different studies to be commensurable, we combine information farther down the spectrum from the raw values. In addition, we do not require a common (commensurable) parameter across studies, so we also skip past combining parameter estimates. But in order to preserve as much information and flexibility as possible, we do not want to reduce to ranks or decisions. Since we are able to estimate all single-gene models in each study, we are in a position to combine the single-gene statistics across studies.

A useful statistic for the purpose of combining is the z-score. Within the GLM framework, this statistic is often straightforward to obtain. In the simplest case, combining tests of single coefficients, one option is to combine the K study standardized coefficients

$$Z_i = \hat{\beta}_i / \mathrm{SE}(\hat{\beta}_i),$$

where i indexes genes. For sufficiently large sample sizes, Z_i are each approximately distributed as standard normal under the null hypothesis H: $\beta = 0$. An alternative which is approximately equivalent is to use the signed square root of the deviance of the likelihood ratio test for one additional parameter[28]:

$$Z_D = \mathrm{sgn}(\hat{\beta})\sqrt{\mathrm{D}}.$$

Where z-statistics are not readily available, either the individual model statistics may be transformed to yield an approximate standard normal or the corresponding p-values may be probit-transformed to yield a z-score.

The single-gene individual study z-scores Z_{ij} are then combined metaanalytically over the K studies using equal weighting by the inverse normal method[7]:

$$\bar{Z}_i = \sum_{k=1}^{K_i} Z_{ik} / \sqrt{K_i}\,, \tag{1}$$

where i indicates genes, k indicates study, and K_i is the number of datasets in which gene i is present (that is, any platform missing the gene is ignored). The resulting $\bar{Z}$ should be (approximately) distributed as standard normal ($N(0, 1)$), and can be ranked according to size or p-value.

We note that this procedure is not necessarily optimal for every situation. For example, if all test statistics from the individual studies have a common distribution that is not Gaussian, then a specialized procedure which is more powerful can generally be constructed. In addition, we use equal weighting for each study. It may instead be desired to use different weights, based on sample size, variance, or quality measures, for instance. However, optimal weights will depend on the alternatives of interest, which may also vary between studies. We have found this equal weight form to be quite useful for large-scale, automated combining and exploratory analysis of studies, for which the models may be very heterogeneous.

6.6. *Multiple Testing*

An issue in large-scale genomic studies is the multiplicity problem when testing thousands of null hypotheses. Although often ignored in meta-analyses of genomic data, adjustment of p-values is needed to provide a realistic assessment of significance for each gene.

We carry out p-value adjustment based on the final combined statistics $\bar{Z}_i$. Among the most utilized adjustments are the family-wise error rate (FWER)-controlling Bonferroni correction, not recommended due to overly conservative p-values but is very quick to compute, and FDR-adjusted p-values.[29] We typically adjust using either FDR or max$T$,[17] another FWER-controlling adjustment which is less conservative than Bonferroni correction as it takes between-gene correlations into account.

To compute max$T$, the joint null distribution of test statistics is estimated by bootstrapping. Bootstrap replicates are obtained for each individual study and then modeled and analyzed as described above, resulting in a new set of combined z-scores, $\bar{Z}_i^*$. Then, for each bootstrap replicate, the maximum value of $\bar{Z}_i^*$ is chosen, yielding a null distribution of max $\bar{Z}^*$, from which the final adjusted p-values are obtained.

7. Breast Cancer Examples

We present two applications to breast cancer gene expression data integration, focusing on methodology rather than biological interpretation.

We collected publicly available breast cancer survival datasets from repositories such as GEO and ArrayExpress, as well as from journal articles, selecting those produced on whole-genome microarrays with medium to large sample sizes (Tables 1 and 2). Small numbers of nonmalignant samples (normal breast tissue or fibroadenoma) were present in some datasets. Almost all malignant tumors were invasive ductal carcinoma.

Since multiple publications sometimes reuse the same patients, we created datasets with unique patients by merging some publication-based datasets or removing redundant patients. We used processed gene expression values ($\log_2$ expression or ratio) as provided by the original studies without further normalization.

Hybridization probes were remapped to Entrez GeneID[51] through sequence alignment against the well-curated subset of the RefSeq mRNA sequence database. This mapping procedure is conservative, but it ensures high-quality cross-platform matching. Within a study, multiple probes of the same genes were made unique by choosing the most variable probe across samples to represent the gene.

Only 1963 genes were present in all platforms. To avoid discarding useful information about many genes, we performed meta-analyses on the union of all 17198 genes. Summary statistics of absent genes were considered as missing values.

Pooling patients from heterogeneous datasets to treat them as if they were from a single cohort may result in false associations. Therefore, we stratified all analyses by dataset and combined only summary statistics (such as *z*-scores of regression models).[7] This approach also circumvents the problem of combining potentially incommensurable expression measures from different microarray datasets. The *z*-scores are not affected by arbitrary shifting or scaling of the expression data matrix of each dataset.

Table 1. Publicly available breast cancer survival datasets.

Dataset symbol	No. of arrays	Institution	References	Platform	Data source	No. of GeneIDs
NKI	337	Nederlands Kanker Instituut	30,31	Agilent	author website	13120
EMC	286	Erasmus Medical Center	32	Affy U133A	GEO:GSE2034	11837
UPP	249	Karolinksa Institute (Uppsala)	33,34	Affy U133A,B	GEO:GSE4922	15684
STOCK	159	Karolinska Institute (Stockholm)	34,35	Affy U133A,B	GEO:GSE1456	15684
DUKE	171	Duke University	36,37	Affy U95Av2	author website	8149
UCSF	161+8	UC San Francisco	38	cDNA	author website	6178
UNC	143+10	University of Carolina	39	Agilent HuA1	author website	13784
NCH	135	Nottingham City Hospital	40	Agilent HuA1	AE:E-UCON-1	13784
STNO	115+7	Stanford + Norwegian Radium Hosp.	41,42	cDNA	author website	5614
JRH1	99	John Radcliffe Hospital	43	cDNA	journal website	4112
JRH2	61	John Radcliffe Hospital	44	Affy U133A	GEO:GSE2990	11837
MGH	60	Massachusetts General Hospital	45	Agilent	GEO:GSE1379	11421
Total	2530	= 2505 carcinomas + 25 nonmalignant breast tissues		No. of the union of all GeneIDs:		17198
				No. of GeneIDs common to genomic platforms:		1963

Abbreviations: No., number; GEO:, Gene Expression Omnibus accession; AE:, ArrayExpress accession; Affy, Affymetrix.

Table 2. Publicly available breast cancer gene signature datasets.

		Number of genes	
Signature symbol	**Reference**	**Original Probes**	**Mapped to GeneID**
ONC-16	46	16	16
NKI-70	30	70	52
EMC-76	32	60+16	48+12
NCH-70	40	70	69
CON-52	47	52	50
p53–32	33	32	19
CSR	48	512	457
IGS	49	—	—
GGI-128	44	128	98
CCYC	50	NA	126

7.1. *Example I: Breast Cancer Survival*

A Cox model of metastasis-free survival was fitted separately for each gene (no covariates were included in this example). For individual i and gene j, we model the instantaneous failure rate, or hazard function $h(t)$, as a function of the expression level x_{ij} with the Cox proportional hazard model:

$$h(t) = h_0(t) \exp(\beta_j x_{ij}). \tag{2}$$

For each gene, there is an estimated coefficient $\hat{\beta}_j$. These estimated coefficients are standardized and then combined across studies to obtain $\bar{Z}_j$.

Figure 1 shows scatter plots of single-gene z-scores (standardized Cox model regression coefficients) from the two largest studies, NKI and EMC. These scatter plots illustrate three rules for determining the significance of combined results: combined Z, Fisher p-value combination, and the Venn diagram rule requiring genes to be selected by both studies. Equal-significance contours for each rule are given for four significance levels ($\alpha = 0.01$, 0.001, 0.0001, and 0.00001). One-sided tests for only large values of the relevant statistic are depicted; the largest negative values are tested similarly.

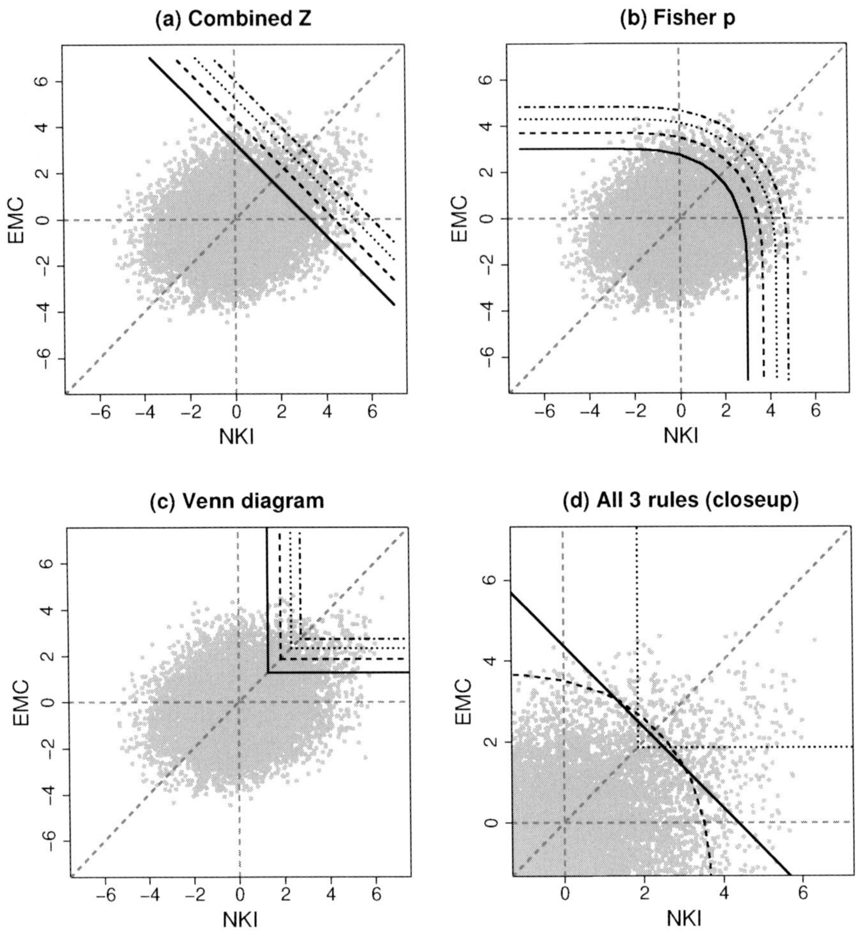

Fig. 1. Contours of equal values at significance levels $\alpha = 0.01$ (solid line), 0.001 (dashed line), 0.0001 (dotted line), and 0.00001 (dotted-dashed line): (a) combined Z, (b) Fisher combined p, and (c) Venn diagram. (d) All three rules ($\alpha = 0.001$).

In Fig. 1(a); the rule for combined Z (i.e. $\bar{Z}$) with equal weighting is shown. Contours for this rule are straight lines in this two-dimensional case, and are more generally hyperplanes. Equal weighting implies a slope of -1; use of other weights just changes the angle of the line. Figure 1(b) gives the contours for the Fisher combined p-statistic, while Fig. 1(c) shows contours for the Venn diagram rule. Figure 1(d) shows a closeup, providing a sense of how all three rules are related ($\alpha = 0.001$ only). The Venn diagram rule gives up much power by stringent within-study requirements. While the Fisher rule picks up genes that provide moderate to strong evidence in both studies, it also calls significant genes with very strong positive evidence in one study yet moderate to strong negative evidence in the other (points in quadrants II and IV). This is essentially because these points are closer to quadrant I (the alternative hypothesis) than to the origin (the null). The test based on $\bar{Z}$ does not suffer so much from this drawback, and picks up some power in the range of roughly equal moderate evidence in both studies.

Table 3 gives the z-scores from individual studies for genes with $\bar{Z} > 8$ and $\bar{Z} < -7$. Genes not measured in an individual study have blank entries. The table also shows the number of overall top and bottom genes occurring in the top and bottom 100 genes in each individual study. The utility of combining results is clearly demonstrated here.

7.2. *Example II: Breast Cancer Gene Signatures*

In the previous example, we focused on combining information on the effect on survival of expression of single genes. Here, we show how meta-analysis can also be used to integrate patterns of gene expression, or gene "signatures". Signatures can be useful for tumor subtyping, diagnosis, or prognosis.

Disparate breast cancer gene expression signatures have been proposed, with little agreement in the constituent genes.[30,32,33,37,40,44–46,52] We show how our meta-analytical approach can be used to uncover relationships that are consistent in a large collection of public datasets, and are thus unlikely to be artifacts of specific cohorts or microarray platforms.

We introduce the concept of "coexpression modules", comprehensive lists of genes with highly correlated expression, which we use to analyze gene expression signatures.

Table 3. Individual study z-scores for top ($\bar{Z} > 8$) and bottom ($\bar{Z} < -7$) genes.

Gene ID	$\bar{Z}$	NKI	DUKE	UCSF	STNO	JRH1	MGH	UPP	STOCK	EMC	UNC	JRH2
AURKA	9.67	6.33	1.09	2.33	3.05	1.83	1.56	3.38	3.28	4.52	3.55	1.16
CCNB2	9.17	5.56	3.95				1.17	3.67	4.18	3.64	2.70	1.05
MELK	8.82	4.51	4.10			2.77		3.64	3.84	3.31	2.11	0.66
MYBL2	8.79	4.94	3.20	0.56	3.38	2.73	1.23	4.37	3.02	2.61	3.01	0.11
BUB1	8.70	4.43	1.15	1.24	3.65	2.63	0.79	2.88	4.24	3.37	2.78	1.69
AURKB	8.47	5.01	4.12	−0.12	3.56	2.09		3.44	3.71	1.15	3.00	0.84
RACGAP1	8.47	5.48					0.48	4.24	3.76	4.91	1.99	1.56
CENPA	8.40	5.75	2.43	2.35				3.41	3.70	2.84	2.19	1.09
DDX39	8.35	5.49	3.29				1.09	3.53	4.49	2.71	1.15	1.89
UBE2C	8.32	5.63	3.56	1.15	2.07	0.66		3.68	3.48	3.43	1.70	0.94
FEN1	8.15	5.31	1.43	0.81	1.92	1.99		4.49	3.28	2.47	3.05	1.00
DLG7	8.13	4.31	2.64	0.88	3.14	1.27		3.18	3.96	3.75	1.81	0.77
DKFZp762E1312	8.12	6.10					1.68	4.00	3.72	2.52	2.73	0.74
TRIP13	8.02	4.97	3.11	0.53	2.90	0.71		4.33	3.79	1.34	2.68	1.01
# in top 100		38	16	3	11	7	0	26	15	16	6	2
CRIM1	−7.25	−4.43	−1.39	−1.11	−2.02	−0.73	−2.97	−3.14	−5.06	−2.09	−1.70	0.63
PTGER3	−7.22	−4.44	−2.38	−0.30	−0.95	−1.18	−3.46	−2.91	−2.38	−2.66	−1.70	−1.59
LRIG1	−7.07	−4.45	−2.53		−2.50		−3.03	−0.34	−3.69	−2.31	−2.33	−0.01
# in bottom 100		3	0	0	1	0	1	1	2	0	0	0

7.2.1. *Prototype-based coexpression module analysis*

To identify coexpression modules associated with specific biological processes, we have devised a supervised approach where a small number of "prototype" genes, each representing an important disease process, are selected based on biological knowledge about breast cancer[53] and previous results of expression studies. Each prototype forms the core of a coexpression module. Expression values of these prototype genes are then used simultaneously as explanatory variables in a regression model to group other genes according to their coexpression with the respective prototype. The modules are created by adding genes based on the association of expression with prototype expression.

For breast cancer, we have identified five key processes: estrogen receptor signaling, ERBB2 amplification, proliferation, invasion, and immune response. We represent these processes with prototype genes, respectively, ESR1, ERBB2, AURKA (aurora-related kinase 1; also known as STK6 or STK15), PLAU (urokinase-type plasminogen activator; uPA), and STAT1 (signal transducer and activator of transcription 1). Other choices of well-known genes for the prototypes do not affect the overall conclusions.

To identify genes associated with each prototype, we use the following meta-analysis scheme:

(1) within a dataset and for each gene separately, fit a multiple regression which models expression as a function of prototype expression;
(2) carry out a t-test for each coefficient, yielding an approximate z-score;
(3) combine z-scores across studies as above using the inverse normal method; and
(4) select for each prototype the genes most strongly associated.

7.2.2. *Model for identifying coexpression modules*

The expression levels of the prototype genes on the $\log_2$ scale are used as explanatory variables in a multiple regression with Gaussian error, using the following equation (gene symbols stand for their log expression and coefficients are omitted for clarity):

$$Y_i = \mathrm{ESR1} + \mathrm{ERBB2} + \mathrm{AURKA} + \mathrm{PLAU} + \mathrm{STAT1}, \qquad (3)$$

where the response variable Y_i is the expression of gene i. This model is fitted separately for each gene i in the array. The association between gene i and prototype j conditional on all other prototypes is tested using the t-statistic for each coefficient. Because the t-statistics for different datasets have different degrees of freedom, we put them all on the same scale by transforming to the corresponding cumulative probabilities and then to z-scores using the inverse standard normal cumulative distribution function.

The linear model in Eq. (3) is fitted separately to each gene in each dataset, and the z-scores are combined meta-analytically using the inverse normal method [Eq. (1)].[7] Genes with large values of $\bar{Z}$ are assigned to the module for which the association is highest. A stringent criterion for $|\bar{Z}|$ is used to maintain interpretability and to keep the modules to a manageable size.

7.2.3. *Coexpression patterns*

Coexpression patterns of those genes assigned to modules are shown in heat maps for two of the datasets (NKI and EMC) in Fig. 2, which also shows survival information for each patient. There are three major subgroupings of samples, corresponding to combinations of conventional markers. The tumors (columns) were sorted first according to breast cancer subtypes — (1) basal-like ER–/ERBB2–, (2) ERBB2+, and (3) luminal ER+/ERBB2– — and then, within each subtype, according to the average expression of proliferation genes in the AURKA module. For clarity, the five modules are horizontally separated in the figure.

Each module contains highly correlated or anticorrelated genes, as shown by the vertical color patterns. The annotation of the modules shows that they correspond well to the expected biological processes. The strong banding patterns show that the modules corresponding to estrogen

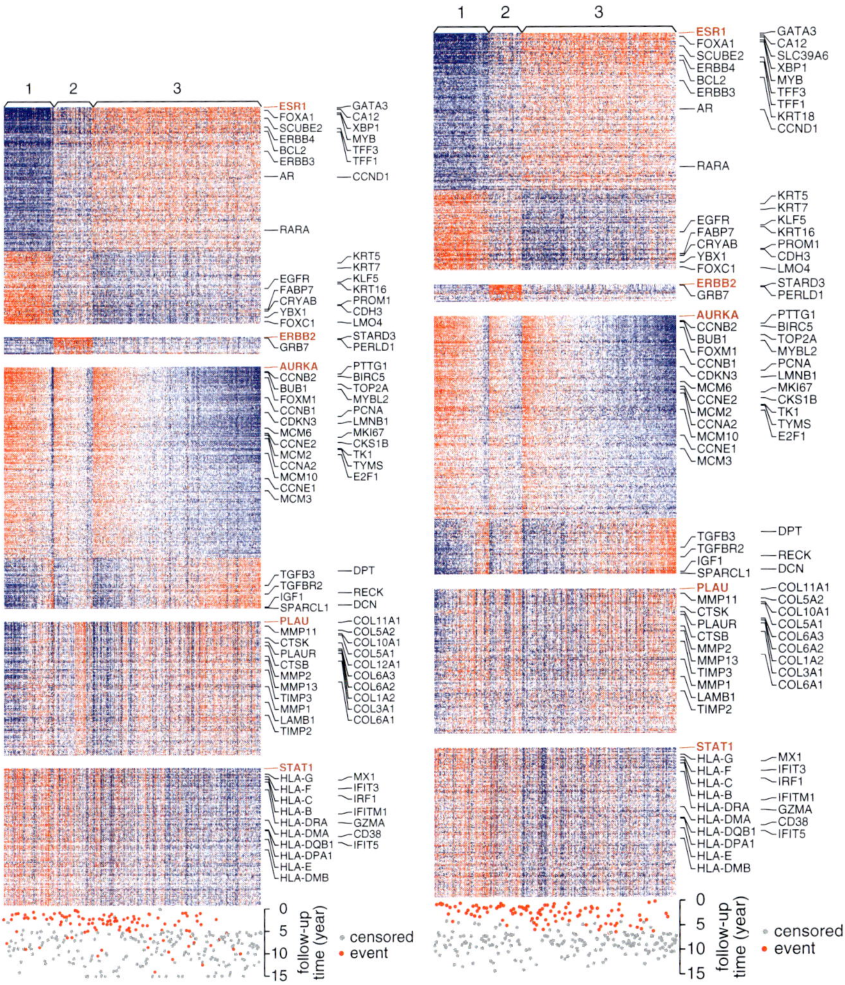

(a) **NKI** (b) **EMC**

Fig. 2. Coexpression module heat maps for the (a) NKI and (b) EMC datasets. The expression level from high to low is coded with a red-white-blue color gradient.

receptor signaling, ERBB2 amplification, and proliferation (ESR1, ERBB2, and AURKA) discriminate well between the three subgroups. The modules for invasion and immune response (PLAU and STAT1) do not show a distinct pattern.

The correlated expression measures in a module provide redundant information about the module's overall expression in a tumor. They can be summarized into a single number by averaging. We called the resulting value a "module score". In addition to examining genes in the modules, we can study properties of the module scores, for example, their effectiveness in tumor subtyping or prognostic value. We have found that this type of examination can produce not only effective tumor classifications or predictions, but also new biological insights.

Focusing for now on AURKA, we look for an association with survival. For each gene, we combine the z-scores from Cox regression [Eq. (2)] according to Eq. (1), and consider these as a function of the combined z-score for the AURKA regression coefficient [Eq. (3)].

Figure 3(a) shows single-gene z-scores for the AURKA coefficient within the NKI and EMC signature datasets. The genes corresponding to each signature are highlighted in the plot, showing that the genes comprising the signature are largely different and also span the range of Z-AURKA.

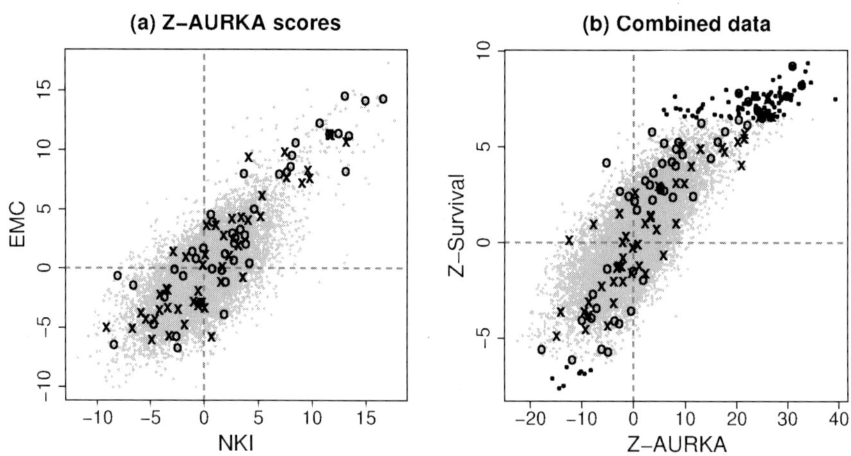

Fig. 3. Gene signature prognostic performance for NKI (o), EMC (x), and all datasets combined (•). (a) Within-dataset AURKA z-scores for NKI and EMC; (b) overall $\bar{Z}$ for survival vs. $\bar{Z}$-AURKA.

In Fig. 3(b), which shows the overall combined scores $\bar{Z}$, we can see that there is a strong positive correlation between survival and Z-AURKA ($\rho = 0.73$). Genes corresponding to the NKI and EMC signatures are again highlighted. Also highlighted are the integrated signature genes, defined by choosing those genes most strongly associated with survival based on the combined Cox regression, $|\bar{Z}| \geq 6.5$. It is notable that many NKI and EMC signature genes are clustered near the middle of the survival distribution, indicating that they are not predictive of survival across studies.

8. Conclusion

Meta-analytical approaches are valuable not only for increasing the sample size (thus decreasing sampling error artifacts), but also for reducing the contribution of strong but nonreproducible associations caused by platform or cohort-specific biases.

Here, we have outlined a general procedure for combining information across independent genomic studies. Our procedure relies on the availability and careful curation of primary data from the original studies. The strategy produces a straightforward combination of z-scores, which can be used to rank genes according to evidence for the study question of interest. We have assumed sufficiently large sample sizes so that the test statistics should each be approximately normally distributed. This assumption will be satisfied for many clinical studies, but is less likely to hold in the case of experimental studies. Studies with smaller sample sizes may be included as well for a suitably transformed statistic.

Because we use all genes measured in each study rather than focusing on only those genes found to be associated with outcome, conclusions based on the combined analysis should be less subject to the problem of publication bias. In addition, the original data handling and analysis are less likely to impact conclusions of meta-analysis based on the data rather than with the original data summaries.

We have also assumed that each study is independent, but it is not difficult to envision scenarios where independence is violated. For example, it is increasingly common to obtain longitudinal data on the same set of patients. Hartung[54] has shown that, by modifying the weights of the inverse normal method, dependent statistics can be accommodated. Thus, our procedure should readily extend to this more general case.

We have found the concept of coexpression modules to be a versatile tool for biologically unifying disparate results and producing an integrated gene signature. Although coexpression does not imply direct physical interactions, the highly correlated genes in a module can be considered surrogate markers of one another and of the same underlying transcriptional process. Thus, coexpression is more appropriate for understanding the equivalence of signatures than for functional annotations of the genes. Coexpression modules can also be used to dissect signatures, revealing the parts that are essential. The strong correlation of expression within a module allows summarizing the module's overall expression by simple averaging. These module scores concisely and robustly characterize a tumor by a handful of quantitative measures with straightforward interpretation.

In summary, we provide a unified, flexible, and extensible framework for integrated analysis of heterogeneous genomic datasets. We have demonstrated here how this framework can be used to unify results of previous gene expression studies in breast cancer. These methods are practical and widely applicable to a variety of technologies and biological investigations.

Acknowledgments

This work was supported by the European Commission Framework Programme VI (FP6-LSHC-CT-2004-503426) and by the Swiss National Science Foundation through the National Centres for Competence in Research in Plant Survival and Molecular Oncology.

References

1. Edgar R, Domrachev M, Lash AE. (2002) Gene Expression Omnibus: NCBI gene expression and hybridization array data repository. *Nucleic Acids Res* **30**: 207–10.

2. Barrett T, Suzek TO, Troup DB *et al.* (2005) NCBI GEO: mining millions of expression profiles — database and tools. *Nucleic Acids Res* **33**: D562–6.
3. Parkinson H, Sarkans U, Shojatalab M *et al.* (2005) ArrayExpress — a public repository for microarray gene expression data at the EBI. *Nucleic Acids Res* **33**: D553–5.
4. Sutton AJ, Abrams KR, Jones DR *et al.* (2000) *Methods for Meta-analysis in Medical Research.* New York, NY: John Wiley & Sons.
5. Chalmers I. (1993) The Cochrane Collaboration: preparing, maintaining, and disseminating systematic reviews of the effects of health care. *Ann NY Acad Sci* **703**: 156–65.
6. Punt CJA, Buyse M, Köhne CH *et al.* (2007) Endpoints in adjuvant treatment trials: a systematic review of the literature in colon cancer and proposed definitions for future trials. *J Natl Cancer Inst* **99**: 998–1003.
7. Hedges LV, Olkin I. (1985) *Statistical Methods for Meta-analysis.* New York, NY: Academic Press.
8. Altman DG, Deeks JJ. (2002) Meta-analysis, Simpson's paradox and the number needed to treat. *BMC Med Res Methodol* **2**: 3.
9. Benito M, Parker J, Du Q *et al.* (2004) Adjustment of systematic microarray data biases. *Bioinformatics* **20**: 105–14.
10. Goldstein DR, Delorenzi M, Luthi-Carter R, Sengstag T. (2009) Meta-analysis of microarray studies. In: Guerra R, Goldstein DR (eds.). *Meta-analysis and Combining Information in Genetics.* Boca Raton, FL: Chapman-Hall/CRC Press.
11. Choi JK, Yu U, Kim S, Yoo OJ. (2003) Combining multiple microarray studies and modeling interstudy variation. *Bioinformatics* **19**(Suppl 1): i84–90.
12. Cooper HM, Hedges LV. (1994) *The Handbook of Research Synthesis.* New York, NY: Russell Sage Foundation.
13. Stevens JR, Doerge RW. (2005) Combining Affymetrix microarray results. *BMC Bioinformatics* **6**: 57.
14. Grützmann R, Boriss H, Ammerpohl O *et al.* (2005) Meta-analysis of microarray data on pancreatic cancer defines a set of commonly dysregulated genes. *Oncogene* **24**: 5079–88.
15. Ghosh D, Barette TR, Rhodes D, Chinnaiyan AM. (2003) Statistical issues and methods for meta-analysis of microarray data: a case study in prostate cancer. *Funct Integr Genomics* **3**: 180–8.
16. Kulinskaya E, Morgenthaler S, Staudte RG. (2008) *A Guide to Calibrating and Combining Statistical Evidence.* New York, NY: Wiley-Interscience.
17. Westfall PH, Young SS. (1993) *Resampling-based Multiple Testing: Examples and Methods for p-Value Adjustment.* New York, NY: Wiley.
18. Yekutieli D, Benjamini Y. (1999) Resampling-based false discovery rate controlling multiple test procedures for correlated test statistics. *J Stat Plan Inference* **82**: 171–96.

19. Fisher RA. (1932) *Statistical Methods for Research Workers*, 4th ed. Oxford, UK: Oxford University Press.
20. Rhodes DR, Barrette TR, Rubin MA *et al.* (2002) Meta-analysis of microarrays: interstudy validation of gene expression profiles reveals pathway dysregulation in prostate cancer. *Cancer Res* **62**: 4427–33.
21. Wise LH, Lanchbury JS, Lewis CM. (1999) Meta-analysis of genome searches. *Ann Hum Genet* **63**: 263–72.
22. Breitling R, Armengaud P, Amtmann A, Herzyk P. (2004) Rank products: a simple, yet powerful, new method to detect differentially regulated genes in replicated microarray experiments. *FEBS Lett* **573**: 83–92.
23. Cahan P, Ahmad AM, Burke H *et al.* (2005) List of lists-annotated (LOLA): a database for annotation and comparison of published microarray gene lists. *Gene* **360**: 78–82.
24. Pirooznia M, Nagarajan V, Deng Y. (2007) GeneVenn — a web application for comparing gene lists using Venn diagrams. *Bioinformation* **10**: 420–2.
25. Wilkinson B. (1951) A statisical consideration in psychological research. *Psychol Bull* **48**: 156–8.
26. Birnbaum A. (1954) Combining independent tests of significance. *J Am Stat Assoc* **49**: 559–74.
27. Koziol JA, Perlman MD. (1978) Combining independent chi-squared tests. *J Am Stat Assoc* **73**: 753–63.
28. McCullagh P, Nelder JA. (1989) *Generalized Linear Models*, 2nd ed. London, UK: Chapman & Hall.
29. Benjamini Y, Hochberg Y. (1995) Controlling the false discovery rate: a practical and powerful approach to multiple testing. *J R Stat Soc Ser B* **57**: 289–300.
30. van't Veer LJ, Dai H, van de Vijver MJ *et al.* (2002) Gene expression profiling predicts clinical outcome of breast cancer. *Nature* **415**: 530–6.
31. van de Vijver MJ, He YD, van't Veer LJ *et al.* (2002) A gene-expression signature as a predictor of survival in breast cancer. *N Engl J Med* **347**: 1999–2009.
32. Wang Y, Klijn JGM, Zhang Y *et al.* (2005) Gene-expression profiles to predict distant metastasis of lymph-node-negative primary breast cancer. *Lancet* **365**: 671–9.
33. Miller LD, Smeds J, George J *et al.* (2005) An expression signature for p53 status in human breast cancer predicts mutation status, transcriptional effects and patient survival. *Proc Natl Acad Sci USA* **102**: 13550–5.
34. Calza S, Hall P, Auer G *et al.* (2006) Intrinsic molecular signature of breast cancer in a population-based cohort of 412 patients. *Breast Cancer Res* **8**: R34.
35. Pawitan Y, Bjöhle J, Amler L *et al.* (2005) Gene expression profiling spares early breast cancer patients from adjuvant therapy: derived and validated in two population-based cohorts. *Breast Cancer Res* **7**: R953–64.
36. Huang E, Cheng SH, Dressman H *et al.* (2003) Gene expression predictors of breast cancer outcomes. *Lancet* **361**: 1590–6.

37. Bild AH, Yao G, Chang JT *et al.* (2006) Oncogenic pathway signatures in human cancers as a guide to targeted therapies. *Nature* **439**: 353–7.
38. Korkola JE, DeVries S, Fridlyand J *et al.* (2003) Differentiation of lobular versus ductal breast carcinomas by expression microarray analysis. *Cancer Res* **63**: 7167–75.
39. Hu Z, Fan C, Oh DS *et al.* (2006) The molecular portraits of breast tumors are conserved across microarray platforms. *BMC Genomics* 7: 96.
40. Naderi A, Teschendorff AE, Barbosa-Morais NL *et al.* (2006) A gene-expression signature to predict survival in breast cancer across independent data sets. *Oncogene* **26**: 1507–16.
41. Sorlie T, Perou CM, Tibshirani R *et al.* (2001) Gene expression patterns of breast carcinomas distinguish tumor subclasses with clinical implications. *Proc Natl Acad Sci USA* **98**: 10869–74.
42. Sorlie T, Tibshirani R, Parker J *et al.* (2003) Repeated observation of breast tumor subtypes in independent gene expression data sets. *Proc Natl Acad Sci USA* **100**: 8418–23.
43. Sotiriou C, Neo SY, McShane L *et al.* (2003) Breast cancer classification and prognosis based on gene expression profiles from a population-based study. *Proc Natl Acad Sci USA* **100**: 10393–8.
44. Sotiriou C, Wirapati P, Loi S *et al.* (2006) Gene expression profiling in breast cancer: understanding the molecular basis of histologic grade to improve prognosis. *J Natl Cancer Inst* **98**: 262–72.
45. Ma XJ, Wang Z, Ryan PD *et al.* (2004) A two-gene expression ratio predicts clinical outcome in breast cancer patients treated with tamoxifen. *Cancer Cell* **5**: 607–16.
46. Paik S, Shak S, Tang G *et al.* (2004) A multigene assay to predict recurrence of tamoxifentreated, node-negative breast cancer. *N Engl J Med* **351**: 2817–26.
47. Teschendorff AE, Naderi A, Barbosa-Morais NL *et al.* (2006) A consensus prognostic gene expression classifier for ER positive breast cancer. *Genome Biol* 7: R101.
48. Chang HY, Sneddon JB, Alizadeh AA *et al.* (2004) Gene expression signature of fibroblast serum response predicts human cancer progression similarities between tumors and wounds. *PLoS Biol* **2**: 206–14.
49. Liu R, Wang X, Chen GY *et al.* (2007) The prognostic role of a gene signature from tumorigenic breast-cancer cells. *New Engl J Med* **356**: 217–26.
50. Whitfield ML, Sherlock G, Saldanha AJ *et al.* (2002) Identification of genes periodically expressed in the human cell cycle and their expression in tumors. *Mol Biol Cell* **13**: 1977–2000.
51. Maglott D, Ostell J, Pruitt KD, Tatusova T. (2005) Entrez Gene: genecentered information at NCBI. *Nucleic Acids Res* **33**: D53–8.

52. Chang HY, Nuyten DSA, Sneddon JB *et al.* (2005) Robustness, scalability and integration of wound-response gene expression signature in predicting breast cancer survival. *Proc Natl Acad Sci USA* **102**: 3738–43.
53. Esteva FJ, Hortobagyi GN. (2004) Prognostic molecular markers in early breast cancer. *Breast Cancer Res* **6**: 109–18.
54. Hartung J. (1999) A note on combining dependent tests of significance. *Biom J* **41**: 849–55.

Chapter 5

Computational Biology of Small Regulatory RNAs

Mihaela Zavolan and Lukasz Jaskiewicz

1. Introduction

The discovery of the let-7 microRNA (miRNA) and the realization that small RNA regulators are conserved over large evolutionary distances[1,2] prompted a vast number of studies aiming to uncover these molecules across species and to characterize their function. Specialized protocols for isolation of miRNAs[3–5] lead to the identification of large numbers of mammalian miRNAs whose function — in contrast to that of the founder of the class, the lin-4 miRNA[6] — has yet to be discovered. Nonetheless, initial efforts aimed to generate catalogs of miRNAs in various species. miRNAs associate with Argonaute members of the PAZ/PIWI domain (PPD) family of proteins to form RNA-induced silencing complex (RISC)-like ribonucleotide particles,[7,8] which have also been observed during silencing that occurs upon injection of small interfering RNAs (siRNAs). The first members of the PPD family to be characterized were Argonaute-1 and Zwille in *Arabidopsis thaliana*.[9,10] The genome of the worm *Caenorhabditis elegans* encodes a total of 27 PPD proteins; the mouse genome, 7; and the human genome, 8.

It appears that PPD proteins evolved to perform highly specialized functions that vary widely between species. In humans, for instance four PPD proteins (Ago-1 through Ago-4) bind miRNAs, but only Ago-2 has the catalytic activity required for the mRNA cleavage,[7,8] typically observed with siRNAs. The other PPD family members are specifically expressed in germline and stem cells,[11,12] which prompted researchers to investigate whether these proteins too are associated with small regulatory RNAs. Cloning and sequencing of libraries constructed from immunoprecipitates of PIWI proteins indeed led to the discovery of a second class of small regulatory RNAs, the Piwi-interacting RNAs (piRNAs),[13–15] which so far have been implicated in transposon silencing.[16–18]

In the following, we give an overview of the biogenesis and function of small RNAs, and the computational tools that have been developed to support the research on these topics. The focus will be mostly on miRNAs, which up to this point have been better characterized than other small regulatory RNAs.

2. Identification of Small Regulatory RNAs

The first miRNA was discovered using the so-called "forward genetics" approach, which aims to identify mutations that produce specific phenotypes. Frequently, mutagens are used to generate a panel of mutants that are screened for the phenotype of interest; the genes responsible for the phenotype are then identified with well-established molecular biology techniques. An immediate advantage of this approach is that once the gene is identified, information about its function is already available in the form of the phenotype of the mutant. On the other hand, only genes responsible for dramatic, easily observable phenotypes can be identified this way.

The vast majority of miRNAs known today are studied using "reverse genetics" approaches, in which one starts from the gene sequence rather than from the phenotype in order to characterize gene function. The classical approach to miRNA (and generally to regulatory RNA) identification has been cloning and sequencing. The RNA population within a selected size range (obtained by fractionation) or binding a specific protein (obtained from immunoprecipitates of that protein) is ligated to short RNA adapters, amplified by reverse transcription–polymerase chain reaction (RT-PCR), concatamerized, and sequenced. Such an approach was used in the first studies aiming to identify fly, worm, and human miRNAs,[3,19,20] and even in the identification of piRNAs.[13] Recent developments in sequencing technologies[21,22] allow one to obtain millions of small RNAs from a single sample, making the task of small RNA identification much easier. These technologies have been used in the discovery of U21 RNAs in worm, piRNAs in mammals,[14] and several miRNAs that appear to have evolved rapidly in flies.[23] In parallel, approaches that start with computational miRNA gene prediction followed by microarray-based validation have also been developed and used for miRNA gene discovery.[24]

Both types of methods (cloning and sequencing, and microarrays) have also been used to determine the expression profile of miRNAs across tissues. The largest database of miRNA expression profiles obtained through cloning was developed by Landgraf *et al.*,[25] and can be accessed at http://www.mirz.unibas.ch/smiRNAdb/. Expression profiling is more commonly done using microarrays, with many different platforms — using modified and unmodified DNA, RNA, and locked-nucleic acid (LNA) probes[26–32] — having been developed.

3. Classes of Small Regulatory RNAs

3.1. *miRNAs*

miRNAs are noncoding RNAs that regulate gene expression in a sequence-specific manner (reviewed by Bartel[33]). They are generated from the genome-encoded precursor hairpins[3] by the sequential action of two complexes containing RNase III-type nucleases: Drosha in the nucleus,[34] with its partner Pasha[35]; and Dicer in the cytoplasm,[36] with its partner known as Loquacious in *Drosophila melanogaster*[37] and the human immunodeficiency virus (HIV) transactivating response RNA-binding protein (TRBP) in humans.[38,39]

The founding member of the miRNA family, lin-4, was discovered in nematode *C. elegans* through a genetic screen for defects in the temporal control of postembryonic development.[40] Mutations in lin-4 disrupt the temporal regulation of larval development, causing the first larval stage-specific cell-division pattern (L1) to reiterate at later developmental stages.[40,41] The opposite developmental phenotype is observed in worms that are deficient in lin-14.[42] While most genes identified from mutagenesis screens encode proteins, lin-4 encodes a 22-nucleotide non-coding RNA with partial complementarity to evolutionarily conserved sites located in the 3′ untranslated region (3′ UTR) of the lin-14 gene,[6,43] which encodes a nuclear protein involved in the L1-to-L2 transition of larval development. The discovery of lin-4 and its target-specific translational inhibition pointed to a new mechanism of gene regulation during development, yet until the discovery of the let-7 miRNA this mechanism was thought to be specific to worms. The let-7 miRNA also encodes a temporally regulated 21-nucleotide small RNA, which controls the developmental transition from the last larval stage (L4) to the adult stage.[1] Similar to lin-4, let-7 has partial complementarity to its targets (among which are lin-41 and hbl-1[44,45]), whose translation it inhibits. Both let-7 and its target lin-41 are evolutionarily conserved throughout metazoans, with homologs that were detected in mollusks, sea urchins, flies, mice, and humans.[2] This degree of conservation strongly pointed to a general role of small RNAs in developmental regulation. Orthologs of lin-4 were also later identified in flies and mammals.[4,46] Many subsequent studies contributed to the catalog of 5234 mature miRNA forms (564 from human) that are currently present in the miRBase repository.[47] While initially the focus was on identification of deeply conserved miRNAs,[48] it has recently been proposed that many miRNAs of recent evolutionary origin exist,[24] although, at least in this particular study, the clustered miRNAs expressed in the placenta share intriguing sequence similarities with more deeply conserved miRNAs.

miRNAs have also been found in the genomes of several DNA viruses, with the Epstein–Barr virus (EBV) being the first found to encode miRNAs.[5] Computational and experimental analysis further revealed miRNAs in other *Herpesviridae* such as the γ-herpesvirus Kaposi's sarcoma-associated virus (KSHV), the γ-herpesvirus murine herpesvirus 68 (MHV68), and the β-herpesvirus human cytomegalovirus (HCMV).[49–52] miRNA gene prediction provides no evidence that other herpesviruses, like the α-herpesvirus HHV3 (varicella-zoster virus) or β-herpesviruses HHV6 and HHV7, encode miRNAs.[49] Besides *Herpesviridae*, miRNAs have been found in the genomes of other viruses, particularly polyomaviruses. Computational prediction first indicated the presence of a miRNA in the BK virus,[49] and a study of the Simian virus 40 (SV40)[53] identified one pre-miRNA expressed from the opposite strand of the T antigen. RNA viruses like yellow fever virus, human immunodeficiency virus (HIV), and hepatitis C virus (HCV) appear to lack miRNAs.[49] As is the case with most miRNAs, the functions of viral miRNAs are only beginning to be characterized. The available data point to the role of viral miRNA in regulating viral gene expression. For instance, the EBV miRNA miR-BART2 is expressed from the complementary strand of the BALF5 transcript, which encodes EBV DNA polymerase, suggesting that the miRNA is involved in the cleavage and downregulation of the viral polymerase.[49] This function of miR-BART2 has been very recently demonstrated.[54] Similarly, the SV40 miRNA was shown to downregulate early viral mRNAs by cleavage, leading to reduced T antigen expression and lower interferon-γ, and allowing the virus to evade the immune response.[53] Finally, the KSHV-encoded miR-K12-10 is located in the open reading frame (ORF) of the kaposin gene, which triggers cell transformation, so processing of the kaposin ORF would downregulatate the kaposin expression.[49,51]

3.2. *Piwi-Interacting RNAs*

The novel class of Piwi-interacting associated small RNAs (piRNAs) has only recently been discovered. They are generated by a Dicer-independent mechanism, are longer than the miRNAs (24–30 nt); and associate with members of the Piwi subfamily of Argonaute proteins.[13–15] Piwi proteins are required for male and female fertility,[55] and are involved in the silencing of the transposons and repeat elements in *Drosophila*.[56] In fact, a cloning study initially identified small RNAs derived from repeat elements in *Drosophila* and called them repeat-associated small interfering RNAs (rasiRNAs)[57]; these were later found to associate with Piwi proteins[16–18] and, for this reason, were renamed piRNAs. In mammals, piRNAs are needed for spermatogenesis and appear responsible for the stability of a subset of mRNAs.[58,59] Various subsets appear to exist, associating with different Piwi proteins in a developmental stage-specific manner.[13,60] The presence of Piwi protein in zebrafish has also been demonstrated; Ziwi, the Piwi homolog in fish, binds piRNAs that derive from both single-stranded transcripts and repetitive elements.[61]

4. Biogenesis of Small Regulatory RNAs

4.1. *miRNA Biogenesis*

Thousands of miRNAs have now been identified in various organisms,[62] and the principles of biogenesis that have been inferred for lin-4 and let-7 miRNAs generally apply to other miRNAs. Typically transcribed as long precursors by the RNA polymerase II,[63] miRNAs form double-stranded RNA (dsRNA) structures that are initially processed in the nucleus by the Drosha–Pasha complex to release 50–70-nucleotide-long pre-miRNAs.[34] These are transported through the exportin-5 system[64,65] to the cytoplasm, where the Dicer–TRBP complex releases the 21–23-nucleotide-long dsRNA. The strand whose 5′ end stability in the duplex is minimal is then incorporated into the RISC complex,[66] whose composition has only been fully characterized in *D. melanogaster*.[66–68] Within RISC, the miRNA acts as a guide for RNA-directed translational inhibition and mRNA degradation. The other strand, generally called miRNA*, is degraded.

4.2. *Biogenesis of piRNAs*

piRNA biogenesis is best understood in *Drosophila*. Analysis of the piRNAs interacting with the three members of the Piwi family of proteins (Piwi, Aubergine (Aub), and Ago-3) revealed that a significant subset of piRNAs associated with Ago-3 are derived from the sense strand of retrotransposons, whereas the antisense strand was predominantly associated with Piwi and Aub.[17,18] The observation that many Aub-associated small RNAs have an adenosine at position 10 led to a model in which Ago-3 is cleaving a dsRNA precursor, generating the 5′ end of rasiRNAs, but this model does not explain the strand selection exhibited by the Piwi proteins. The proposed mechanism thus couples piRNA biogenesis with their function, and implicates Piwi proteins in the silencing of repetitive elements. The biogenesis of piRNAs in mammals remains to be elucidated, although one study suggests that piRNAs also come from repeat elements and may be produced by a mechanism similar to that in flies.[60]

5. Function of Small Regulatory RNAs

Many independent experiments showed that miRNAs can reduce mRNA translation either by triggering endonucleolytic cleavage of the mRNA or by promoting repression of translation. In contrast to plants, in which most known miRNAs hybridize almost perfectly with their targets and direct endonucleolytic cleavage,[69] most metazoan miRNAs have only partial complementarity to their targets and trigger translational repression (reviewed by He and Hannon[70]). The binding of miRNA to mRNA may be facilitated by other RNA-binding proteins such as GW182/TNRC6,[71–73] which are components of the so-called P bodies,[74] and may be modulated under specific conditions by RNA-binding proteins such as HuR[75] and the Dead-end protein.[76] The function of Ago proteins in the RISC complex appears to be manyfold. The crystal structures of Argonaute proteins from *Pyrococcus furiosus*[77] and *Achaeoglobus fulgidus*[78] suggested that the PIWI domain of these proteins, which bear striking similarity to RNase H, is responsible for Slicer activity, performing the small RNA-directed endonucleolytic cleavage of mRNA in RISC. X-ray and nuclear magnetic resonance (NMR) studies of Argonaute PAZ domains, both free and complexed with RNA, revealed that the domain specifically recognizes the two-nucleotide 3′ overhang of the duplex or the 3-OH end of a single-stranded RNA.[8,79] Finally, it has been recently demonstated that the human Ago-2 directly binds the m^7G cap of mRNA targets, most likely preventing the recruitment of eIF4E and blocking initiation of translation.[80]

Although translational repression is believed to be the main mechanism of miRNA activity in animals, some amount of miRNA-induced degradation has been demonstrated.[81,82] In a specific system, namely early development of zebrafish, miRNA-induced mRNA deadenylation and degradation are the main mechanisms behind the clearance of maternal mRNAs.[83,84] To what extent miRNA binding is followed by mRNA degradation in mammals, and whether translational inhibition and mRNA degradation are miRNA- and target-specific, remains to be clarified.

As mentioned above, the function of piRNAs in silencing repeat elements may be directly coupled with their biogenesis. Whether they may also play a role in transcriptional gene silencing (TGS) remains to be further investigated. In flies, Aub and Piwi mutants with low levels of rasiRNAs[85] have been shown to lose the histone 3 lysine 9 methylation and HP1 protein binding that normally are the mark of silenced chromatin loci.[56] In mouse, knockouts of Mili and Miwi-2 show DNA demethylation at the retrotransposon loci concomitant with transposon expression.[60,86] These studies support the potential function of Piwi proteins and piRNAs in TGS mechanisms.

6. MicroRNA Gene Prediction

6.1. *What Does a miRNA Precursor Look Like?*

The discovery of the first miRNA that is strongly conserved in animals, let-7,[1,2] prompted computational biologists to design methods for genome-wide prediction of miRNA genes. The spectrum of RNA molecules that are processed by Drosha and Dicer is still unclear, but a few features that appear to be important for processing have been discovered and incorporated in various combinations in miRNA gene prediction tools.

6.1.1. *Stable hairpin precursors*

The study of Lee *et al.*[6] indicated that mature miRNAs are processed from stem-loop precursor structures; and for 3868 of the 4115 animal miRNAs that are present in the 10.1 release of the miRNA Registry (http://microrna.sanger.ac.uk/sequences/), one can indeed predict a simple stem-loop as the minimal free energy structure. Stem-loops are, however, very frequently predicted in whole genomes. For instance, Lim *et al.*[81] predicted secondary structures in regions of the human genome that were conserved in mouse. Selecting stem-loops with at least 25 base pairs and with a free energy of folding of at least 25 kcal/mol gave rise to approximately 800 000 stem-loops. This gives a frequency of approximately 1 conserved stem-loop every 7500 nucleotides. One expects that only a small proportion of these represent miRNA precursors, and that additional properties, beyond a stem-loop precursor structure, distinguish miRNA precursors from other stem-loop-forming sequences.

One important difference was reported by Bonnet *et al.*[87] By comparing the free energy of folding predicted for miRNA precursors with that predicted for variants whose sequence has been randomized while preserving the mononucleotide or dinucleotide composition, they found that miRNA precursors have significantly lower free energy. This indicates an evolutionary pressure on the miRNA precursor sequence, such as to form stable secondary structures that are presumably necessary for miRNA biogenesis. Related observations, i.e. that the free energy of folding normalized by the sequence length[88–90] and additionally by the G/C content[89,90] is higher for miRNA precursors compared to other stem-loops, have also been reported. Moreover, miRNA precursors appear to be robust with respect to mutation,[91] meaning that many of their single-point mutant neighbors fold into the same secondary structure, but this does not hold for other sequences that have the same minimum free energy folds as the miRNA precursors.

6.1.2. *Structural constraints*

A second important feature of the secondary structure of miRNA precursors is the relative symmetry of the internal loops.[48,92,93] That is, internal loops tend to involve the same number of nucleotides on the 5′ as on the 3′ side of the loop. A very recent study of the predicted three-dimensional structures of miRNA precursors suggests that the reason behind these initial observations is that these symmetrical loops with particular nucleotide-nucleotide mismatches do not disturb the docking surface of the Dicer processing enzyme.[94] Mismatched nucleotides either interact with a geometry isosteric to Watson–Crick base pairs, or stack inside the helix, or form bulges behind the docking surface of Dicer, without hindering the binding. In contrast to siRNAs, both strands of which can generally be incorporated into the RISC complex, miRNA processing typically results in only one of the 5′ or 3′ arms of the hairpin being incorporated into the Argonaute protein. This asymmetry appears to be imposed through another structural constraint, namely that the pairing of the 5′ end of the miRNA within the miRNA–miRNA* duplex is less stable than the pairing at the 5′ end of the miRNA*.

6.1.3. *Position-dependent selection strength*

The pattern of evolutionary conservation along the miRNA precursor was one of the first features to be incorporated in miRNA prediction methods. Lai *et al.*[92] studied the alignments of 24 orthologous pre-miRNAs in *D. melanogaster* and *D. pseudoobscura* and found that the mature form exhibits the maximum degree of conservation, followed by the complementary strand and then by the loop sequence, which shows the mimumum amount of conservation. Only precursors obeying this order in the number of mutations observed between *D. melanogaster* and *D. pseudoobscura* were considered as candidate miRNA precursors. Numerous other studies used this feature for miRNA gene prediction. Berezikov *et al.*[95] made the additional requirement that the putative miRNA precursor is flanked by regions of even lower conservation than the loop to predict miRNAs that are conserved between primates and rodents.

6.1.4. *Sequence composition*

There is little evidence of sequence specificity in miRNA processing, and miRNA precursors are not especially G/C-poor or G/C-rich. Mature miRNAs tend, however, to start with a U nucleotide. Of the 619 human mature miRNA forms present in miRBase, 235 (38%) start with U, even though the U content of these miRNAs is only 28%.

6.1.5. *miRNA gene prediction methods*

Many methods for miRNA gene prediction have been proposed, but they can be categorized based on a relatively small number of parameters. First, there are methods that aim to predict miRNA genes based solely on the information present in these genes, as opposed to also using information that has so far emerged about the miRNA–target interaction. Second, there are methods designed to predict miRNAs in a given genome and methods that are directly or indirectly based on evolutionary selection of miRNA precursors. Ideally, if we had an accurate description of Drosha and Dicer substrates, one would expect to be able to recognize them in individual genomes. Because such an accurate description is lacking, most miRNA gene prediction methods aim to identify miRNAs that are conserved in some number of species, or use the property that the sequences of miRNA precursors were selected during evolution (e.g. to stably fold into stem-loop structures) while random stem-loops were not. Third, miRNA gene prediction methods differ, of course, in the precise features that they use to distinguish between miRNA precursors and other types of stem-loops. Finally, these features are either used in filters that discard unlikely miRNA precursors, or used to compute scores indicative of the likelihood with which a stem-loop will be recognized and processed as a miRNA precursor.

One of the first methods for miRNA gene prediction, miRseeker, was developed by Lai *et al.*,[92] who focused on fruit fly miRNAs. The 24 fly miRNAs that were known at the time were used to gauge the quality of the prediction method. The procedure was to first identify intronic and intergenic regions in the *D. melanogaster* genome that aligned well (regions of 100 nucleotides with no more than 13% gaps and 15% mismatches) to the genome of *D. pseudoobscura*. These regions were submitted to RNA secondary structure prediction, and stem-loops with at least 23 base pairs and a free energy of folding $\Delta G \leq -23$ kcal/mol were scored with a model that penalized internal loops of increasing size, particularly if they were asymmetrical. The orthologous regions of the *D. melanogaster* top candidates in *D. pseudoobscura* were then evaluated in a similar fashion. Finally, the 600 candidates with the best average score in the two genomes were submitted to a filter of their pattern of evolutionary conservation. This was based on the analysis of the 24 reference miRNAs, which indicated that mutations are least tolerated in the region of the mature miRNA, followed by the complementary strand and finally by the loop region. In the end, the authors suggested that about

110 miRNA genes are present in the drosophilid genomes, and that 18 of the 24 miRNAs of the reference set were present among the top 124 predictions of miRseeker.

A second method designed for the identification of miRNAs conserved between species is MirScan, which was applied to mature miRNA gene prediction in worms and in vertebrates. Regions from a reference genome (*Caenorhabditis elegans* and *Homo sapiens*) that were not part of protein-coding genes or repetitive elements and were conserved in another species (*Caenorhabditis briggsae* and *Mus musculus*, respectively) were analyzed for the potential of forming secondary structure roughly resembling that of miRNA precursors. This meant stem-loop structures with a predicted $\Delta G \leq -25$ kcal/mol. Regions that passed these initial filters were then aligned with ClustalW,[96] and the alignment, as well as the alidot-predicted common secondary structure[97] were scored using the MirScan scoring model. The model takes into account the following features:

- miRNA base pairing — the sum of the base pairing probabilities for the pairs involving the 21-nucleotide-long miRNA candidate;
- base pairing outside of the miRNA — the sum of the base pairing probabilities for the pairs in the same helix, but not involving the 21-nucleotide-long miRNA candidate;
- 5′ conservation — the number of bases among the first 10 of the putative miRNA that are conserved in the alignment between the two species;
- 3′ conservation — the number of bases among the last 11 of the putative miRNA that are conserved in the alignment between the two species;
- bulge symmetry — the difference in the number of nucleotides involved in bulges or mismatches in the region encoding the putative miRNA and its complementary sequence;
- distance from the loop — the number of base pairs between the loop of the stem-loop and the closest end of the mature miRNA candidate; and
- initial pentamer score — given a weight matrix model of this region constructed from a reference set of known miRNAs.

The score of a candidate miRNA x is defined as $S(x) = \sum_{i=1}^{7} s_i(x)$, where $s_i(x)$ is the contribution of feature i to the score and is computed as $s_i(x) = \log_2 (f_i(x)/b_i(x))$. $f_i(x)$ is an estimate of the frequency of the value of the ith feature of candidate x in the training set of reference miRNAs, while $b_i(x)$ is a similar estimate computed over a background set of stem-loops. For the reference set of miRNAs, which was very small, these estimates were obtained by smoothing the empirical frequency distributions. In combination with large-scale cloning, this method raised the number of validated *C. elegans* miRNAs to 88,[98] while the study in human led to an estimated upper bound of 255 human miRNA genes.[48] This latter estimate was obtained considering candidates that are conserved up to fish.

The first method that could in principle predict miRNAs in individual genomes was ProMir,[99] which employed a hidden Markov model. The structures of putative miRNA precursors were predicted using programs from the Vienna package.[100] Describing the predicted hybrid formed between the 5′ and the 3′ arms of a stem-loop in the terminology of sequence alignments, one can then distinguish the following states:

- match (M) — one of the possible base pairs (A–U, U–A, G–C, C–G, G–U, U–G);
- mismatch (N) — one of the remaining (not matched) base pairs;
- insertion (I) — base on the 5′ arm is bulged (A-, C-, G-, U-); or
- deletion (D) — base on the 3′ arm is bulged (-A, -C, -G, -U).

To be able to predict the precise location of the start and end of the mature miRNA in the stem-loop, the hidden states have an additional qualifier, namely *true* (inside the duplex that involves the mature miRNA) or *false* (outside of this duplex): The transition probabilities between states therefore depend on this qualifier, as well as on the type of states (M, N, I, D). By cross-validation, the authors showed that ProMir[99] achieves a sensitivity of 73% and a specificity of 96%. Although this specificity seems very good, applying the algorithm on the scale of the human genome will generate presumably thousands of false-positive predictions. Therefore, in order to predict miRNA in the human genome, several filters have been implemented. First, evidence of expression in the form of expressed sequence tag (EST) data was required. Second, the Randfold algorithm[87] was used to evaluate whether the pre-miRNA sequence formed an unually stable structure. Third, multi-genome alignments were used to determine whether the candidates had the expected pattern of conservation (highest in the mature miRNA region, then the complementary region, the loop, and finally the flanking regions).

A second method to attempt miRNA gene prediction in a single genome was developed by Pfeffer *et al.*[49] It was motivated by the discovery of miRNAs encoded by the Epstein–Barr virus,[5] which raised the question of what other viruses may encode miRNAs. Viral genomes are considerably smaller than animal genomes, yet the viral miRNA biogenesis proceeds along the same path as the biogenesis of host miRNAs. Thus, a model trained on host miRNAs can be expected to perform well in predicting viral miRNAs, while generating a very small number of false positives. Pfeffer *et al.*[49] used a support vector machine (SVM) framework in their prediction method. The procedure started with the identification of "robust" stem-loops in viral genomes; these were defined as those stem-loops that are not sensitive to the precise sequence boundary of the region which is submitted to the secondary structure prediction program. The rationale behind this choice was that the miRNA stem-loop is necessary in multiple biogenesis steps, from pri-miRNA and pre-miRNA processing to nuclear export of pre-miRNAs and the final release of the miRNA duplex by the Dicer enzyme. Features of the robust human miRNA stem-loops as well as robust stem-loops predicted from noncoding transcripts, mRNAs, and genomic regions were used to train a SVM. One subset of features captured information about the entire stem-loop:

- free energy of folding;
- length of the stem;
- length of the hairpin loop;
- length of the longest perfect stem;
- number of nucleotides in symmetrical loops;
- number of nucleotides in asymmetrical loops;

- average distance between internal loops;
- average size of symmetrical loops;
- average size of asymmetrical loops;
- proportion of A/C/G/U nucleotides in the stem; and
- proportion of A–U/C–G/G–U base pairs in the stem.

A second subset captured information about the longest symmetrical region in the stem, i.e. the longest region devoid of bulges or asymmetrical loops, and information about the longest region with relaxed symmetry, i.e. the region in which the difference in the number of unpaired nucleotides on the 5′ and the 3′ arm of the stem was below a given threshold:

- length;
- distance from the hairpin loop;
- number of nucleotides involved in internal loops (calculated separately for symmetrical and asymmetrical loops in the case of the region with relaxed symmetry);
- proportion of A/C/G/U nucleotides; and
- proportion of A–U/C–G/G–U base pairs.

Finally, another subset of features captured information determined from sliding windows of the length of the mature miRNA (22 nucleotides):

- maximum number of base pairs;
- minimum number of nucleotides in asymmetrical loops; and
- minimum asymmetry over the internal loops in this region.

The model obtained this way was used to score robust stems extracted from a large panel of animal pathogenic viruses. miRNAs were predicted in viruses of the herpes and polyoma families, with overall sensitivities and specificities around 50%.[49]

Finally, a fundamentally different approach to miRNA gene prediction, moving away from biogenesis requirements and focusing on the effector arm of the miRNA pathway, was taken by Xie *et al.*[101] Since various studies indicated that seven to eight nucleotides at the 5′ end of the miRNA are most important for miRNA function,[102–104] Xie *et al.*[101] identified 8mers that were strongly conserved in the 3′ UTRs of protein-coding transcripts, and searched for regions in the genome that were reverse complementary to the 3′ UTR sequences and could form stable stem-loops. This procedure generated a set of 129 novel predicted miRNAs.

Still more miRNA prediction programs are available, particularly those based on SVMs. Of these, some of the most distinct ones are the triplet-SVM classifier,[105] which has as features the frequencies of triplets representing the secondary structure of three contiguous nucleotides; RNAmicro,[106] which uses a small number of descriptors downstream of a conservation filter, among them a structure conservation index; and the Microprocessor SVM,[107] which includes a number of features aiming to capture the specificity of Drosha processing.

7. MicroRNA Target Prediction

7.1. *What Does a miRNA Target Site Look Like?*

In contrast to lin-4, the founder of the miRNA class, which was discovered by forward genetics and for which the phenotype was known at the time when the gene was discovered, the vast majority of miRNAs that have been identified experimentally or predicted computationally do not yet have an associated function. This generated great interest in computational predictions of miRNA targets and, not surprisingly, several methods were already proposed very shortly after miRNA gene prediction methods were published. Perhaps similar to the situation of miRNA genes, the main roadblock is that it is still unclear what functional miRNA target sites look like.

7.2. *The miRNA Seed Region*

One of the most predictive features that has so far been described for miRNA target sites is the perfect complementarity with the first seven to eight nucleotides at the 5′ end of the miRNA, a region which has been called the "nucleus"[108] or "seed".[81] Lai[109] initially observed that sequence elements found in the 3′ UTRs of transcripts encoding Notch pathway basic helix-loop-helix (bHLH) repressors and Bearded family proteins in *D. melanogaster* are complementary to the 5′ end of various miRNAs. A subsequent experimental study in which mutations were introduced individually in each of the nucleotides of the miRNA binding site of a reporter transcript[103] confirmed that mutations in nucleotides complementary to positions 2–8 of the miRNA are most deleterious for the miRNA-induced translation inhibition. On the computational side, Lewis *et al.*[110] analyzed the conservation of subsequences of human 3′ UTRs that are complementary to miRNA 5′ ends and found that they are considerably more conserved than random subsequences of the same length, indicating that miRNAs have many mRNA targets that are conserved in evolution. These computational studies do not imply that perfect complementarity to the 5′ end of the miRNA is indispensable for function[111]; rather, they indicate that the accuracy of prediction of seed-complementary sites should be higher than the accuracy of prediction of sites with imperfect pairing in the 5′ region of the miRNA.[81,112]

7.3. *Structural Determinants*

Given that the targets of miRNAs are mRNAs, it was expected that this interaction is governed by RNA–RNA interaction rules, which is why the first miRNA target prediction programs[108,110,113] took into account the hybridization energy between miRNA and putative target sites. Additional indication that the hybridization between miRNAs and target sites is important for target recognition came from the experimental study of Brennecke *et al.*,[103] who defined three types of miRNA target sites; one of these is the "3′-compensatory" type, in which relatively poor complementarity of the 5′ end of the miRNAs to the target sites appears to be compensated by strong complementarity of the 3′ end. More recently, Lewis *et al.*[104] showed that a better specificity of miRNA target prediction can be achieved by considering only the evolutionary conservation of miRNA-complementary motifs. The role of structural determinants is currently under debate. Some studies[114–116] do indicate that the secondary structure formation and, in particular, the accessibility of the target site play a detectable role in the efficacy of target site recognition, yet the magnitude of this effect relative to sequence determinants[117] is yet to be determined.

7.4. *Other Determinants*

With the availability of sufficiently accurate target prediction methods as well as relatively large data sets of mRNA expression profiles obtained after miRNA and siRNA transfection,[118–120] it became possible to ask the question of what other features may contribute to the efficacy of miRNA target sites. The relative position of the target site with respect to the 3′ UTR boundaries was one of the first such determinants discovered[112,117,121]: target sites that are close, yet not adjacent, to the stop codon and the poly(A) tail are more strongly conserved and appear to be more efficient in triggering mRNA degradation. The conservation profile of 3′ UTRs follows a similar pattern (not shown), suggesting that miRNAs exert an important pressure on the evolution of 3′ UTRs, as argued by Stark *et al.*[122] An interesting question is whether this positional bias is a reflection of spatial constraints on the RISC interaction with other complexes involved in translation silencing and mRNA degradation. Alternatively, the regions close to 3′ UTR boundaries may provide the optimal sequence and structure environment for the recruitment of components necessary for silencing. Grimson *et al.*[117] showed that the efficacy of target sites correlates with the A/U-richness of their environments, and Robins and Press[123] argued that miRNA targets generally have A/U-rich 3′ UTRs. Thus, the local sequence composition, either through the recruitment of A/U-rich sequence binding proteins or indirectly, through the secondary structure, is a determinant of miRNA target site efficacy.

Apart from the efficacy of individual sites, many authors have made the observation that the presence of multiple target sites for an individual miRNA in the 3′ UTR of a given transcript is more indicative of a functional miRNA–target interaction than the presence of single sites. Interestingly, some of the best known miRNA targets, such as the hunchback-like 1[45] and the dauer formation family member 12 transcripts, targeted by let-7 in *C. elegans*, do indeed have multiple putative miRNA target sites. These observations lead to the speculation that miRNAs may act in a "cooperative"[124] or "coordinated"[125] manner. The mechanism(s) behind this cooperativity remains to be established. Nonetheless, incorporating these features into miRNA target prediction programs increases the accuracy of target prediction,[117] even though the prediction of mRNA changes upon miRNA overexpression remains a challenging task.

7.5. *miRNA Target Prediction Methods*

Initially, miRNA target prediction methods attempted to identify 3′ UTRs that could form extensive hybrids with miRNAs. To improve their accuracy, these methods also filtered their predictions using evolutionary conservation.

The first genome-wide target prediction methods to be published were those of Stark *et al.*,[113] dealing with miRNA target prediction in *D. melanogaster*, and TargetScan,[110] applied to the prediction of mammalian miRNA targets. The method of Stark *et al.*[113] used the alignment tool HMMer[126] to search for 3′ UTR segments that were complementary (allowing for G-U pairs) to the first eight nucleotides at the 5′ end of miRNAs. These segments were extended to the length of the miRNA plus another five nucleotides, and Mfold[127] was then used to predict the free energy of folding between the putative site and the miRNA. The score of the site was the *Z*-value of its value relative to 10 000 random target sequences. The score of the 3′ UTR was the number of sites with a *Z*-value ≥ 3.

TargetScan used a fairly similar approach. Initially, 3′ UTR sequences were searched for the presence of segments that had perfect Watson–Crick complementarity to nucleotides 2–8 of miRNAs, a region which was referred to as the "seed". The complementarity was then extended in both directions, allowing G–U pairs, but stopping at mismatches. Finally, the 3′ UTR regions obtained in the previous steps were extended to 35 nucleotides upstream of the "seed match", the RNAfold program[128] was used to predict the hybridization of the 3′ end of the miRNA to the putative target site, and a final hybridization energy between the miRNA and the putative target site constructed in this piecemeal way was obtained using the RNAeval[128] algorithm. The score of a 3′ UTR was computed as $Z = \sum_{k=1}^{n} e^{-G_k/T}$, where n is the number of seed matches in the 3′ UTR, G_k is the energy of miRNA:target site interaction for site k, and T is a free parameter. Predictions were obtained by setting a cut-off on Z ($Z_i \geq Z_C$, with Z_i being the score of the 3′ UTR in species i) and on R, the rank of the prediction ($R_i \leq R_C$, with R_i being the rank of the prediction in species i).

The third method proposed for miRNA target prediction was miRanda,[124] initially applied to *D. melanogaster* target prediction. Unlike TargetScan, which used an RNA secondary structure prediction program based on energy minimization to predict hybrids between miRNAs and target sites,[110] miRanda started from the premise that, in the context of the RISC complex, the parameters of RNA–RNA interactions will change. Therefore, miRanda used user-defined scores as opposed to energetic parameters to predict hybrids (the scores of A–U, G–C, and G–U interactions were 5, 5, and 2, respectively), and individual positions in the miRNA were weighted differently in their contribution to the total score. Mismatches were penalized with −3, and gaps were scored using an affine model.

Also initially applied to the miRNA target prediction in *D. melanogaster* was the algorithm of Rajewsky and Socci,[108] which was based on the identification of good matches between the miRNA 5′ end and putative target sites, followed by further analysis of the overall hybridization energy between the miRNA and the target site. These authors further made the argument that the 5′ end of the miRNA implements an important kinetic component of miRNA–mRNA interaction.

Since their initial inception, these target prediction methods have changed (and converged) substantially.[104,125,129,130] In particular, evolutionary conservation and close-to-perfect complementarity to the 5′ end of miRNAs have now been incorporated in virtually all target prediction programs, while much less emphasis is currently placed on the energy of miRNA–target site hybridization.

Given that miRNA–mRNA hybridization was believed to play such an important role in miRNA target recognition, and that none of the then-available RNA secondary structure prediction programs directly calculated the hybridization energy of two RNA molecules, Rehmsmeier *et al.*[131] developed RNAhybrid, a fast algorithm for predicting miRNA/target duplexes. In essence, RNAhybrid uses the energy parameters from Mathews *et al.*,[132] which are also used by Mfold and RNAfold secondary structure prediction programs, to compute the minimum free energy hybrid between an miRNA and a target. Intramolecular (within-miRNA or within-mRNA) pairs are not allowed. Important contributions from the RNAhybrid package are the programs used to evaluate the significance of the results. For instance, to account for the variable length of 3′ UTRs, the authors use — instead of the free energy of hybridization — normalized energies defined as $e_n = \frac{e}{\log(mn)}$, with e being the free energy, m the length of the target sequence, and n the length of the miRNA. The statistical significance of the normalized free energies is computed using extreme value statistics, and the significance of multiple binding sites is computed assuming a Poisson distribution of the number of miRNA target sites per 3′ UTR. Finally, the evolutionary conservation of putative miRNA sites is taken into account by computing joint p-values for the sites $P\,[Z_1 \geq e_1, Z_2 \geq e_2,..., Z_k \geq e_k] = (\max\{P\,[Z_1 \geq e_1], P\,[Z_2 \geq e_2],..., P\,[Z_k \geq e_k]\})^{k_{\text{eff}}}$, with $e_1, e_2, ..., e_k$ being the normalized free energies of hybridization of the orthologous sites in the k species, and k_{eff} being the effective number of sequences, taking into account the phylogenetic distance between the species. k_{eff} is fitted using random miRNAs.

While RNAhybrid aimed to fill the need for a program that predicts hybrids between two RNA molecules, another miRNA target prediction program, ElMMo,[112] aimed to rigorously treat the comparative genomic information for miRNA target prediction. With the exception of RNAhybrid,[131] all of the previously proposed programs for miRNA target prediction that used evolutionary conservation treated each species independently, regardless of the relative position of the species in the phylogenetic tree. In ElMMo, the authors treated conservation in a Bayesian framework that explicitly took the phylogeny of the species into account, and explicitly dealt with the possibility that a conserved site is conserved by chance in a subset of the species while being conserved due to its being functional and under selection in another subset of species. Thus, ElMMo infers how likely it is that a given putative site with a given conservation pattern across species is functional and has been maintained by selection in one or more species, as opposed to being maintained in the absence of selection pressure in any of the species. Additionally, ElMMo infers the strength of selection pressure on target sites for a given miRNA along each branch of the evolutionary tree, thus enabling one to analyze the dynamics of miRNA target site evolution.

As was the case with miRNA gene prediction, new methods for miRNA target prediction continue to emerge. Among those that use the most distinct set of features are

- TargetBoost[133] — uses an adaptation of a genetic programming algorithm to define motifs that presumably capture the characteristics of interaction between miRNAs and targets;
- miTarget[134] — uses an SVM based on local and global structural and thermodynamic features;
- GenMir[135] — uses miRNA and mRNA expression profiles to identify sets of genes whose expression changes from sample to sample in a way that is consistent with the miRNA expression;
- a linear model that takes into account features computed over the context of target sites: A/U content, base pairing with nucleotides 13–17 of the miRNA, and distance to the nearest 3′ UTR boundary[117]; and
- Pita[136] — predicts target sites based on the energy of interaction between miRNA and target sites, taking into account both the energetic cost of unfolding the target site and the energy gained through the hybridization of the miRNA with the target.

8. Conclusions

The abundance and the functions of non-protein-coding RNAs in regulating gene expression have only just begun to be explored. Computational approaches play a prominent role in this field, yet they face the problem that, with the relatively poor understanding of the constraints on particular types of molecules or interactions, accurate, predictive models are difficult to construct. Nonetheless, computational biology of small regulatory RNAs is a field in which combined computational and experimental approaches are common, and are generally successful.

The open problems have to a large extent remained unchanged: what is a miRNA, and what parameters determine the efficacy of a miRNA target site in mRNA degradation or translational repression? As answers to these questions start to emerge, the focus is shifting to the structure of the microRNA nucleoprotein (miRNP) code. Such a code is believed to control gene expression at the posttranscriptional level, similar to the transcription regulatory code implemented at the DNA level.[137] At this point, this analogy has relatively little experimental basis because the equivalents at the RNA level of the multi-molecular complexes implementing the combinatorial transcription control have not been characterized. Biochemical characterization of such complexes and identification of binding sites of miRNA- and RNA-binding proteins using, again, high-throughput data and computational analyses will start uncovering this code in the very near future.

References

1. Reinhart B, Slack F, Basson M *et al.* (2000) The 21-nucleotide let-7 RNA regulates developmental timing in *Caenorhabditis elegans*. *Nature* **403**: 901–6.
2. Pasquinelli A, Reinhart B, Slack F *et al.* (2000) Conservation of the sequence and temporal expression of let-7 heterochronic regulatory RNA. *Nature* **408**: 86–9.
3. Lagos-Quintana M, Rauhut R, Lendeckel W, Tuschl T. (2001) Identification of novel genes coding for small expressed RNAs. *Science* **294**: 853–8.
4. Lagos-Quintana M, Rauhut R, Yalcin A *et al.* (2002) Identification of tissue-specific microRNAs from mouse. *Curr Biol* **12**: 735–9.
5. Pfeffer S, Zavolan M, Grässer F *et al.* (2004) Small RNA profiling of virus-infected human cells identifies viral microRNAs. *Science* **304**: 734–6.
6. Lee R, Feinbaum R, Ambros V. (1993) The *C. elegans* heterochronic gene lin-4 encodes small RNAs with antisense complementarity to lin-14. *Cell* **75**: 843–54.
7. Meister G, Landthaler M, Patkaniowska A *et al.* (2004) Human Argonaute2 mediates RNA cleavage targeted by miRNAs and siRNAs. *Mol Cell* **15**: 185–97.
8. Liu J, Carmell M, Rivas F *et al.* (2004) Argonaute2 is the catalytic engine of mammalian RNAi. *Science* **305**: 1437–41.
9. Bohmert K, Camus I, Bellini C *et al.* (1998) AGO1 defines a novel locus of *Arabidopsis* controlling leaf development. *EMBO J* **17**: 170–80.

10. Moussian B, Schoof H, Haecker A *et al.* (1998) Role of the ZWILLE gene in the regulation of central shoot meristem cell fate during *Arabidopsis* embryogenesis. *EMBO J* **17**: 1799–809.
11. Reinke V, Smith H, Nance J *et al.* (2000) A global profile of germline gene expression in *C. elegans. Mol Cell* **6**: 605–16.
12. Cox D, Chao A, Lin H. (2000) *piwi* encodes a nucleoplasmic factor whose activity modulates the number and division rate of germline stem cells. *Development* **127**: 503–14.
13. Aravin A, Gaidatzis D, Pfeffer S *et al.* (2006) A novel class of small RNAs bind to MILI protein in mouse testes. *Nature* **442**: 203–7.
14. Girard A, Sachidanandam R, Hannon G, Carmell M. (2006) A germline-specific class of small RNAs binds mammalian Piwi proteins. *Nature* **442**: 199–202.
15. Lau N, Seto A, Kim J *et al.* (2006) Characterization of the piRNA complex from rat testes. *Science* **313**: 363–7.
16. Saito K, Nishida K, Mori T *et al.* (2006) Specific association of Piwi with rasiRNAs derived from retrotransposon and heterochromatic regions in the *Drosophila* genome. *Genes Dev* **20**: 2214–22.
17. Brennecke J, Aravin A, Stark A *et al.* (2007) Discrete small RNA-generating loci as master regulators of transposon activity in *Drosophila. Cell* **128**: 1089–103.
18. Gunawardane L, Saito K, Nishida K *et al.* (2007) A slicer-mediated mechanism for repeat-associated siRNA 5′ end formation in *Drosophila. Science* **315**: 1587–90.
19. Lau N, Lim L, Weinstein E, Bartel D. (2001) An abundant class of tiny RNAs with probable regulatory roles in *Caenorhabditis elegans. Science* **294**: 858–62.
20. Lee R, Ambros V. (2001) An extensive class of small RNAs in *Caenorhabditis elegans. Science* **294**: 797–9.
21. Bennett S. (2004) Solexa Ltd. *Pharmacogenomics* **5**: 433–8.
22. Margulies M, Egholm M, Altman W *et al.* (2005) Genome sequencing in microfabricated high-density picolitre reactors. *Nature* **437**: 376–80.
23. Lu J, Shen Y, Wu Q *et al.* (2008) The birth and death of microRNA genes in *Drosophila. Nat Genet* **40**: 351–5.
24. Bentwich I, Avniel A, Karov Y *et al.* (2005) Identification of hundreds of conserved and nonconserved human microRNAs. *Nat Genet* **37**: 766–70.
25. Landgraf P, Rusu M, Sheridan R *et al.* (2007) A mammalian microRNA expression atlas based on small RNA library sequencing. *Cell* **129**: 1401–14.
26. Calin G, Liu C, Sevignani C *et al.* (2004) MicroRNA profiling reveals distinct signatures in B cell chronic lymphocytic leukemias. *Proc Natl Acad Sci USA* **101**: 11755–60.
27. Miska E, Alvarez-Saavedra E, Townsend M *et al.* (2004) Microarray analysis of microRNA expression in the developing mammalian brain. *Genome Biol* **5**: R68.

28. Babak T, Zhang W, Morris Q *et al.* (2004) Probing microRNAs with microarrays: tissue specificity and functional inference. *RNA* **10**: 1813–19.
29. Barad O, Meiri E, Avniel A *et al.* (2004) MicroRNA expression detected by oligonucleotide microarrays: system establishment and expression profiling in human tissues. *Genome Res* **14**: 2486–94.
30. Liang R, Li W, Li Y *et al.* (2005) An oligonucleotide microarray for microRNA expression analysis based on labeling RNA with quantum dot and nanogold probe. *Nucleic Acids Res* **33**: e17.
31. Wienholds E, Kloosterman W, Miska E *et al.* (2005) MicroRNA expression in zebrafish embryonic development. *Science* **309**: 310–1.
32. Beuvink I, Kolb F, Budach W *et al.* (2007) A novel microarray approach reveals new tissue-specific signatures of known and predicted mammalian microRNAs. *Nucleic Acids Res* **35**: e52.
33. Bartel D. (2004) MicroRNAs: genomics, biogenesis, mechanism, and function. *Cell* **116**: 281–97.
34. Lee Y, Ahn C, Han J *et al.* (2003) The nuclear RNase III Drosha initiates microRNA processing. *Nature* **425**: 415–29.
35. Denli A, Tops B, Plasterk R *et al.* (2004) Processing of primary microRNAs by the Microprocessor complex. *Nature* **432**: 231–5.
36. Bernstein E, Caudy A, Hammond S, Hannon G. (2001) Role for a bidentate ribonuclease in the initiation step of RNA interference. *Nature* **409**: 363–6.
37. Förstemann K, Tomari Y, Du T *et al.* (2005) Normal microRNA maturation and germ-line stem cell maintenance requires Loquacious, a double-stranded RNA-binding domain protein. *PLoS Biol* **3**: e236.
38. Chendrimada T, Gregory R, Kumaraswamy E *et al.* (2005) TRBP recruits the Dicer complex to Ago2 for microRNA processing and gene silencing. *Cell* **436**: 740–4.
39. Haase A, Jaskiewicz L, Zhang H *et al.* (2005) TRBP, a regulator of cellular PKR and HIV-1 virus expression, interacts with Dicer and functions in RNA silencing. *EMBO Rep* **6**: 961–7.
40. Liu Z, Ambros V. (1989) Heterochronic genes control the stage-specific initiation and expression of the dauer larva developmental program in *Caenorhabditis elegans. Genes Dev* **3**: 2039–49.
41. Chalfie M, Horvitz H, Sulston J. (1981) Mutations that lead to reiterations in the cell lineages of *C. elegans. Cell* **24**: 59–69.
42. Ambros V, Horvitz H. (1984) Heterochronic mutants of the nematode *Caenorhabditis elegans. Science* **226**: 409–16.
43. Wightman B, Ha I, Ruvkun G. (1993) Posttranscriptional regulation of the heterochronic gene lin-14 by lin-4 mediates temporal pattern formation in *C. elegans. Cell* **75**: 855–62.
44. Abrahante J, Daul A, Li M *et al.* (2003) The *Caenorhabditis elegans* hunchback-like gene lin-57/hbl-1 controls developmental time and is regulated by microRNAs. *Dev Cell* **4**: 625–37.

45. Lin S, Johnson S, Abraham M *et al.* (2003) The *C. elegans* hunchback homolog, hbl-1, controls temporal patterning and is a probable microRNA target. *Dev Cell* **4**: 639–50.
46. Sempere L, Dubrovsky E, Dubrovskaya V *et al.* (2002) The expression of the let-7 small regulatory RNA is controlled by ecdysone during metamorphosis in *Drosophila melanogaster*. *Dev Biol* **244**: 170–79.
47. Griffiths-Jones S, Saini H, van Dongen S, Enright A. (2008) miRBase: tools for microRNA genomics. *Nucleic Acids Res* **36**: D154–8.
48. Lim L, Glasner M, Yekta S *et al.* (2003) Vertebrate microRNA genes. *Science* **299**: 1540.
49. Pfeffer S, Sewer A, Lagos-Quintana M *et al.* (2005) Identification of microRNAs of the herpesvirus family. *Nat Methods* **2**: 269–76.
50. Samols M, Hu J, Skalsky R, Renne R. (2005) Cloning and identification of a microRNA cluster within the latency-associated region of Kaposi's sarcoma-associated herpesvirus. *J Virol* **79**: 9301–5.
51. Cai X, Lu S, Zhang Z *et al.* (2005) Kaposi's sarcoma-associated herpesvirus expresses an array of viral microRNAs in latently infected cells. *Proc Natl Acad Sci USA* **102**: 5570–5.
52. Grey F, Antoniewicz A, Allen E *et al.* (2005) Identification and characterization of human cytomegalovirus-encoded microRNAs. *J Virol* **79**: 12095–9.
53. Sullivan C, Grundhoff A, Tevethia S *et al.* (2005) SV40-encoded microRNAs regulate viral gene expression and reduce susceptibility to cytotoxic T cells. *Nature* **435**: 682–6.
54. Barth S, Pfuhl T, Mamiani A *et al.* (2007) Epstein-Barr virus-encoded microRNA miR-BART2 down-regulates the viral DNA polymerase BALF5. *Nucleic Acids Res* **36**: 666–75.
55. Lin H, Spradling A. (1997) A novel group of pumilio mutations affects the asymmetric division of germline stem cells in the *Drosophila* ovary. *Development* **124**: 2463–76.
56. Pal-Bhadra M, Leibovitch B, Gandhi S *et al.* (2004) Heterochromatic silencing and HP1 localization in *Drosophila* are dependent on the RNAi machinery. *Science* **303**: 669–72.
57. Aravin A, Lagos-Quintana M, Yalcin A *et al.* (2003) The small RNA profile during *Drosophila melanogaster* development. *Dev Cell* **5**: 337–50.
58. Deng W, Lin H. (2002) *miwi*, a murine homolog of *piwi*, encodes a cytoplasmic protein essential for spermatogenesis. *Dev Cell* **2**: 819–30.
59. Kuramochi-Miyagawa S, Kimura T, Ijiri T *et al.* (2004) Mili, a mammalian member of piwi family gene, is essential for spermatogenesis. *Development* **131**: 839–49.
60. Aravin A, Sachidanandam R, Girard A *et al.* (2007) Developmentally regulated piRNA clusters implicate MILI in transposon control. *Science* **316**: 744–7.

61. Houwing S, Kamminga L, Berezikov E *et al.* (2007) A role for Piwi and piRNAs in germ cell maintenance and transposon silencing in Zebrafish. *Cell* **129**: 69–82.
62. Griffiths-Jones S, Grocock R, van Dongen S *et al.* (2006) miRBase: microRNA sequences, targets and gene nomenclature. *Nucleic Acids Res* **34**: D140–4.
63. Lee Y, Kim M, Han J *et al.* (2004) MicroRNA genes are transcribed by RNA polymerase II. *EMBO J* **23**: 4051–60.
64. Yi R, Qin Y, Macara I, Cullen B. (2003) Exportin-5 mediates the nuclear export of pre-microRNAs and short hairpin RNAs. *Genes Dev* **17**: 3011–6.
65. Lund E, Güttinger S, Calado A *et al.* (2004) Nuclear export of microRNA precursors. *Science* **303**: 95–8.
66. Tomari Y, Matranga C, Haley B *et al.* (2004) A protein sensor for siRNA asymmetry. *Science* **306**: 1377–80.
67. Pham J, Pellino J, Lee Y *et al.* (2004) A Dicer-2-dependent 80S complex cleaves targeted mRNAs during RNAi in *Drosophila*. *Cell* **117**: 83–94.
68. Tomari Y, Du T, Haley B *et al.* (2004) RISC assembly defects in the *Drosophila* RNAi mutant armitage. *Cell* **116**: 831–41.
69. Llave C, Xie Z, Kasschau K, Carrington J. (2002) Cleavage of Scarecrow-like mRNA targets directed by a class of *Arabidopsis* miRNA. *Science* **297**: 2053–6.
70. He L, Hannon G. (2004) MicroRNAs: small RNAs with a big role in gene regulation. *Nat Rev Genet* **5**: 522–31.
71. Jakymiw A, Lian S, Eystathioy T *et al.* (2005) Disruption of GW bodies impairs mammalian RNA interference. *Nat Cell Biol* 7: 1267–74.
72. Liu J, Valencia-Sanchez M, Hannon G, Parker R. (2005) MicroRNA-dependent localization of targeted mRNAs to mammalian P-bodies. *Nat Cell Biol* 7: 719–23.
73. Rehwinkel J, Behm-Ansmant I, Gatfield D, Izaurralde E. (2005) A crucial role for GW182 and the DCP1:DCP2 decapping complex in miRNA-mediated gene silencing. *RNA* **11**: 1640–7.
74. Eystathioy T, Chan E, Tenenbaum S *et al.* (2002) A phosphorylated cytoplasmic autoantigen, GW182, associates with a unique population of human mRNAs within novel cytoplasmic speckles. *Mol Biol Cell* **13**: 1338–51.
75. Bhattacharyya S, Habermacher R, Martine U *et al.* (2006) Relief of microRNA-mediated translational repression in human cells subjected to stress. *Cell* **125**: 1111–24.
76. Kedde M, Strasser M, Boldajipour B *et al.* (2007) RNA-binding protein Dnd1 inhibits microRNA access to target mRNA. *Cell* **131**: 1273–86.
77. Song J, Smith S, Hannon G, Joshua-Tor L. (2004) Crystal structure of Argonaute and its implications for RISC slicer activity. *Science* **305**: 1434–7.
78. Parker J, Roe S, Barford D. (2004) Crystal structure of a PIWI protein suggests mechanisms for siRNA recognition and slicer activity. *EMBO J* **23**: 4727–37.

79. Lingel A, Simon B, Izaurralde E, Sattler M. (2004) Nucleic acid 3′-end recognition by the Argonaute2 PAZ domain. *Nat Struct Mol Biol* **11**: 576–7.
80. Kiriakidou M, Tan G, Lamprinaki S *et al.* (2007) An mRNA m^7G cap binding-like motif within human Ago2 represses translation. *Cell* **129**: 1141–51.
81. Lim L, Lau N, Garrett-Engele P *et al.* (2005) Microarray analysis shows that some microRNAs downregulate large numbers of target mRNAs. *Nature* **433**: 769–73.
82. Krutzfeldt J, Rajewsky N, Braich R *et al.* (2005) Silencing of microRNAs *in vivo* with 'antagomirs'. *Nature* **438**: 685–9.
83. Giraldez A, Cinalli R, Glasner M *et al.* (2005) MicroRNAs regulate brain morphogenesis in zebrafish. *Science* **308**: 833–8.
84. Giraldez A, Mishima Y, Rihel J *et al.* (2006) Zebrafish MiR-430 promotes deadenylation and clearance of maternal mRNAs. *Science* **312**: 75–9.
85. Vagin V, Sigova A, Li C *et al.* (2006) A distinct small RNA pathway silences selfish genetic elements in the germline. *Science* **313**: 320–4.
86. Carmell M, Girard A, van de Kant H *et al.* (2007) MIWI2 is essential for spermatogenesis and repression of transposons in the mouse male germline. *Dev Cell* **12**: 503–14.
87. Bonnet E, Wuyts J, Rouz P, Van de Peer Y. (2004) Evidence that microRNA precursors, unlike other non-coding RNAs, have lower folding free energies than random sequences. *Bioinformatics* **20**: 2911–7.
88. Freyhult E, Gardner P, Moulton V. (2005) A comparison of RNA folding measures. *BMC Bioinformatics* **6**: 241.
89. Zhang B, Pan X, Cox S *et al.* (2006) Evidence that miRNAs are different from other RNAs. *Cell Mol Life Sci* **63**: 246–54.
90. Ng Kwang Loong S, Mishra S. (2007) Unique folding of precursor microRNAs: quantitative evidence and implications for *de novo* identification. *RNA* **13**: 170–87.
91. Borenstein E, Ruppin E. (2006) Direct evolution of genetic robustness in microRNA. *Proc Natl Acad Sci USA* **103**: 6593–8.
92. Lai E, Tomancak P, Williams R, Rubin G. (2003) Computational identification of *Drosophila* microRNA genes. *Genome Biol* **4**: R42.
93. Sewer A, Paul N, Landgraf P *et al.* (2005) Identification of clustered microRNAs using an *ab initio* prediction method. *BMC Bioinformatics* **6**: 267.
94. Parisien M, Major F. (2008) The MC-Fold and MC-Sym pipeline infers RNA structure from sequence data. *Nature* **452**: 51–5.
95. Berezikov E, Guryev V, van de Belt J *et al.* (2005) Phylogenetic shadowing and computational identification of human microRNA genes. *Cell* **120**: 21–4.
96. Thompson J, Higgins D, Gibson T. (1994) CLUSTAL W: improving the sensitivity of progressive multiple sequence alignment through sequence weighting, position-specific gap penalties and weight matrix choice. *Nucleic Acids Res* **22**: 4673–80.

97. Hofacker I, Stadler P. (1999) Automatic detection of conserved base pairing patterns in RNA virus genomes. *Comput Chem* **23**: 401–14.
98. Lim L, Lau N, Weinstein E *et al.* (2003) The microRNAs of *Caenorhabditis elegans*. *Genes Dev* **17**: 991–1008.
99. Nam J, Shin K, Han J *et al.* (2005) Human microRNA prediction through a probabilistic co-learning model of sequence and structure. *Nucleic Acids Res* **33**: 3570–81.
100. Hofacker I. (2003) Vienna RNA secondary structure server. *Nucleic Acids Res* **31**: 3429–31.
101. Xie X, Lu J, Kulbokas E *et al.* (2005) Systematic discovery of regulatory motifs in human promoters and 3′ UTRs by comparison of several mammals. *Nature* **434**: 338–45.
102. Doench J, Sharp P. (2004) Specificity of microRNA target selection in translational repression. *Genes Dev* **18**: 504–11.
103. Brennecke J, Stark A, Russell R, Cohen S. (2005) Principles of microRNA-target recognition. *PLoS Biol* **3**: e85.
104. Lewis B, Burge C, Bartel D. (2005) Conserved seed pairing, often fianked by adenosines, indicates that thousands of human genes are microRNA targets. *Cell* **120**: 15–20.
105. Xue C, Li F, He T *et al.* (2005) Classification of real and pseudo microRNA precursors using local structure-sequence features and support vector machine. *BMC Bioinformatics* **6**: 310.
106. Hertel J, Stadler P. (2006) Hairpins in a haystack: recognizing microRNA precursors in comparative genomics data. *Bioinformatics* **22**: e197–202.
107. Helvik S, Snove O, Saetrom P. (2007) Reliable prediction of Drosha processing sites improves microRNA gene prediction. *Bioinformatics* **23**: 142–9.
108. Rajewsky N, Socci N. (2004) Computational identification of microRNA targets. *Dev Biol* **267**: 529–35.
109. Lai E. (2002) Micro RNAs are complementary to 3′ UTR sequence motifs that mediate negative post-transcriptional regulation. *Nat Genet* **30**: 363–4.
110. Lewis B, Shih I, Jones-Rhoades M *et al.* (2003) Prediction of mammalian microRNA targets. *Cell* **115**: 787–98.
111. Didiano D, Hobert O. (2006) Perfect seed pairing is not a generally reliable predictor for miRNA–target interactions. *Nat Struct Mol Biol* **13**: 849–51.
112. Gaidatzis D, van Nimwegen E, Hausser J, Zavolan M. (2007) Inference of miRNA targets using evolutionary conservation and pathway analysis. *BMC Bioinformatics* **8**: 69.
113. Stark A, Brennecke J, Russell R, Cohen S. (2003) Identification of *Drosophila* microRNA targets. *PLoS Biol* **1**: e60.
114. Long D, Lee R, Williams P *et al.* (2007) Potent effect of target structure on microRNA function. *Nat Struct Mol Biol* **14**: 287–94.

115. Vermeulen A, Robertson B, Dalby A *et al.* (2007) Double-stranded regions are essential design components of potent inhibitors of RISC function. *RNA* **13**: 723–30.
116. Ameres S, Martinez J, Schroeder R. (2007) Molecular basis for target RNA recognition and cleavage by human RISC. *Cell* **130**: 101–12.
117. Grimson A, Farh K, Johnston W *et al.* (2007) MicroRNA targeting specificity in mammals: determinants beyond seed pairing. *Mol Cell* **27**: 91–105.
118. Linsley P, Schelter J, Burchard J *et al.* (2007) Transcripts targeted by the microRNA-16 family cooperatively regulate cell cycle progression. *Mol Cell Biol* **27**: 2240–52.
119. Jackson A, Burchard J, Schelter J *et al.* (2006) Widespread siRNA "off-target" transcript silencing mediated by seed region sequence complementarity. *RNA* **12**: 1179–87.
120. Schwarz D, Ding H, Kennington L *et al.* (2006) Designing siRNA that distinguish between genes that differ by a single nucleotide. *PLoS Genet* **2**: e140.
121. Majoros W, Ohler U. (2007) Spatial preferences of microRNA targets in 3′ untranslated regions. *BMC Genomics* **8**: 152.
122. Stark A, Brennecke J, Bushati N *et al.* (2005) Animal microRNAs confer robustness to gene expression and have a significant impact on 3′ UTR evolution. *Cell* **123**: 1133–46.
123. Robins H, Press W. (2005) Human microRNAs target a functionally distinct population of genes with AT-rich 3′ UTRs. *Proc Natl Acad Sci USA* **102**: 15557–62.
124. Enright A, John B, Gaul U *et al.* (2003) MicroRNA targets in *Drosophila*. *Genome Biol* **5**: R1.
125. Krek A, Grun D, Poy M *et al.* (2005) Combinatorial microRNA target predictions. *Nat Genet* **37**: 495–500.
126. Eddy S. (1996) Hidden Markov models. *Curr Opin Struct Biol* **6**: 361–5.
127. Zuker M, Mathews D, Turner D. (1999) Algorithms and thermodynamics for RNA secondary structure prediction. In: Barciszewski J, Clark B (eds.). *A Practical Guide in RNA Biochemistry and Biotechnology*. Dordrecht, The Netherlands: Kluwer Academic Publishers, pp. 11–43.
128. Hofacker I, Fontana W, Stadler P *et al.* (1994) Fast folding and comparison of RNA secondary structures. *Monatsh Chem* **125**: 167–88.
129. John B, Enright A, Aravin A *et al.* (2004) Human microRNA targets. *PLos Biol* **2**: e363.
130. Lall S, Grü D, Krek A *et al.* (2006) A genome-wide map of conserved microRNA targets in *C. elegans*. *Curr Biol* **16**: 460–71.
131. Rehmsmeier M, Steffen P, Hochsmann M, Giegerich R. (2004) Fast and effective prediction of microrNA/target duplexes. *RNA* **10**: 1507–17.

132. Mathews D, Sabina J, Zuker M, Turner D. (1999) Expanded sequence dependence of thermodynamic parameters improves prediction of RNA secondary structure. *J Mol Biol* **288**: 911–40.
133. Saetrom O, Snove O, Saetrom P. (2005) Weighted sequence motifs as an improved seeding step in microRNA target prediction algorithms. *RNA* **11**: 995–1003.
134. Kim S, Nam J, Rhee J *et al.* (2006) miTarget: microRNA target gene prediction using a support vector machine. *BMC Bioinformatics* **7**: 411.
135. Huang J, Babak T, Corson T *et al.* (2007) Using expression profiling data to identify human microRNA targets. *Nat Methods* **4**: 1045–9.
136. Kertesz M, Iovino N, Unnerstall U *et al.* (2007) The role of site accessibility in microRNA target recognition. *Nat Genet* **39**: 1278–84.
137. Hobert O. (2008) Gene regulation by transcription factors and microRNAs. *Science* **319**: 1785–6.

Section II

PROTEINS AND PROTEOMES

Chapter 6

UniProtKB/Swiss-Prot Manual and Automated Annotation of Complete Proteomes: The *Dictyostelium discoideum* Case Study

Amos Bairoch and Lydie Lane

1. Introduction

1.1. *What is UniProtKB?*

The Swiss Institute of Bioinformatics (SIB), the European Bioinformatics Institute (EBI), and the US-based Protein Information Resource (PIR) group at Georgetown University Medical Center and the National Biomedical Research Foundation united 6 years ago to form the Universal Protein Resource (UniProt) Consortium (www.uniprot.org), which aims at providing a stable, high-quality, comprehensive, and authoritative resource for protein sequences and functional information.

The core component provided by the UniProt Consortium is the UniProt Knowledgebase (UniProtKB), which is composed of two sections: Swiss-Prot and TrEMBL.[1] Swiss-Prot contains nonredundant, fully annotated records; while TrEMBL contains the computer-annotated translations of coding sequences (CDSs) proposed by submitters for every nucleotide sequence incorporated in the public nucleic acid databases. Taken together, Swiss-Prot and TrEMBL cover all proteins identified so far, whether characterized or only inferred from nucleotide sequences. However, the manual annotation effort carried out by Swiss-Prot is mainly concentrated on proteins from model organisms to ensure the presence of high-quality annotation for representative members of most protein families. What makes Swiss-Prot unique is that it strives to present a high-quality synthetic view of the knowledge that is available on a particular protein.[2] This requires the expertise of a dedicated set of curators who extract, from the scientific literature, information that will be summarized and integrated into the knowledgebase. Curators also make use of state-of-the-art bioinformatics sequence analysis tools and, together with information gleaned from the scientific community, aim to help bench scientists through accurate protein annotation. Currently, Swiss-Prot contains about 400 000 reviewed (annotated) entries; while TrEMBL contains more than 6 million unreviewed, computer-annotated entries.

1.2. *What is Dictyostelium discoideum?*

Dictyostelium discoideum is one of the few organisms selected by the National Institutes of Health (NIH) as a suitable model organism for functional analysis of sequenced genes. It is a unicellular ameba during most of its life cycle, but starvation induces a unique developmental program during which individual cells aggregate by a chemotactic response to a cAMP gradient to form a multicellular entity. A morphogenetic process involving cell migration and cellular morphogenesis then transforms a simple mound of cells into a slug called pseudoplasmodium, establishing a relatively simple developmental pattern. This structure further develops into a fruiting body consisting of multiple terminally differentiated cell types, including spores and stalk cells.[3] These unique features of *D. discoideum* — together with its easy and efficient genetic manipulation by gene targeting, gene replacement, insertional mutagenesis, and suppressor screens — have made it a popular experimental system for the study of cytokinesis, motility, phagocytosis, chemotaxis, cell sorting, pattern formation, and cell-type determination. Many of these cellular behaviors and biochemical mechanisms are either absent or less accessible to experimental studies in other model organisms.

Sequencing and analysis of the genome of *D. discoideum* was completed in 2005.[4] This genome consists of six chromosomes with a total of 34 million base pairs of DNA. It is predicted to harbor about 13 000 protein-coding genes — a number comparable with that of the fruit fly, *Drosophila*. The introns and intergenic regions are usually more than 90% AT-rich, making them easily distinguishable from the GC-rich protein-coding regions and facilitating the prediction of open reading frames (ORFs).

The predicted proteome supports the hypothesis that *Dictyostelium* represents an early branch in the eukaryotic tree of life that diverged after the split between animals, plants, and fungi, with Dictyostelium and other amebae more closely related to animals. Therefore, many *Dictyostelium* proteins are more similar to their human orthologs than they are to those of the yeast *Saccharomyces cerevisiae.*

1.3. *The Dictyostelium discoideum Proteome at DictyBase*

dictyBase (dictybase.org), created in June 2003, is the model organism database (MOD) for D. discoideum.[5] It houses and maintains the genome sequence, and integrates biological knowledge obtained from experimental studies. Currently, 5372 gene models have already been manually checked and annotated, and 7515 genes have a product description assigned either manually or automatically, representing 40% and 60% of the estimated number of genes, respectively.

1.4. *The Dictyostelium Annotation Project at UniProt*

Among the large number of species contained within UniProtKB, a subset — targets of large-scale genome sequencing and mapping projects — has been selected. Dedicated annotation programs have been initiated around these priority organisms to concentrate annotation efforts on these species. The Human Proteomics Initiative (HPI) aims to annotate all known human protein sequences and orthologous sequences from other mammals such as other primates, mouse, rat, cow, pig, and rabbit. The Plant Proteome Annotation Project (PPAP) focuses on the annotation of plant-specific proteins with efforts directed primarily towards *Arabidopsis thaliana* and *Oryza sativa*. The Fungal Proteome Annotation Project (FPAP) focuses on the annotation of complete fungal proteomes, concentrating on *Saccharomyces cerevisiae* and *Schizosaccharomyces pombe* as well as other model species such as *Neurospora crassa* and the human pathogens *Candida albicans* and *Emericella nidulans*. Bacterial and archaeal model organisms are annotated as part of the HAMAP (High-quality Automated and Manual Annotation of microbial Proteomes) project, which identifies proteins that are part of well-conserved families and semiautomatically annotates them based on manually created family rules. Manual annotation efforts are also directed towards nonmammalian vertebrates (e.g. zebrafish and *Xenopus*) and invertebrate species (e.g. *Drosophila melanogaster* and *Caenorhabditis elegans*).

Until very recently, *D. discoideum* was not one of UniProt's priorities; but when we decided to explore how a generalist data resource such as UniProt could most effectively work in synergy with an MOD (such as dictyBase), *D. discoideum* appeared as an optimal test case. Furthermore, to work with the community of researchers trying to understand the particular and peculiar lifestyle of this fascinating organism seemed an enticing project. This community is quite unique in its sociological aspects. Its collaborative and altruistic spirit makes a refreshing contrast to the more competitive atmosphere which is pervasive in larger communities working with organisms that are deemed to be more important to the life sciences research universe.

We therefore decided to work in close collaboration with the dictyBase team to fully annotate all of the *D. discoideum* proteins and, subsequently, bring them into Swiss-Prot. We bootstrapped such an effort by organizing, in Geneva in February 2008, an annotation jamboree that brought together Swiss-Prot annotators from the SIB, EBI, and PIR with dictyBase curators and some *D. discoideum* researchers. Not only was the jamboree a complete success, but it also led us — as will be later pointed out — to the unforeseen conclusion that *D. discoideum* is a good prototype for establishing rules to propagate annotation to many uncharacterized eukaryotic organisms whose genome is or will soon be available.

2. Annotation Pipeline of the *Dictyostelium discoideum* Proteome in Uniprotkb

2.1. *Creating a Complete Proteome Set Across Swiss-Prot and TrEMBL*

Not all organisms whose genome has been sequenced are included in the complete proteome sets of UniProtKB. We consider as a "complete proteome" those sets of proteins that originate from genomes which have been fully sequenced and for which good-quality gene prediction models are available. These criteria are fulfilled for *Dictyostelium discoideum* and for an increasing number of other eukaryotic organisms such as *Drosophila melanogaster*, *Caenorhabditis elegans*, *Arabidopsis thaliana*, and *Saccharomyces cerevisiae*, as well as for a plethora of bacterial and archaeal species. The up-to-date list of available complete proteomes can be retrieved at http://www.uniprot.org/taxonomy/?query=complete%3ayes/.

Defining and deciding which proteins belong to a complete proteome set is often not a trivial task. Prior to our annotation jamboree, TrEMBL contained three types of protein entries for *D. discoideum*:

(1) entries derived from the predicted gene models of the submitted genomic sequence;
(2) entries derived from the predicted gene models of a preliminary sequence of chromosome 2, as carried out by one member of the genome sequencing consortium in 2002[6]; and
(3) entries derived from full-length cDNAs or from genomic DNA segments which have been submitted over the last 20 years by individual *Dictyostelium* laboratories.

To add a bit of complexity to this situation, not all proteins were linked to the same taxonomic node. The proteins from the genome sequence were stored as originating from the taxonomic identifier (TaxID) 352472, which corresponds to *Dictyostelium discoideum* strain AX4 (the specific strain selected to be sequenced by the genome consortium); while the two other categories of entries were said to originate from TaxID 44689, which corresponds to the generic *Dictyostelium discoideum*. In agreement with the dictyBase staff, we decided to consolidate all *D. discoideum* UniProtKB entries at the level of TaxID 44689.

As a result, we have a complete proteome set in UniProtKB, and the keyword "complete proteome" was added to the relevant entries. This allows users to download a complete, nonredundant set of proteins from a given organism by specifying the TaxID of the species of interest with the concomitant presence of the "complete proteome" keyword. Hence, 13 018 *D. discoideum* UniProtKB entries (out of 14 179) are now tagged with the "complete proteome" keyword (release 14.0 of July 22, 2008). They can be retrieved at the following URL: http://www.uniprot.org/uniprot/?query=organism%3Adictyostelium+AND+keyword%3A%22complete+proteome.

2.2. *UniProtKB/Swiss-Prot Annotation*

The UniProtKB annotation process encompasses the annotation not only of the sequence itself, but also of sequence features, gene and protein nomenclature, and protein functional properties. Basic steps are done in all of these areas for the whole UniProtKB, but an entry will be integrated into Swiss-Prot only after it has been completely annotated and manually reviewed in all of these annotation aspects. TrEMBL records only partially fulfill the criteria outlined below, and are not manually checked.

Some 6500 publications specifically related to *D. discoideum* are currently referenced in PubMed and constitute the basic material for manual annotation, both at dictyBase and at UniProt. Systematic BLAST analysis[7] allows one to find the closest annotated protein and thus classify the protein as a member of a given family. This step is very important to ensure consistency of annotation between organisms within a family. Since the *D. discoideum* proteome is highly enriched in asparagine stretches, the BLAST tool had to be configured to filter low-complexity regions. In cases where BLAST does not allow one to unambiguously identify the family or subfamily to which a protein belongs, further analysis is performed using phylogenetic tools such as the PhyloFacts database, the team of which helped us to select candidates for priority annotation during the annotation jamboree.[8]

2.2.1. *Sequence annotation*

Since UniProtKB is a protein-centric resource, it is critical to get as much information as possible in order to obtain the correct, biologically significant protein sequence(s). This step requires a number of distinct, yet related annotation activities:

- capturing sequencing conflicts to display, in the entry, the protein sequence the most likely to be correct;
- correcting incorrect gene models (using multiple alignments to detect potential problems);
- determining the most probable initiation codon (using multiple alignments);
- validating models using full-length cDNAs or, in some cases, expressed sequence tag (EST) sequences;
- validating protein sequences using high-quality mass spectrometry (MS) or, more rarely, Edman sequencing data;
- detecting/predicting rare sequence-modifying biological events such as the presence of nonstandard amino acids like selenocysteines; and
- annotating splice variants using published reports, full-length cDNAs, and bioinformatics predictions from trusted sources.

For *D. discoideum*, unfortunately only a few full-length cDNAs are available to confirm gene models, and in most cases sequencing at the protein level is not available. However, we often correct exon or initiation codon prediction using comparative sequence analysis, as illustrated in Figs. 1 and 2.

In addition to the sequence itself, UniProtKB provides the protein existence (PE) line, which is an "evidence level" for the *in vivo* existence of a protein, regardless of the accuracy or correctness of the sequence displayed. This is particularly important for model organisms such as *D. discoideum*, for which a large proportion of the available sequences are pure predictions from the genome. To gain this "level 1" ("existence at protein level"), a protein has to be clearly observed experimentally. The corresponding entry should contain a characterization paper, Edman sequencing information, clear identification by MS, or an X-ray or nuclear magnetic resonance (NMR) structure, such as the entry displayed in Fig. 3.

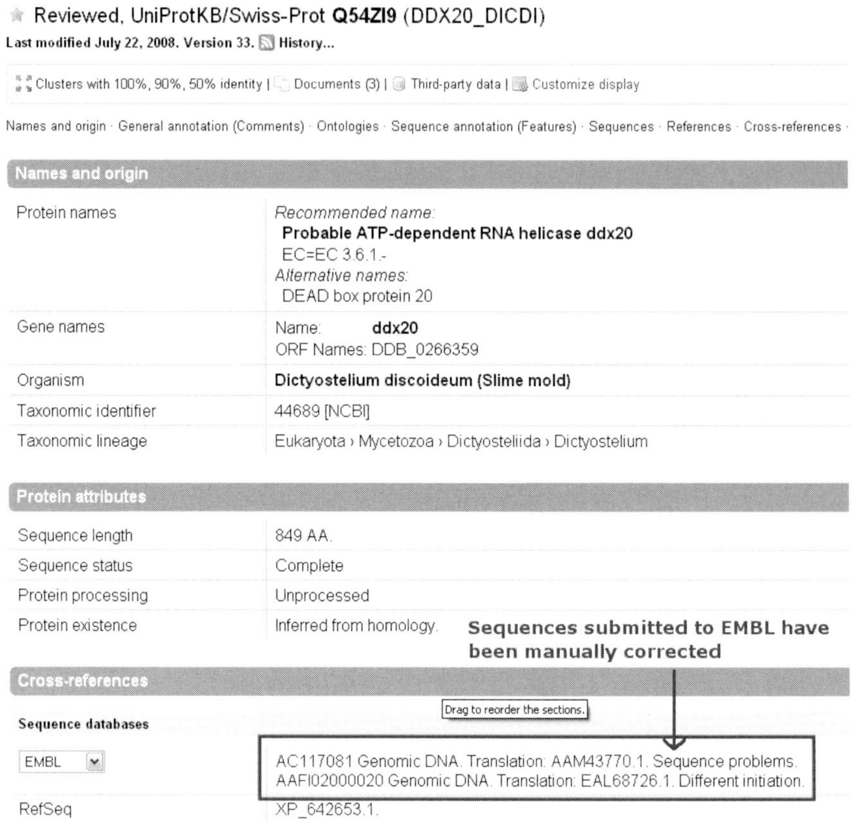

Fig. 1. Part of the UniProtKB entry Q54ZI9 displayed at www.uniprot.org/. The manually added tags to the EMBL cross-references show that the originally submitted sequences have been corrected.

2.2.2. *Sequence feature annotation*

Providing correct protein sequence information also implies being as comprehensive as possible in terms of the representation of posttranslational modifications (PTMs). This requires the following:

- annotation of processing events (cleavage of initiator methionines, signal sequences, mitochondrial transit peptides, propeptides, etc.);
- annotation of cross-linking events (mainly disulfide bonds); and

Fig. 2. Part of the UniProtKB entry Q55EX3 displayed at www.uniprot.org/. The sequence annotation field mentions that the sequence has been corrected to include a selenocysteine at position 112.

Fig. 3. Part of the UniProtKB entry P10733 displayed at www.uniprot.org/. This protein has been extensively characterized in the cited papers. This is reflected in the protein existence tag set "at protein level".

- annotation of all other types of PTMs that generally result in the addition of a simple (e.g. acetyl, methyl, or phosphate groups) or complex (e.g. carbohydrates and lipids) chemical entity to the protein chain.

Information on PTMs can be obtained by

- using results of published low- or high-throughput proteomics studies;
- using some specific high-quality prediction tools (e.g. for signal sequence, transit peptide, N-glycosylation, etc.); or
- propagation from already annotated orthologous proteins. This process must be carried out with the utmost care, since it is important to avoid propagating species- or phylum-specific PTMs outside of their realm.

Figure 4 provides an example of entry with a predicted PTM feature.

It is also of paramount importance to correctly and fully represent the domain structure of proteins as well as to report relevant sites and motifs.

- Annotation of topological domains (transmembrane, extracellular regions, etc.) is made on the basis of
  - published experimental topological data;
  - transmembrane prediction tools;
  - results of some PTM prediction tools that offer insight into the topology of the protein, such as GPI-anchor prediction, signal or transit peptide prediction, etc.; or
  - similarity to close orthologs, complemented by a high-level manual check to carefully estimate the specific biological context of some topological information.
- Annotation of specific validated domains and important sites (active sites, metal-binding sites, etc.) is made on the basis of
  - information derived from three-dimensional (3D) structures through software-assisted data mining of the relevant Protein Data Bank (PDB)[9] entries;
  - InterPro[10] and ProRule,[11] a domain annotation rule system that we developed to help in the annotation of important sequence

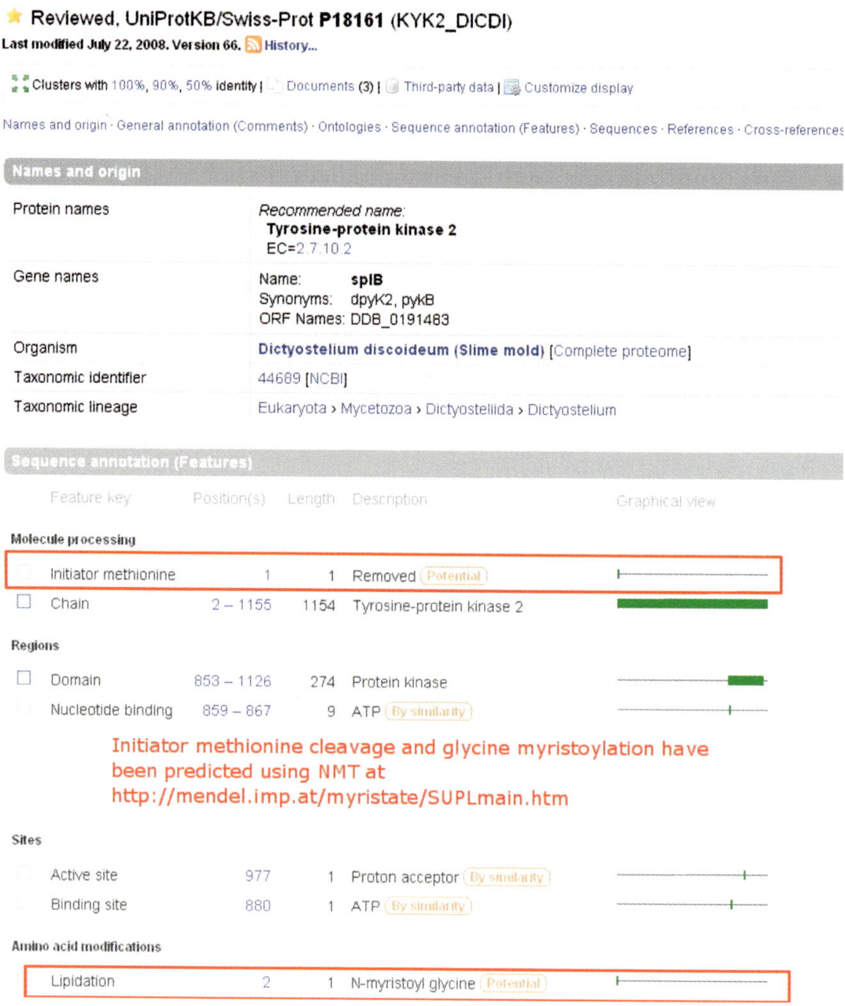

Fig. 4. Part of the UniProtKB entry P18161 displayed at www.uniprot.org/. The potential *N*-myristoylation detected through the use of the NMT predictor Web tool is described in the sequence annotation field.

features at the domain level. We have currently annotated almost 1000 different protein domains;

- published results of mutagenesis deletion; or
- similarity to close orthologs.

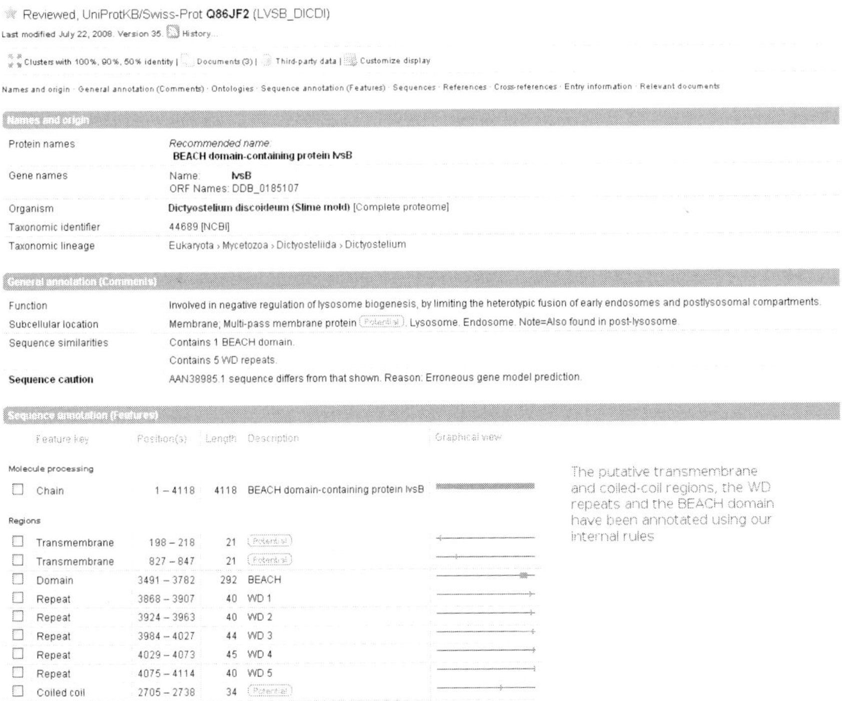

Fig. 5. Part of the UniProtKB entry Q86JF2 displayed at www.uniprot.org/. The different topological and functional domains detected through our annotation pipeline are described in the sequence annotation field.

Figure 5 is an example of a protein harboring transmembrane domains, WD repeats, and a BEACH domain.

2.2.3. *Nomenclature annotation*

Associating a protein with a recommended name and a gene symbol is also a crucial mission of the UniProt Consortium. This manually intensive task is a very useful service to the community since it helps in the process of standardizing shared nomenclature, which is often inadequate in the life sciences field.

Cross-references to dictyBase are added automatically to TrEMBL entries at each release. At the same time, the gene name provided by dictyBase is added as the primary name in the gene name line. Thus, even before an entry is manually curated, it already possesses a link to dictyBase and includes the dictyBase-approved gene symbol. During manual Swiss-Prot annotation of *D. discoideum* entries, we

- check and use recommended gene symbols from dictyBase when available. These gene symbols conform to the nomenclature rules described at http://dictybase.org/Dicty_Info/nomenclature_guidelines.html/. Genes that have been studied by *Dictyostelium* geneticists generally follow the naming conventions derived by Demerec *et al.*[12]: three lowercase letters followed by an uppercase letter (e.g. carA, dmtA, fttB). This is a naming convention which was originally developed for bacterial genes. However, many recent gene symbols in *Dictyostelium* are derived from those of the cognate genes in human or mouse and, in some cases, in *S. cerevisiae*, reflecting the pivotal roles currently played by these organisms as models;
- are proactive in proposing new names for entries lacking a proper gene symbol. As noted above, for the sake of having consistent gene symbols over a wide range of orthologs, we try, whenever appropriate, to select names which are as close as possible to the names of their human or yeast counterparts;
- capture as many alternative protein names from literature reports and external resources; and
- actively send feedback to dictyBase to clean up inconsistencies, propose updates to gene symbols, and, in general, try to make sure that they share the same symbols whenever adequate.

When appropriate, assigning a protein to a family is also part of the naming task. For proteins that are not yet characterized, we have created a system of temporary family names, the so-called "Uncharacterized Protein Family" (UPF) nomenclature. Such an annotation is displayed in Fig. 6.

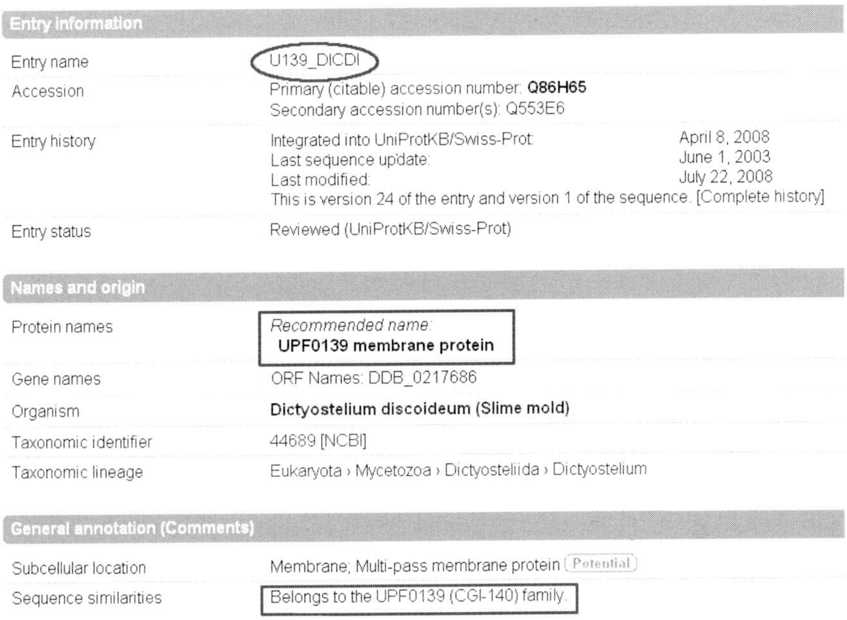

Entry information	
Entry name	U139_DICDI
Accession	Primary (citable) accession number: **Q86H65** Secondary accession number(s): Q553E6
Entry history	Integrated into UniProtKB/Swiss-Prot: April 8, 2008 Last sequence update: June 1, 2003 Last modified: July 22, 2008 This is version 24 of the entry and version 1 of the sequence. [Complete history]
Entry status	Reviewed (UniProtKB/Swiss-Prot)

Names and origin	
Protein names	*Recommended name:* **UPF0139 membrane protein**
Gene names	ORF Names: DDB_0217686
Organism	**Dictyostelium discoideum (Slime mold)**
Taxonomic identifier	44689 [NCBI]
Taxonomic lineage	Eukaryota › Mycetozoa › Dictyosteliida › Dictyostelium

General annotation (Comments)	
Subcellular location	Membrane; Multi-pass membrane protein (Potential)
Sequence similarities	Belongs to the UPF0139 (CGI-140) family.

Fig. 6. Part of the UniProtKB entry Q86H65 displayed at www.uniprot.org/. This protein has clear orthologs in other species, but none have been characterized yet. This makes it belong to the "Uncharacterized Protein Family", which is mentioned in the general annotation field.

2.2.4. *Functional annotation*

The holy grail of annotating a proteome is to be able to know the role of each of the protein entities. This goal is far from being reached, even in model organisms that have been the target of the largest number of concerted or individual research initiatives (i.e. *Saccharomyces cerevisiae* and *Escherichia coli*). Therefore, the process of providing functional annotation is a never-ending process that is time-consuming, yet nevertheless a priority for the UniProt Consortium. Although Gene Ontology (GO) terms can be used to summarize some functional aspects of a protein,[13] additional textual descriptions often remain essential for a comprehensive overview.

Fig. 7. Part of the UniProtKB entry O76329 displayed at www.uniprot.org/. This protein has been extensively characterized. A detailed summary of the data available is given in the general annotation field.

As shown in Fig. 7, the general annotation (also known as comment (CC) lines) of *D. discoideum* Swiss-Prot entries mainly provide information relative to the following:

- functions and biological roles;
- potential cofactors and activity regulators;

- subcellular locations;
- protein/protein interactions; and
- expressions — tissue specificity and developmental stage.

To populate Swiss-Prot entries with this type of information, we

- use published experimental reports. This is by far the bulk of the manual functional annotation process. It necessitates the full scientific expertise of the annotators to ensure the quality of the textual representation of the role and function of the protein;
- use and actively request information provided by the scientists who have carried out functional characterization studies;
- use prediction tools to infer information from the full set of topological, domain, and PTM information; and
- use sequence comparison tools to infer information from orthologs and, when relevant, paralogs.

The UniProtKB keywords are useful in providing a very high-level summary of some of the information displayed in a UniProtKB entry. We maintain a keyword controlled vocabulary and ensure that keywords are consistently used in the relevant entries. This is done partially in a programmatic manner, but also requires annotator judgment skills.

Unfortunately, there is still a large set of proteins that do not have any homologs and that lack InterPro domain predictions. For these proteins, inferring a function is a challenging, if not impossible, task. Those so-called "ORFan" proteins are semiautomatically annotated using Anabelle, our sequence analysis workbench. The result of this process is generally restricted to the annotation of predicted topological information (signal sequences, transmembrane domains) as well as the annotation of regions of compositional biases (coiled-coil domains, runs of particular amino acids).

2.3. *Why are DictyBase and UniProtkb Complementary?*

dictyBase, as with other MODs, is genome-centric. In addition to the complete genome sequence, groups performing high-throughput experiments such as large-scale mutagenesis and microarray-based gene expression studies deposit their data in dictyBase to be integrated and distributed to the research community. Phenotypic data are represented using a controlled vocabulary that describes the consequences of mutations reported for that gene. This type of information is not available in UniProtKB, since its mission is to concentrate on information obtained at the protein level. Conversely, most MODs, including dictyBase, rely on UniProtKB to provide protein-specific information. In addition, UniProtKB provides family annotation across a wide spectrum of organisms, which, by definition, is not part of the mission of MODs.

3. Conclusion: Status and Perspective

3.1. *Status of Dictyostelium Annotation in Swiss-Prot*

Release 14.0 of UniProt contains 2479 manually reviewed *D. discoideum* entries. In terms of their protein existence level, the breakdown is as follows:

Evidence at protein level	368
Evidence at transcript level	302
Inferred from homology	1611
Predicted	194
Uncertain	6

3.2. *Future of Dictyostelium Annotation in Swiss-Prot*

We plan to achieve the complete manual annotation of the *D. discoideum*, proteome in the Swiss-Prot section of UniProtKB within 2 years (i.e. by mid-2010). To ensure that we achieve this goal, we intend to make full use of the synergy between the annotation efforts of dictyBase and those of UniProt. We also plan to speed up the work by soliciting help from the *Dictyostelium* research community. To increase the percentage of proteins that have been confirmed at the protein level, we hope to convince a number of proteomics facilities to run *D. discoideum* cellular extracts through their protein identification pipelines. We will also import

EST-based information from dictyBase to significantly increase the proportion of proteins confirmed at the transcript level.

The genome sequence of a highly related organism, *Dictyostelium purpureum*, is expected to be released in the near future. Once dictyBase has run gene prediction tools on the available sequence, we will be able to annotate the two proteome sets in parallel with a negligible amount of annotation time overhead.

3.3 *How may the Annotation of Dictyostelium Help to Annotate other Eukaryotic Proteomes?*

Proteins from organisms other than model organisms are generally predicted from genome sequences and are not the target of experimental characterization studies. To annotate these proteins with as much information as possible, we use an automated annotation pipeline based on a carefully developed integrated rule system. Annotation rules are critically dependent on having at least one and preferably more "template" entries that have gone through the process of knowledge-based annotation.

A careful study of the *D. discoideum* proteins annotated in Swiss-Prot shows that a large number of them belong to conserved families covering a wider-than-previously-thought taxonomic range and generally having a single copy in each genome (few in-paralogs and between-species paralogs). In those cases, *D. discoideum* can be used as the "seed" organism to implement eukaryotic-wide UniRules. In the near future, we plan to create about 1500–2000 eukaryotic automatic annotation rules that will be used to annotate TrEMBL entries, which will, following manual checks, be entered into Swiss-Prot.

Acknowledgments

The authors want to thank Lydie Bougueleret, Pascale Gaudet, Janet James, and Patricia Palagi for critical reading of the manuscript, as well as Ron Appel for his patience and understanding in the face of our complete lack of abidance to the planned date at which we were supposed to submit the text of this chapter. UniProt annotation activities at the SIB are supported by the Swiss Federal Government through the Federal Office of Education and Science, by the National Institutes of Health (NIH) grant 2 U01 HG02712-04, and by the European Commission contract FELICS (021902).

References

1. UniProt Consortium. (2008) The Universal Protein Resource (UniProt). *Nucleic Acids Res* **36**: D190–5.
2. Bairoch A, Boeckmann B, Ferro S, Gasteiger E. (2004) Swiss-Prot: juggling between evolution and stability. *Brief Bioinform* **5**: 39–55.
3. Thomason P, Traynor D, Kay R. (1999) Taking the plunge. Terminal differentiation in *Dictyostelium*. *Trends Genet* **15**: 15–19.
4. Eichinger L, Pachebat JA, Glockner G *et al.* (2005) The genome of the social amoeba *Dictyostelium discoideum*. *Nature* **435**: 43–57.
5. Chisholm RL, Gaudet P, Just EM *et al.* (2006) dictyBase, the model organism database for *Dictyostelium discoideum*. *Nucleic Acids Res* **34**: D423–7.
6. Glockner G, Eichinger L, Szafranski K *et al.* (2002) Sequence and analysis of chromosome 2 of *Dictyostelium discoideum*. *Nature* **418**: 79–85.
7. Altschul SF, Gish W, Miller W *et al.* (1990) Basic local alignment search tool. *J Mol Biol* **215**: 403–10.
8. Brown DP, Krishnamurthy N, Sjolander K. (2007) Automated protein subfamily identification and classification. *PLoS Comput Biol* **3**: e160.
9. Berman H, Henrick K, Nakamura H, Markley JL. (2007) The worldwide Protein Data Bank (wwPDB): ensuring a single, uniform archive of PDB data. *Nucleic Acids Res* **35**: D301–3.
10. Mulder NJ, Apweiler R. (2008) The InterPro database and tools for protein domain analysis. *Curr Protoc Bioinformatics* (Chapter 2): Unit 2.7.
11. de Castro E, Sigrist CJ, Gattiker A *et al.* (2006) ScanProsite: detection of PROSITE signature matches and ProRule-associated functional and structural residues in proteins. *Nucleic Acids Res* **34**: W362–5.
12. Demerec M, Adelberg EA, Clark AJ, Hartman PE. (1966) A proposal for a uniform nomenclature in bacterial genetics. *Genetics* **54**: 61–76.
13. Hill DP, Smith B, McAndrews-Hill MS, Blake JA. (2008) Gene Ontology annotations: what they mean and where they come from. *BMC Bioinformatics* **9**(Suppl 5): S2.

Chapter 7

Analytical Bioinformatics for Proteomics

Patricia M. Palagi and Frédérique Lisacek

1. Introduction

Proteomics was originally defined as the combination of protein separation techniques, such as two-dimensional gel electrophoresis (2-DE) or liquid chromatography (LC), with mass spectrometry (MS). Analysis of these experimental data encompasses the identification and quantification of proteins, as well as the determination of their localization, modifications, interactions, activities, and, ultimately, function. Proteomes — complete sets of proteins expressed by a genome, cell, tissue, or organism — are now routinely compared under different conditions (time, temperature, pH, pathological conditions, etc.). This comparison is both qualitative and quantitative, and it depends heavily on computers.

The Proteome Informatics Group (PIG) at the Swiss Institute of Bioinformatics (SIB) focuses its activities on the development of software tools and databases for proteomics, and targets research projects in three domains: proteome imaging, protein identification and characterization with MS data, and proteomics knowledge integration and databases. As shown in Fig. 1, these domains are chronologically integrated in the proteomics workflow. Some of the derived projects have been ongoing since 1984, long before the SIB was created in 1998. The state of the art of the three topics is detailed in this chapter, whilst the contributions of the PIG are highlighted.

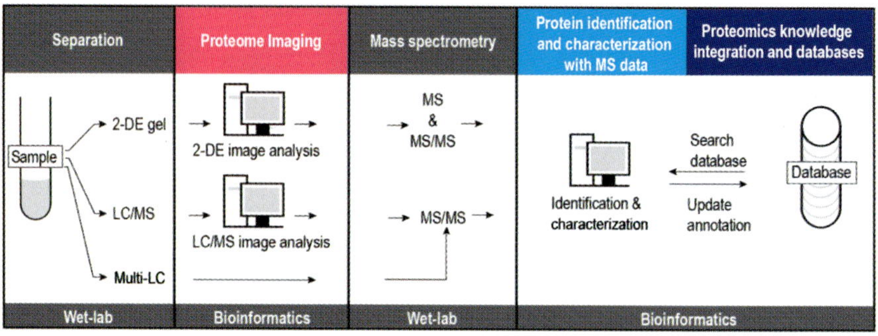

Fig. 1. This proteomics workflow highlights the chronological steps alternating wet-lab and dry-lab (bioinformatics) operations. Each bioinformatics step matches one of the three domains covered by the PIG. Their corresponding colors are used in Fig. 3 to emphasize the bioinformatics developments specific to each domain.

Proteome informatics currently faces at least three important challenges. Firstly, a majority of mass spectra collected in high-throughput experiments still cannot find confident molecular identification. Explanations for this are twofold: they involve, on the one hand, the quality of data and, on the other hand, the efficiency of mass spectra data-matching procedures. The quality of the separation, the presence of artifacts/contaminants/posttranslational modifications (PTMs), and mass calibration status need assessing to ensure fruitful interpretation of the data. This issue can be tackled by visually outputting two-dimensional (2-D) LC-MS data and exploiting image intensities and contrasts for monitoring data. More information from tandem mass spectrometry (MS/MS) can be retrieved using workflows that combine several complementary software tools. A platform supporting such workflows can address the question of efficient MS/MS data-matching procedures. In this case, efficiency is envisaged as increasing both the precision and the speed of data processing. Sections 2 and 3 of this chapter show how we implement these strategies.

Secondly, the large volume of data from proteomic experiments needs to be integrated and exploited. As a first step, these datasets require a standardized description and format. Data standardization efforts undertaken up to now are expected to increasingly bear on future software and database development. Most proteomics journals have actually defined guidelines for publishing proteomics data that comply with standards. Publicly accessible standardized proteomic data resources will grow and spread. Automated procedures for the design and update of such databases will therefore be essential to keep up with the production and with the increasing need to retrieve and mine the data. Section 4 of this chapter emphasizes these aspects.

Thirdly, protein expression is yet to be adequately measured and quantified for the extraction of reliable differential quantitative data from MS experiments. In this context, comparative tools will remain essential for data interpretation. 2-DE gels now share the proteomics scene with other high-resolution separation techniques (e.g. LC), which when combined with MS, produces large, highly correlated datasets. Exploiting the correlations that exist within these datasets allows for the extraction of information that cannot be visible when exclusively analyzing individual spectra. The 2-D visual representation of LC-MS data helps in tracking biomarkers and comparing peptide profiles. Adding tools for the quantitative analysis of MS data from isotope-labeled or label-free complex mixtures is the next requisite. Incidentally, a current bottleneck is the huge size of data files that would warrant a universal format for compressed raw spectrum storage. We are also committed to solving this critical point, as mentioned throughout the text.

2. Proteome Imaging

Protein separation techniques exploit the diversity of proteins, including their size, shape, electrical charge, molecular weight, hydrophobicity, and predisposition to interact with other proteins. A number of technologies performing large-, medium-, or small-scale separation of complex protein mixtures have been investigated, such as capillary and gel electrophoresis, microchannel, protein chips, LC, and high-performance liquid chromatography (HPLC). Among these separation technologies, gel electrophoresis and LC coupled to mass spectrometers are well suited to display proteomes in a form that is amenable to human vision and computer analysis; this can subsequently facilitate the comparison of two or more samples and assist protein identification.

2.1. 2-DE Gel Imaging

2-DE gels can simultaneously resolve thousands of proteins separated by their molecular weight and isoelectric point. The 2-DE gel patterns provide an important research tool for quantitative analysis and comparative proteomics. The possibility of detecting protein expression changes associated with diseases and treatments or of finding therapeutic molecular targets has been, among many other applications, a major incentive to the development of specialized software systems for 2-DE gel image analysis since 1975.

In general, 2-DE gel image analysis packages have the same basic operations and functionalities necessary to carry out a complete gel study, which should end with highlighting differentially expressed spots in populations of gels. Dowsey *et al.*[1] give a comprehensive review of computational techniques and algorithm details. Besides a variety of ways of displaying and viewing gel images (some different viewpoints of gel images are illustrated in Fig. 2), the major functions of software systems for 2-DE gel image analysis are (a) the detection and quantification of protein spots on the gels; (b) the matching of corresponding spots across the gels; and (c) the localization of significant protein expression changes. Any other additional features, such as data management and database integration, may or may not be included. Functions (a) and (b) have to be successfully executed prior to carrying out function (c).

The optimal and reproducible definition of the spot borders, as a consequence of spot quantitation, depends mostly on gel experimental details as well as on the quality of focusing and polymerization. In order to eliminate or reduce the impact of overlapping or weak spots, detection algorithms included in the packages generally comprise filtering steps to automatically remove streak artifacts and noise spikes[2] or a segmentation process based on the analysis of the gray levels.[3] Spot detection algorithms also produce quantitative information on the protein spots, such as the spots' area, optical density (maximum intensity value in the area), volume (integration of all intensity values over the area), and relative measures of these values that partially compensate for variations in sample load or staining, and as a consequence provide better reproducibility of data analysis and results.

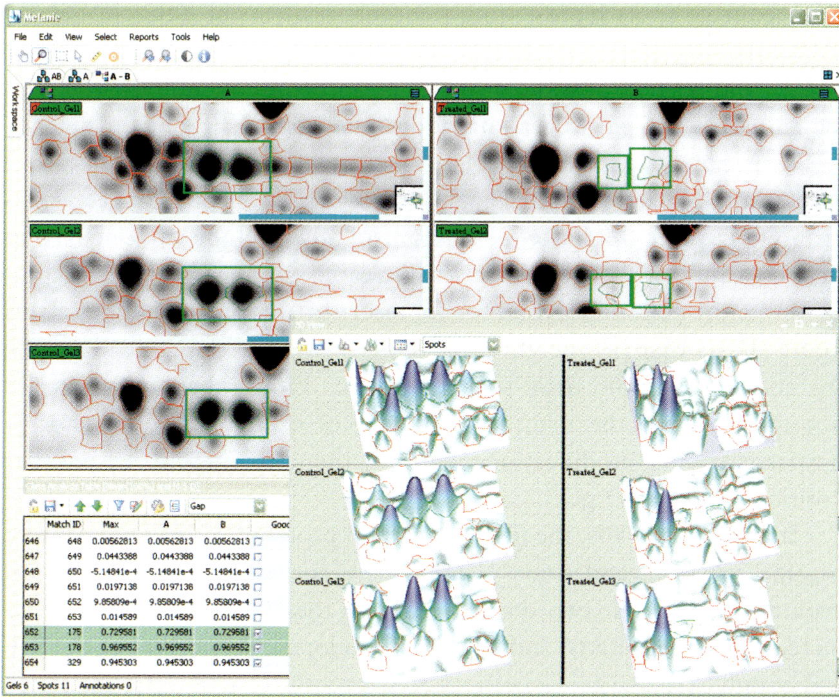

Fig. 2. Differential image analysis of two groups of gels with the Melanie package. Each group has 3 replicate gels. The left column shows three control gels, and the right column shows the corresponding regions on three treated gels. The three-dimensional (3-D) window (bottom right) highlights two neighboring spots present in the control sample, but absent from the drug-treated sample.

Matching gel images is another critical process, whether based on the previous detection of spots[4] or on the intensities of the regions before the detection of spots.[5] Gel matching involves the comparison of the spatial distribution of spots of a gel image taken as a reference image (e.g. control) with a second gel image (e.g. disease). Expectable variations induced by gel running conditions and gel scanning should not interfere with the assessment of changes in protein expression. Needless to say, the subsequent steps on gel image analysis will be erroneous if spots representing the same protein are not correctly matched or if spots representing different proteins are mistakenly matched together. To minimize this risk, some tools propose the initialization of a few pairs of spots representing the same proteins in different gels — a landmarking step. These landmarks are then used to warp the gel images and correct possible distortions, and consequently improve the matching quality. However, this additional step has the inconvenience of being time-consuming and labor-intensive for end-users. Figure 2 is an example of a differential analysis between two groups of gels. Each group has three replicate gels. Control gels belong to one group (left column), while gels from drug-treated samples are in a second group (right column). Notably, the insert in the bottom right of this figure shows that two neighboring spots present in the control sample are absent from the drug-treated sample. Either these proteins are not expressed in the drug-treated samples or they have undergone posttranslational modifications, causing them to migrate to another position on the 2-D gel.

In the early 1980s, the first packages of proteomics imaging software became available, and some of them have survived the computational evolution of the last two decades. Among these are PDQuest™ (commercialized by BioRad)[6] and Melanie[7] developed by the PIG at the SIB and commercialized by GeneBio S.A. Currently, several other dedicated software packages are also commercialized, such as DeCyder (from GE Healthcare) and Progenesis (from Nonlinear Dynamics).

Most current packages include two-dimensional difference gel electrophoresis (2-D DIGE)[a] analysis. In this fluorescent technique for protein labeling,[8] each sample is labeled with a fluorescent dye (Cy2, Cy3, or Cy5) prior to electrophoresis and then the samples are coseparated on the same 2-D gel. Scanning the gel at a specific wavelength for each dye reveals the different proteomes. These images are then overlaid using the above-mentioned software and the differences in abundance of specific protein spots can be detected. One advantage of this technique is that variation in spot location due to gel-specific experimental factors is the same for each sample within a single DIGE gel; consequently, the relative amount of a protein in a gel in one sample compared to another is unaffected. From the bioinformatics point of view, only the spot detection function requires some adaptation when the software processes

[a] Also called multiplex experiments.

DIGE gels. Since the same proteins are localized in the same *x*- and *y*-coordinates on the gels, the spot detection procedure is the same for the codetected gels. The matching step is thus straightforward and less subject to error.

Free tools available through the Internet are an alternative to commercial systems, although they are usually limited to simple operations on a small number of gels. The free tool Flicker was created to visually compare 2-DE gels.[9,10] It has a Java applet and a stand-alone version to be installed on the user's computer. Both versions allow the visual comparison of two gels either side by side or superimposed. When gels are in the overlay mode, "flickering" both images affects protein spot intensities so that expression is discernable. GelScape[11] is a Web-based tool to display gel images and at the same time a database to house gels and their annotations. The gels uploaded in GelScape can also be compared with the other available gels in this database. The free Melanie Viewer has the usual visualization operations of the full version and most of the analysis procedures as well; however, the analysis is restricted to a small number of proteins and only from gels that have already been analyzed by a full version. The viewer version of PDQuest™ also gives the possibility of elementary visualization of gel images.

2.2. *LC/MS Imaging for Label-Free Quantitation*

An alternative, though common, proteomics workflow combines the separation of proteins and peptides by LC followed by direct analysis by MS. This workflow has initiated the design of new bioinformatics tools for proteomics that are complementary to 2-DE gel image analysis. In the proteomics imaging of LC/MS studies, data are also represented in two dimensions, i.e. the elution time and m/z, and they can be visualized and analyzed as images. LC/MS image analysis, albeit being a recent proteomics field (the first articles were only published in 2002[12,13]), shows promising applications in differential proteome analysis by comparing several label-free proteome sets and detecting significant quantitative differences, and in discovering specific proteins. These expectations certainly explain the sudden emergence of a number of packages within a short period of time.

Filtering is one of the basic operations, and functionalities are necessary to carry out a complete LC/MS study. Usually, this operation removes peaks with weaker intensities (e.g. background noise) or high spikes constant in time (e.g. chemical noise such as column contaminants) so as to reduce the complexity of spectra and facilitate peak detection. Peak detection involves looking for the monoisotopic peaks (also called the deisotoping procedure) and determining the states of the ion charge. Finally, isotopic peaks of the same corresponding mass value are clustered into one single peak signal. In fact, this procedure selects the peaks of interest from the enormous quantity of data.

Ideally, the same molecules analyzed in the same LC/MS platform would have the same retention time, molecular weight, and signal intensity. However, due to experimental variations, this is not always the case. While m/z values depend on mass accuracy and the resolution of the mass spectrometer, the retention times largely depend on the analytical method used. Peaks from the same compound or peptide match fairly close in m/z values, but the retention times between the runs can vary significantly. The peak alignment operation corrects these variations and finds corresponding peaks across different LC/MS runs. Once runs are aligned, they can be compared and statistically analyzed in order to find differentially expressed proteins and peptides, and to quantify these differences.

Similarly to 2-DE gel analysis software, due to the large amount of data generated through LC/MS experiments, LC/MS image analysis software does not usually run through Web interfaces or on a remote basis. All tools are available as stand-alone versions that have to be installed and run locally.

MSight®,[14] a software developed by the PIG, is a free downloadable software available through the ExPASy[15] server. Its interface and functionalities are based on the Melanie gel image analysis system (mentioned in the previous section). It runs on the latest Windows™ operating systems and accepts data generated from the majority of mass spectrometers such as those supplied by Bruker, Waters, and ABI-SCIEX. It also supports the mzXML[16] and mzData[17] formats. It is worth noting that the mzXML team (http://sashimi.sourceforge.net/) provides various converter tools to generate mzXML files from mass spectrometer native acquisition files. MSight has the advantage of a user-friendly interface, which eases navigation through large volumes of data. Several visualization tools allow one to discriminate peptide or protein from noise or to perform differential analysis. Peak detection and peak alignment have been integrated in its latest version, as has the semiautomatic analysis of LC-MS datasets. The procedures for quantitative differential proteome analysis are currently under development. MSight can also be considered a resourceful tool for data quality control.

MZmine[18] and MapQuant[19] are open-source software packages for LC/MS analysis written in Java and ANSI C, respectively. In MZmine, several spectral filters are implemented to correct the raw data files, such as smoothing for noise filtering of the mass spectra. Other methods are also implemented, such as peak detection, peak alignment, and normalization of multiple data files. Despite a user-friendly interface, the tool misses statistical analysis procedures to quantify differences in a comparative study.

Other tools, such as XCMS,[20] SpecArray,[21] msInspect,[22] and OpenMS,[23] automatically detect potential peptide features directly from the raw data and extract the corresponding quantitative information without the support of image analysis. Ion intensities are simply integrated over time for measuring the total ion abundance for any peptide ion within a LC-MS experiment. Some programs in the SpecArray software suite[21] share functionalities with image analysis. In a sequential mode, the Pep3D program generates 2-D gel-type images from LC/MS data; the mzXML2dat program extracts high-quality data by cleaning the spurious noise and creating centroid MS spectra; the PepList program extracts a list of peptide features from LC/MS data, such as the monoisotopic masses, the charges, and the retention times of the MS spectra; the PepMatch program aligns peptide features of multiple samples; and finally, the PepArray program generates an array of peptide information. For each selected peptide present in each sample, this program generates its normalized abundance value and its retention time. These final arrays can be exported to a clustering tool and then be further analyzed, i.e. to find quantitative differences in LC/MS samples.

3. Protein Identification and Characterization with MS Data

Protein identification plays a central role in the investigation of proteomes, whenever using 2-DE gels or LC for sample separation followed by MS analysis. Typically, the spots of a 2-DE gel are enzymatically digested and the resulting peptide masses are measured, producing an MS spectrum, also called "peptide mass fingerprinting" (PMF). Peptides can also be isolated and fragmented within the mass spectrometer, leading to MS/MS spectra, also called "Peptide Fragmentation Fingerprinting" (PFFs).

3.1. *PMF*

In PMF analysis, peak lists extracted from an experimental spectrum are compared with theoretical masses computed from protein sequences stored in databases that were digested *in silico* using the same cleavage specificity of the protease employed in the experiment. Practically, the procedure counts matching experimental and theoretical peptide masses. A protein is selected as a candidate when it contains a threshold-dependent number of peptide hits. Other criteria, such as peptide coverage of the sequence, mass error, etc., contribute to defining a score. The candidate proteins are then sorted according to their scores. The top-ranked proteins are considered as potential identifications of the spectrum.

The key step of the procedure lies in the scoring function, a higher score indicating a higher likelihood that the corresponding protein is the target one. Many factors are usually taken into account to produce a robust score, such as dissimilarities in the peak positions due to internal or calibration errors or modified amino acids, expected peak intensities, noise, contaminant or missing peaks, and so on. A variety of different scoring schemes have been implemented in various algorithms and integrated in current software. PeptideSearch[24] and PepFrag[25] use a simple score based on the number of common masses between the experimental and theoretical spectra. Pappin *et al.*[26] designed a scoring function for the algorithm called MOWSE, which accounts for the nonuniform distribution of protein and peptide molecular weights in databases. Similar score schemes are exploited in MS-Fit,[27] Mascot,[28] and ProFound.[29] Complete reviews on these and other related scoring functions are given in Refs.[30–33]

Aldente,[34] a PMF tool developed by the PIG and available from the ExPASy server,[15] exploits the Hough transform to determine the mass spectrometer deviation, to realign the experimental masses, and to exclude outliers. This method is particularly insensitive to noise. Other main features of Aldente are

- a tuneable score parametrization, where the user can choose the parameters that will be considered in the score and in what proportion;
- an extensive use of the UniProtKB/Swiss-Prot[35,36] annotations such as protein mature form, PTMs, and alternative splicing, offering a degree of protein characterization as part of the identification procedure; and
- the possibility for the user to define one or more chemical amino acid modifications (besides the usual oxidation of methionine and acrylamide adducts on cysteine residues, alkylation products on cysteine residues, etc.), and to define the contributions of these modifications to the score.

A stand-alone version of Aldente is commercialized as an in-house solution.[b]

3.2. *PFF*

PMF is an efficient approach when searching databases for species of small and fully sequenced genomes. PFF is better adapted to searching in larger databases as a more specific and sensitive identification method when working with complex mixtures of peptides. Indeed, the sequence information given by peptide fragmentation in MS/MS spectra increases the chances of finding true positive hits in database

[b] This version keeps track of the history of submitted jobs for easier job resubmission, manages batch queries, and allows result annotation.

searches. A single peptide (a single MS/MS spectrum) may even correctly identify a protein depending on the number of amino acids in its sequence. However, the downside of PFF identification lies in the high proportion of unassigned MS/MS spectra collected during an experiment. The usual explanations for these nonmatches are the presence of experimental contaminants appearing before or during MS acquisition, bad-quality spectra with noise and unusual fragmentation for certain mass spectrometers, spectra derived from proteins not present in the database or products of alternatively spliced genes not annotated in the database, etc.

Some algorithms have been developed to deal with most of these issues, although none of them currently handle all problems at once. Some have opted to reduce the number and complexity of MS/MS spectra while increasing their quality (e.g. NoDupe[37]), but most available software have adopted a classical search strategy. Most classical search programs split the identification process into two stages. The first stage is aimed at building a list of candidate proteins from confidently identified spectra, and the second one is aimed at matching unidentified spectra against this list with more combinatorial parameters (e.g. taking into account a larger number of modification types or increasing the mass error tolerance). The main idea behind this strategy is to increase the sequence coverage by loosening constraints. Popular programs of that category are Mascot[28]; Sequest[38]; Phenyx,[39] a software platform developed by GeneBio S.A.[c] in collaboration with the PIG; and X!Tandem.[40]

Classical database search algorithms aim to identify the best sequence match given a list of fixed and/or potential modifications in a database for each spectrum analyzed. The advantage of classical search tools is the speed with which larger data sets can be processed in a reasonable amount of time. The drawback is their conceptual limitation of identifying only spectra with expected modifications. Open modification search tools, also known as tag approach or blind search, address this problem. This search strategy is also based on detecting matches

[c] Phenyx runs publicly through a Web interface (http://phenyx.vital-it.ch/); its commercial version is installed locally.

between experimental spectra and sequence entries of a database; but in contrast to classical search tools, they implement algorithms optimized to match spectra with unexpected mass shifts such as PTMs and/or transpeptidations. Four programs are known to a large community: GutenTag,[41] Modiro,[42] Popitam,[43] and InsPecT.[44] Open modification search algorithms are designed to take into account any type of modification that would allow a better match between the spectrum peak pattern and a candidate peptide.

Popitam was developed by the PIG. Original features of Popitam are (a) the sequence-guided tag extraction performed by the simultaneous parsing of the spectrum graph and of an indexed representation of a candidate sequence; and (b) the tag compatibility graph, in which cliques represent possible interpretation scenarios of the spectrum; and (c) optimization of scoring functions using parallel multi-objective genetic programming. Popitam is specially adapted to deal with PTMs, mutations, missed cleavages or half-cleavages, and transpeptidation. This list can be extended to other events, like errors in databases, badly calibrated data or imprecise precursor masses, and in general any event that modifies the sequence or residue masses of a peptide analyzed by MS. Popitam is currently available on the ExPASy server (http://www.expasy.org/tools/popitam/).

3.3. *Identification Platforms — SwissPIT*

Software tools are run several times in order to empirically discover the best parameter settings.[45] When various strategies of MS analysis are used, the results are manually selected and combined. In most situations, though, only one single tool is utilized for protein identification along with a unique parameter setting. Many spectra are thus missed due to inappropriate parameter values, to inadequate filtering, or merely to underperformance of certain scoring schemes for the quality of the spectra at hand. The flexible automation of computer tasks as well as a combination of different workflow strategies are thus necessary to enhance data analysis, to reduce human interaction, and to achieve high-throughput analysis. Therefore, a major goal in MS/MS identification projects is the matching of as many spectra as possible while keeping the false-positive rate as low as possible. The most promising way to simultaneously solve both problems is the combination of several search strategies and identification tools in so-called identification "workflows", which should combine various tools and search strategies to improve MS/MS data analysis by increasing the number of confidently identified peptides.[46,47]

Up to now, very few platforms dedicated to proteomic data processing have been implemented. These platforms aim to automate the identification process so as to reduce data analysis time and to enhance the quality of identification as well as the coverage of matched spectra. The Trans-Proteomic Pipeline (TPP)[46] is an open source platform comprising the suite of tools for MS/MS analysis pointed out in Sec. 2.2. This pipeline allows importing output files from Sequest and comprises various modules, mainly for postprocessing, including result validation, quantification of isotopically labeled samples, and the Pep3D tool for viewing raw LC/MS data and results at the peptide and protein levels. MASPECTRAS[48] also includes several identification tools through importing result files that are input in further processing. The TOPP package,[49] a framework for processing simple MS/MS data, should also be mentioned as its parts can be connected to a pipeline by self-written scripts. In all cases, TPP and MASPECTRAS support a fixed set of identification programs implementing the same search strategy.

Several commercial platforms are also available. For example, Scaffold analyzes Mascot and Sequest results and validates hits by cross-correlation with X!Tandem, filters out uninteresting spectra, and exports high-quality unidentified ones for future analysis. ProteinScape[50] and ProteinLynx Global SERVER also adopt a stepwise approach to proteome study.

The swissPIT[51] platform was designed by the PIG as a workflow-oriented toolbox giving access to a number of MS/MS analysis software tools. The first identification workflow was created using the JOpera engine.[52] Currently, four identification tools (Phenyx, Popitam, X!Tandem, and InsPecT[44]) and two protein datasets (UniProtKB/Swiss-Prot and UniProtKB/TrEMBL) are being tested in the pipeline. The choice of these first four algorithms has been motivated by their popularity and known efficiency through the implementation of various parameterized search strategies, as well as the authors' access to the source code. The computing resources that currently serve the purposes of the swissPIT project are aggregated, controlled, and coordinated by Grid infrastructure, which is based on the ARC middleware.[53] It provides user-level interfaces and abstractions of a seamless, homogeneous distributed system. A high-level Web portal to swissPIT manages the Grid security certificates on behalf of the users, thus sparing the scientists from having to deal with credential delegation and renewal.

4. Proteomics Knowledge Integration and Databases

A proteomic study typically generates a huge list of identified, and sometimes relatively quantified, proteins. However, a list of hundreds, or even thousands, of independent items is actually poorly informative when one aims at selecting a few potentially relevant biomarkers for further validation studies. At this stage, a considerable amount of work still has to be done on two fronts. On the one hand, database search mechanisms need improvement to increase the knowledge on individual proteins and to help sort out the candidates with the highest value[30]; these issues have been described in the previous section. On the other hand, the representation of data and knowledge requires further reflection to create more meaning, as it enriches the interpretation of experimental results.[54] Characterization of the selected proteins is the next stage and is often a very difficult task. It involves not only determining the precise form of each protein (e.g. splice variant, phenotype, posttranslational processing), but also investigating their possible interactions. The latter studies raise a number of issues common to many bioinformatics applications, given that sequences constitute the core data type for all of these applications.

4.1. *Standards for High-Throughput Data*

High-throughput analytical methods have produced an ever-growing flow of new data. This situation has spawned a number of initiatives to address the issue of data storage and data standardization. The transcriptomics community has led by example, introducing a data format termed MIAME (Minimum Information About a Microarray Experiment).

Following the same principle, a MIAPE (Minimum Information About a Proteomics Experiment) format is currently being defined. These formats are set to enable the unambiguous interpretation of the results of an experiment and to allow its reproduction.[d]

The MIAPE project is carried out by a working group in charge of defining standardized formats for proteomics experiments. This working group is one of several, all launched through the Proteomics Standards Initiative (PSI) of the Human Proteome Organisation (HUPO).[16,55] Besides being involved in setting a standardized general proteomics format, the PSI supports other working groups in key areas of proteomics. These include 2-DE, MS, and protein–protein interaction data. The MIAPEGelDB[56] was set up by the PIG to facilitate the submission of required data for the MIAPE Gel Electrophoresis module.

4.2. *Integrative Proteomics Data*

Various independent sources are available for the manual or automated extraction of relevant information related to a sequence or a collection of sequences. However, the reliability of all extracted information needs to be ascertained. The different sources need to be cross-checked, and further digging into mass data is possibly involved. Only then can contextual constraints be reliably expressed for characterizing protein structure, function, modifications, and interactions. These characteristics constitute what is often called "sequence annotation". Well-annotated biological data are clearly valuable for interpreting experimental results.

The move towards integration was initiated years ago by some of the most established and recognized bioinformatics resources, among them UniProt (see Chapter 6), KEGG, and FlyBase, to name just a few. They centralized information from multiple sources, and the resultant knowledge is comparable to that of an encyclopedia.

---

[d] This is the mission of the HUPO Proteomics Standards Initiative (http://www.psidev.info).

A recent shift in data collection and representation is seen in the move from bioinformatic databases to atlases (e.g. EMAGE,[57] Protein Atlas,[58] Novartis Atlas,[59] PeptideAtlas[60]). For example, PeptideAtlas and Protein Atlas have recently been introduced as resources for browsing and querying protein data. In this manner, data exploration may become an indispensable preliminary step towards integration. Interestingly, the navigation between different views in an atlas might help us refine the definitions required for the overall task of biological data integration.

So far, Gene Ontology (GO)[61] remains the most popular initiative for organizing biologically interpretable protein data at molecular and cellular levels. But in GO, knowledge is currently unevenly represented due to the relative novelty of this effort and also to our limited understanding of biology. Ontology design has recently become an active field in bioinformatics,[62] and this trend further justifies the need for the determination of principles that set a more stable basis for data integration.

The strength of integration has been defined to distinguish a portal (loose integration) from solutions based on global data models (tight integration).[63] In the following, integrated resources are presented in order of growing strength of integration.

4.2.1. *Proteomics servers*

ExPASy was created in 1993,[64] and is still developed and maintained at the SIB.[15] ExPASy is dedicated to the federation of databases and tools that are relevant to proteomics studies. The server hosts the following databases, which are essentially developed and maintained by SIB members:

- the UniProt Knowledgebase (UniProtKB; see Chapter 6);
- SWISS-2DPAGE (described in Sec. 4.2.3);
- the PROSITE database of protein domains and families;
- the ENZYME repository of information relative to the nomenclature of enzymes; and

- the SWISS-MODEL Repository, a database of automatically generated 3-D protein structural models.

All databases available on ExPASy are explicitly cross-referenced to other molecular biology databases or resources available on the Internet. Approximately 30 implicit links to other additional resources are created on demand when certain views of UniProtKB are generated. This concept is targeted at data collections that do not have their own system of unique identifiers, but can be referenced via identifiers such as accession numbers or gene names. GeneCards is an example of a database implicitly linked to UniProtKB; it only shares an identifier with UniProtKB (the HUGO-approved gene name). Implicit links are a specific feature of ExPASy and are not available on other Web servers or in the UniProtKB data files downloadable by File Transfer Protocol (FTP). They significantly enhance database interoperability and strengthen the role of UniProtKB as a central hub for the interconnection of molecular biology resources.

ExPASy has an extensive collection of software tools. Some of them are targeted toward access and display of the databases listed above. Others can be used for data analysis, such as the prediction of protein sequence features or the processing of proteomics data originating from 2-D PAGE and MS experiments. ExPASy tools that are specifically designed for MS data analysis perform computations and predictions while using annotations documented in the UniProtKB feature tables (e.g. the input form of PMF engine Aldente includes the optional selection of a range of protein features). This assists the detection of possible splice variants, PTMs, or protein processing.

The European Bioinformatics Institute (EBI) also develops and maintains a number of proteomics databases and federates a very wide spectrum of protein-related databases; however, software tools available at the EBI are not specifically targeted at MS data analysis. Conversely, the Institute for Systems Biology (ISB) offers a wide panel of software tools for proteomics studies, but does not develop or maintain a wide variety of protein-related databases with the exception of PeptideAtlas. ExPASy thus remains the server that best covers the diversity of requirements for proteomics researchers as well as the window into PIG and SIB developments and databases.

4.2.2. *Proteomics repositories*

The HUPO PSI is also closely related to the development of the PRoteomics IDEntifications database (PRIDE). PRIDE was put together to provide the proteomics community with a public repository for protein and peptide identifications together with the evidence supporting these identifications.[65] PRIDE is a centralized, standards-compliant, public data repository for proteomics data. Each entry in the database contains the proteins identified, a list of peptides used to make identifications, the tissue, the experiment conducted, the conditions of the experiment, any PTMs to those peptides, and links to any publication describing the experiment. PRIDE has a dual objective: it is designed to provide (1) a common data exchange format and repository to support proteomics literature publications, and (2) a reference set of tissue-based identifications for use by the community.

PeptideAtlas[60] was designed to store proteomics data generated by high-throughput methods. It is actually presented as an expandable resource for integration of data from diverse proteomics experiments. This initiative is driven by the assumption that protein expression data can contribute to the annotation of eukaryotic genomes. MS/MS spectra are stored in the PeptideAtlas database. Each statistically validated assignment of a peptide to a mass spectrum is recorded. A range of confidence thresholds is optionally made available for selecting peptides likely to match proteins, and false-positive rates are correspondingly estimated. A database scheme supports different builds of PeptideAtlas, several versions of Ensembl, a range of eukaryotic organisms, and several reference protein sequence sets. PeptideAtlas currently includes data produced at the ISB, which hosts the resource. As it is intended as a data repository, a large set of published data are also available for download.

The Global Proteome Machine (GPM)[66] is another similar approach to storing MS/MS data. The GPM database is used on its own, to provide answers to specific queries, as well as to serve as an index to experimental information stored in XML documents. The underlying schema serves as both an extension and a simplification of the MIAPE idea, for the purpose of validating observed protein coverage and peptide fragmentation data.

4.2.3. *Proteomics integrated resources*

SWISS-2DPAGE is an annotated database developed by the PIG. It assembles data on proteins from a variety of human and mouse biological samples as well as from *Arabidopsis thaliana*, *Dictyostelium discoideum*, *Escherichia coli*, *Saccharomyces cerevisiae*, and *Staphylococcus aureus*. In all cases, proteins have been identified on two-dimensional polyacrylamide gel electrophoresis (2-D PAGE) reference maps. SWISS-2DPAGE provides links between sequence data and protein expression. Most recorded proteins have been identified by one or more methods, including MS, microsequencing, immunoblotting, gel comparison, or amino acid composition.

The SWISS-2DPAGE database was the first 2-DE federated database available on the Internet.[67] Since then, it has been continuously accessible and expanded,[68] and contains close to 40 maps. Various types of information (e.g. genome data, organism-specific data, protein families or domains, polymorphisms, mutations, structures, metabolic pathways) are brought together by cross-linking to other resources such as UniProtKB, PubMed, and other federated 2-DE databases (e.g. HSC-2DPAGE, PHCI-2DPAGE, Siena-2DPAGE).

The WORLD-2DPAGE List (http://world-2dpage.expasy.org/list/) is an index to known federated 2-D PAGE databases as well as to 2-D PAGE related servers and services.[69] It is available from the ExPASy proteomics website and has been continuously updated since 1995. It currently lists up to 60 databases, totalizing nearly 420 gel images. These databases are grouped by species and classified into three categories according to their implementation of some or all of the rules defining a federated 2-DE database;[70] these rules (http://www.expasy.org/ch2d/fed-rules.html) have been proposed to facilitate the navigation between remote databases. In short, a federated 2-DE database has to be accessed minimally through a keyword search and graphical queries, and also has to include active links to other related databases.

In the last few years, the accessibility to gel data has not reached expected levels, and most of the proteomics data are confined in scientific articles or as supplementary material on publishers' websites. In either case, the data are not fully exploitable for analysis or comparison without a lot of preliminary manual work. The first contribution of the PIG to the expansion of the pool of proteomics data available as usable electronic resources was the development of the Make2D-DB package,[71] an open source software that helps build a federated 2-DE database on one's own website (http://world-2dpage.expasy.org/make2ddb/). Such a SWISS-2DPAGE-like database maker provides various text search mechanisms (protein/gene name, protein description, author name, species, etc.), search by experimental data (pI/Mw range or identification methods), as well as an interactive graphical query interface. Moreover, all search tools can be used to query any number of local database(s) and/or similar remote interface(s). The output result clearly states the origin of the resulting hits. In such a case, results are produced as human-readable lists in HTML. These comprehensible lists are particularly adapted for end-users who can navigate through related information (spot data, gel image, protocols, etc.); for instance, it is possible to retrieve the complete list of identified spots for a given gel, species, or identification method. Additionally, advanced users can easily retrieve data as computer-readable lists through logical URLs. Make2D-DB is not only an environment to create query interfaces; it also ensures data reliability and consistency, is compliant with current proteomics standards (http://www.psidev.info/), and includes automated updates from external resources (UniProt Knowledgebase, NCBI taxonomy, etc.). It is also designed to take in much experimental information as possible in order to improve the quality assessment of the database content (link to protocols in MIAPE format or in separate plain text file, links to MS data, Spectra Viewer which highlights identified peaks, etc.).

As a continuation to this project, the PIG launched the World-2DPAGE Portal (http://world-2dpage.expasy.org/portal/) in 2006. This complement is more than just a useful list of existing databases. It currently involves eight SWISS-2DPAGE-like databases that can be simultaneously queried, even though they are running in many European and Asian countries. Globally, it can be seen as a virtual unique database with up to 91 reference maps for 10 species, totalizing nearly 10 300 identified spots, enabling the biggest gel-based proteomics datasets to be accessible from a single interface.

The World-2DPAGE Repository[72] (http://world-2dpage.expasy.org/repository/), recently created as a supplement to the World-2DPAGE Constellation, aims to host gel-based proteomics data with protein identifications published in the literature for laboratories that do not have the means of hosting a database. Data from two publications[73,74] are already accessible, including four 2-DE image maps with nearly 1200 identified spots. Practically, the users are asked to upload gel image(s), spot identification list(s), and MS data (if any); and to give relevant information on protocols, publications, etc. If a dataset was already submitted as MS identification data to the PRIDE repository (http://www.ebi.ac.uk/pride/),[75] then the same file can be reused for submission to the World-2DPAGE Repository without any additional work. The bidirectional cross-references between World-2DPAGE and PRIDE have been set up to offer a smooth navigation between both repositories (between gel and MS data).

The World-2DPAGE Constellation, as well as some of the other PIG developments described in the previous sections, is detailed in Fig. 3 in one single and typical 2-DE analysis workflow.

5. Conclusion

The present chapter has described a set of proteomics tools and analysis platforms developed in our group in the past decade or so. Throughout the years, the common trait of this collection has remained the production of collaborative work with other proteomics groups. On the one hand, the exchange with other dry lab groups led to the integration of, or with, other tools; among them is the long-term relationship with the Swiss-Prot group of SIB, where care is taken to develop tools that consider as much information from UniProtKB/Swiss-Prot and other derived tools as possible. On the other hand, interaction with wet lab groups, in particular with the Geneva Hospital and the University of Geneva Faculty of Medicine, has contributed to the integration of data in dedicated databases and the adaptation and fine-tuning of software.

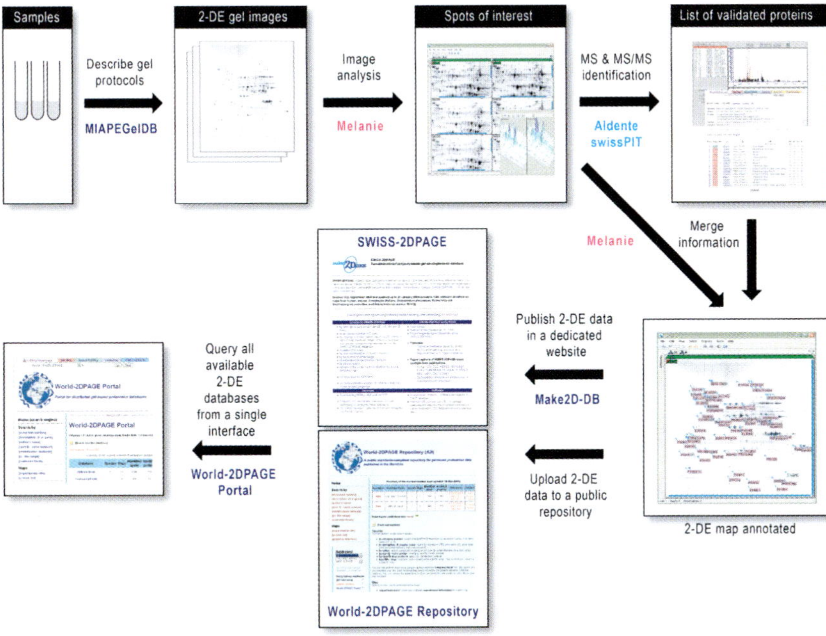

Fig. 3. Integration of PIG developments in a typical 2-DE analysis workflow. Software and database names are colored according to the PIG activity domains shown in Fig. 1.

Acknowledgments

The authors would like to acknowledge their colleagues in the Proteome Informatics Group (PIG) as key contributors to the projects mentioned in this manuscript: Daniel Walther, Gérard Bouchet, and Sébastien Catherinet in the proteome imaging projects; Christine Hoogland, Khaled Mostaguir, Xavier Robin, and Grégoire Rossier in the proteomics knowledge integration and databases projects; and Erik Ahrne, Patricia Hernandez, Céline Hernandez, Yann Mauron, Markus Müller, Roman Mylonas, Andreas Quandt, and Marc Tuloup in projects related to protein identification and characterization with MS. Finally, the authors acknowledge Alexandre Masselot and Pierre-Alain Binz (both affiliated to Geneva Bioinformatics (GeneBio) S.A.) as well as Ron D. Appel (affiliated to the Computer Science Department of the University of Geneva and former leader of the PIG) for their insightful participation and support in all projects.

References

1. Dowsey AW, Dunn MJ, Yang GZ. (2003) The role of bioinformatics in two-dimensional gel electrophoresis. *Proteomics* **3**: 1567–96.
2. Appel RD, Vargas JR, Palagi PM *et al.* (1997) Melanie II — a third-generation software package for analysis of two-dimensional electrophoresis images: II. Algorithms. *Electrophoresis* **18**: 2735–48.
3. Cutler P, Heald G, White IR, Ruan J. (2003) A novel approach to spot detection for two-dimensional gel electrophoresis images using pixel value collection. *Proteomics* **3**: 392–401.
4. Pleissner KP, Hoffmann F, Kriegel K *et al.* (1999) New algorithmic approaches to protein spot detection and pattern matching in two-dimensional electrophoresis gel databases. *Electrophoresis* **20**: 755–65.
5. Smilansky Z. (2001) Automatic registration for images of two-dimensional protein gels. *Electrophoresis* **22**: 1616–26.
6. Garrels JI. (1989) The QUEST system for quantitative analysis of two-dimensional gels. *J Biol Chem* **264**: 5269–82.
7. Appel RD, Hochstrasser DF, Funk M *et al.* (1991) The MELANIE project: from a biopsy to automatic protein map interpretation by computer. *Electrophoresis* **12**: 722–35.
8. Ünlü M, Morgan ME, Minden JS. (1997) Difference gel electrophoresis: a single gel method for detecting changes in protein extracts. *Electrophoresis* **18**: 2071–7.
9. Lemkin PF. (1997) Comparing two-dimensional electrophoretic gel images across the Internet. *Electrophoresis* **18**: 461–70.
10. Lemkin PF, Thornwall G, Evans J. (2005) Comparing 2-D electrophoretic gels across Internet databases. In: Walker J (ed.). *The Protein Protocols Handbook.* Totowa, NJ: Humana Press Inc., pp. 279–305.
11. Young N, Chang Z, Wishart DS. (2004) GelScape: a web-based server for interactively annotating, manipulating, comparing and archiving 1D and 2D gel images. *Bioinformatics* **20**: 976–8.
12. Berger SJ, Lee SW, Anderson GA *et al.* (2002) High-throughput global peptide proteomic analysis by combining stable isotope amino acid labeling and data-dependent multiplexed-MS/MS. *Anal Chem* **74**: 4994–5000.
13. Palmblad M, Ramstrom M, Markides KE *et al.* (2002) Prediction of chromatographic retention and protein identification in liquid chromatography/mass spectrometry. *Anal Chem* **74**: 5826–30.

14. Palagi PM, Walther D, Quadroni M *et al.* (2005) MSight: an image analysis software for liquid chromatography–mass spectrometry. *Proteomics* **5**: 2381–4.
15. Gasteiger E, Gattiker A, Hoogland C *et al.* (2003) ExPASy: the proteomics server for in-depth protein knowledge and analysis. *Nucleic Acids Res* **31**: 3784–8.
16. Pedrioli PG, Eng JK, Hubley R *et al.* (2004) A common open representation of mass spectrometry data and its application to proteomics research. *Nat Biotechnol* **22**: 1459–66.
17. Orchard S, Hermjakob H, Binz PA *et al.* (2005) Further steps towards data standardisation: the Proteomic Standards Initiative HUPO 3rd annual congress, Beijing 25–27th October, 2004. *Proteomics* **5**: 337–9.
18. Katajamaa M, Miettinen J, Oresic M. (2006) MZmine: toolbox for processing and visualization of mass spectrometry based molecular profile data. *Bioinformatics* **22**: 634–6.
19. Leptos KC, Sarracino DA, Jaffe JD *et al.* (2006) MapQuant: open-source software for large-scale protein quantification. *Proteomics* **6**: 1770–2.
20. Smith CA, Want EJ, O'Maille G *et al.* (2006) XCMS: processing mass spectrometry data for metabolite profiling using nonlinear peak alignment, matching, and identification. *Anal Chem* **78**: 779–87.
21. Li XJ, Yi EC, Kemp CJ *et al.* (2005) A software suite for the generation and comparison of peptide arrays from sets of data collected by liquid chromatography-mass spectrometry. *Mol Cell Proteomics* **4**: 1328–40.
22. Bellew M, Coram M, Fitzgibbon M *et al.* (2006) A suite of algorithms for the comprehensive analysis of complex protein mixtures using high-resolution LC-MS. *Bioinformatics* **22**: 1902–9.
23. Gröpl C, Lange E, Reinert K *et al.* (2005) Algorithms for the automated absolute quantification of diagnostic markers in complex proteomics samples. In: *Proceedings of the First Symposium on Computational Life Sciences* (CLS 2005). Lecture Notes in Bioinformatics. Heidelberg, Germany Springer, pp. 151–63.
24. Mann M, Wilm M. (1994) Error-tolerant identification of peptides in sequence databases by peptide sequence tags. *Anal Chem* **66**: 4390–9.
25. Fenyo D, Qin J, Chait BT. (1998) Protein identification using mass spectrometric information. *Electrophoresis* **19**: 998–1005.
26. Pappin DJ, Hojrup P, Bleasby AJ. (1993) Rapid identification of proteins by peptide-mass fingerprinting. *Curr Biol* **3**: 327–32.
27. Clauser KR, Baker P, Burlingame AL. (1999) Role of accurate mass measurement (+/− 10ppm) in protein identification strategies employing MS or MS/MS and database searching. *Anal Chem* **71**(14): 2871–82.
28. Perkins DN, Pappin DDJ, Creasy DM, Cottrell JS. (1999) Probability-based protein identification by searching sequence databases using mass spectrometry data. *Electrophoresis* **20**: 3551–67.

29. Zhang W, Chait BT. (2000) ProFound: an expert system for protein identification using mass spectrometric peptide mapping information. *Anal Chem* **72**: 2482–9.
30. Palagi PM, Hernandez P, Walther D, Appel RD. (2006) Proteome informatics I: bioinformatics tools for processing experimental data. *Proteomics* **6**: 5435–44.
31. Hernandez P, Müller M, Appel RD. (2006) Automated protein identification by tandem mass spectrometry: issues and strategies. *Mass Spectrom Rev* **25**: 235–54.
32. Johnson RS, Davis MT, Taylor JA, Patterson SD. (2005) Informatics for protein identification by mass spectrometry. *Methods* **35**: 223–36.
33. Gras R, Muller M. (2001) Computational aspects of protein identification by mass spectrometry. *Curr Opin Mol Ther* **3**: 526–32.
34. Gasteiger E, Hoogland C, Gattiker A *et al.* (2005) Protein identification and analysis tools on the ExPASy server. In: Walker JM (ed.). *The Proteomics Protocols Handbook.* Totowa, NJ: Humana Press, pp. 571–607.
35. Bairoch A, Apweiler R, Wu CH *et al.* (2005) The Universal Protein Resource (UniProt). *Nucleic Acids Res* **33** (Database Issue): D154–9.
36. Farriol-Mathis N, Garavelli JS, Boeckmann B *et al.* (2004) Annotation of post-translational modifications in the Swiss-Prot knowledge base. *Proteomics* **4**: 1537–50.
37. Tabb DL, MacCoss MJ, Wu CC *et al.* (2003) Similarity among tandem mass spectra from proteomic experiments: detection, significance, and utility. *Anal Chem* **75**: 2470–7.
38. Eng JK, McCormack AL, Yates IJR. (1994) An approach to correlate tandem mass spectral data of peptides with amino acid sequences in a protein database. *J Am Soc Mass Spectrom* **5**: 976–89.
39. Colinge J, Masselot A, Giron M *et al.* (2003) OLAV: towards high-throughput tandem mass spectrometry data identification. *Proteomics* **3**: 1454–63.
40. Craig R, Beavis RC. (2004) TANDEM: matching proteins with tandem mass spectra. *Bioinformatics* **20**: 1466–7.
41. Tabb DL, Saraf A, Yates JR III. (2003) GutenTag: high-throughput sequence tagging via an empirically derived fragmentation model. *Anal Chem* **75**: 6415–21.
42. Schaefer H, Chamrad DC, Marcus K *et al.* (2005) Tryptic transpeptidation products observed in proteome analysis by liquid chromatography–tandem mass spectrometry. *Proteomics* **5**: 846–52.
43. Hernandez P, Gras R, Frey J, Appel RD. (2003) Popitam: towards new heuristic strategies to improve protein identification from tandem mass spectrometry data. *Proteomics* **3**: 870–8.
44. Tanner S, Shu H, Frank A *et al.* (2005) InsPecT: identification of posttranslationally modified peptides from tandem mass spectra. *Anal Chem* **77**: 4626–39.

45. Ossipova E, Fenyo D, Eriksson J. (2006) Optimizing search conditions for the mass fingerprint-based identification of proteins. *Proteomics* **6**: 2079–85.
46. Keller A, Eng J, Zhang N *et al.* (2005) A uniform proteomics MS/MS analysis platform utilizing open XML file formats. *Mol Syst Biol* **1**: Epub Aug 2.
47. Hernandez, P. (2005) Peptide identification by tandem mass spectrometry: a tag-oriented open-modification search method. PhD thesis, University of Geneva, Switzerland.
48. Hartler J, Thallinger GG, Stocker G *et al.* (2007) MASPECTRAS: a platform for management and analysis of proteomics LC-MS/MS data. *BMC Bioinformatics* **8**: 197.
49. Kohlbacher O, Reinert K, Gropl C *et al.* (2007) TOPP — the OpenMS proteomics pipeline. *Bioinformatics* **23**: e191–7.
50. Chamrad DC, Koerting G, Gobom J *et al.* (2003) Interpretation of mass spectrometry data for high-throughput proteomics. *Anal Bioanal Chem* **376**: 1014–22.
51. Quandt A, Hernandez P, Kunzst P *et al.* (2007) Grid-based analysis of tandem mass spectrometry data in clinical proteomics. *Stud Health Technol Inform* **126**: 13–22.
52. Pautasso C, Bausch W, Alonso G. (2006) Autonomic computing for virtual laboratories. In: Kohlas J, Meyer B, Schiper A (eds.). *Dependable Systems: Software, Computing, Networks.* New York, NY: Springer Verlag, pp. 211–230.
53. Smirnova O. (2003) The NorduGrid architecture and middleware for scientific applications. In: Sloot PMA (ed.). *Proceedings of the International Conference on Computer Science* (*ICCS 2003*). Lecture Notes in Computer Science, Vol. 2657. Berlin, Germany: Springer-Verlag, pp. 264–273.
54. Lisacek F, Cohen-Boulakia S, Appel RD. (2006) Proteome informatics II: bioinformatics for comparative proteomics. *Proteomics* **6**: 5445–66.
55. Taylor CF, Paton NW, Garwood KL *et al.* (2003) A systematic approach to modeling, capturing, and disseminating proteomics experimental data. *Nat Biotechnol* **21**: 247–54.
56. Robin X, Hoogland C, Appel RD, Lisacek F. (2008) MIAPEGelDB, a web-based submission tool and public repository for MIAPE gel electrophoresis documents. *J Proteomics* **71**: 249–51.
57. Baldock RA, Bard JB, Burger A *et al.* (2003) EMAP and EMAGE: a framework for understanding spatially organized data. *Neuroinformatics* **1**: 309–25.
58. Uhlen M, Bjorling E, Agaton C *et al.* (2005) A human protein atlas for normal and cancer tissues based on antibody proteomics. *Mol Cell Proteomics* **4**: 1920–32.
59. Su AI, Wiltshire T, Batalov S *et al.* (2004) A gene atlas of the mouse and human protein-encoding transcriptomes. *Proc Natl Acad Sci USA* **101**: 6062–7.
60. Desiere F, Deutsch EW, Nesvizhskii AI *et al.* (2005) Integration with the human genome of peptide sequences obtained by high-throughput mass spectrometry. *Genome Biol* **6**: R9.

61. Harris MA, Clark J, Ireland A *et al.* (2004) The Gene Ontology (GO) database and informatics resource. *Nucleic Acids Res* **32**: D258–61.
62. Schulze-Kremer S. (2002) Ontologies for molecular biology and bioinformatics. *In Silico Biol* **2**: 179–93.
63. Davidson SB, Overton C, Buneman P. (1995) Challenges in integrating biological data sources. *J Comput Biol* **2**: 557–72.
64. Appel RD, Bairoch A, Hochstrasser DF. (1994) A new generation of information retrieval tools for biologists: the example of the ExPASy WWW server. *Trends Biochem Sci* **19**: 258–60.
65. Martens L, Hermjakob H, Jones P *et al.* (2005) PRIDE: the Proteomics Identifications database. *Proteomics* **5**: 3537–45.
66. Craig R, Cortens JP, Beavis RC. (2004) Open source system for analyzing, validating, and storing protein identification data. *J Proteome Res* **3**: 1234–42.
67. Appel RD, Sanchez JC, Bairoch A *et al.* (1993) SWISS-2DPAGE: a database of two-dimensional gel electrophoresis images. *Electrophoresis* **14**: 1232–8.
68. Hoogland C, Mostaguir K, Sanchez JC *et al.* (2004) SWISS-2DPAGE, ten years later. *Proteomics* **4**: 2352–6.
69. Hoogland C, Sanchez JC, Tonella L *et al.* (1999) The SWISS-2DPAGE database: what has changed during the last year. *Nucleic Acids Res* **27**: 289–91.
70. Appel RD, Bairoch A, Sanchez JC *et al.* (1996) Federated two-dimensional electrophoresis database: a simple means of publishing two-dimensional electrophoresis data. *Electrophoresis* **17**: 540–6.
71. Mostaguir K, Hoogland C, Binz PA, Appel RD. (2003) The Make 2D-DB II package: conversion of federated two-dimensional gel electrophoresis databases into a relational format and interconnection of distributed databases. *Proteomics* **3**: 1441–4.
72. Hoogland C, Mostaguir K, Appel RD, Lisacek F. (2008) The World-2DPAGE Constellation to promote and publish gel-based proteomics data through the ExPASy server. *J Proteomics* **71**: 245–8.
73. Li L, Wada M, Yokota A. (2007) Cytoplasmic proteome reference map for a glutamic acid-producing *Corynebacterium glutamicum* ATCC 14067. *Proteomics* 7: 4317–22.
74. Plikat U, Voshol H, Dangendorf Y *et al.* (2007) From proteomics to systems biology of bacterial pathogens: approaches, tools, and applications. *Proteomics* 7: 992–1003.
75. Jones P, Cote RG, Cho SY *et al.* (2008) PRIDE: new developments and new datasets. *Nucleic Acids Res* **36**: D878–83.

Chapter 8

Protein–Protein Interaction Networks: Assembly and Analysis

Christian von Mering

1. Introduction

Protein–protein interactions are essential for all cellular life.[1] Only very few proteins, if any, can function entirely on their own — presumably only in self-contained processes such as constitutive transport or trivial enzymatic reactions. Indeed, the vast majority of proteins are bound by specific protein partners at one point or another during their functional life cycle, either because they need to be controlled and regulated in their activity or because they can only function as part of a larger molecular complex.

Specific and functionally relevant interactions between proteins are often conceptualized in the form of "interaction networks". In interaction networks, the proteins are defined as nodes, and the interactions as edges; the latter can be further subclassified depending on the type and strength of interaction. By convention, interaction networks do not include nonspecific, generic interactions between proteins, such as the interaction between a nascent protein chain and the ribosome, between a freshly produced protein and downstream processing enzymes (chaperones), or between a damaged protein and the protein degradation machinery. Thus, protein–protein interaction networks aim to focus on the set of specific and functionally relevant protein interaction interfaces in a cell, and they capture the majority of direct connections that can occur between proteins.

Of course, current protein–protein interaction networks still represent only a much-simplified view of the true connectivity of proteins inside a cell. For the most part, the networks are not yet capable of describing any spatial or temporal restrictions for protein interactions (i.e. they ignore the fact that not all possible protein–protein interactions actually take place in all cells of an organism and at all time points). The networks are also not annotated with much detail regarding binding energies, interaction stochiometry, or the spatial arrangement of the proteins. They are also vastly incomplete — current knowledge of protein–protein interactions is limited to a few model organisms, and experimental measurements are often done under a limited set of laboratory conditions where many specific interactions may not be forming.

Nevertheless, even the crude protein–protein interaction networks of today represent a very powerful way of integrating and communicating functional information about a cell's proteome. These networks provide an exquisitely intuitive, concise visualization of complicated relationships, and they can be used as a platform to navigate through entire proteomes interactively. Many additional information items can be mapped onto a protein network, and the networks can be used to place most types of experimental data into context. Apart from visualization and data mining, protein interaction networks can also help to address fundamental questions about cell biology and evolution. For example, hierarchical clustering algorithms, when applied to protein networks, can potentially reveal "functional modules" or (sub)complexes forming molecular machines, and can thus help to describe the overall functional organization of a given proteome.[2] Likewise, graph-theoretical analysis of the topology of interaction networks is used as a tool to search for fundamental design principles of evolution.[3]

2. Experimental Protein Interaction Data

Protein–protein interactions cannot yet be reliably predicted *de novo* by computational means, although many promising avenues are using protein structure information or evolutionary signatures from the large number of sequenced genomes.[4] In order to assemble a comprehensive and reliable protein network, experimental interaction data still remain essential. This, however, does not mean that computational predictions would have no value *per se* — indeed, it is often advisable to include them, for example, to augment and filter experimental data with predicted interactions, much like experimental data can be filtered with previous knowledge and with interactions determined in other organisms.

When working with experimental protein–protein interaction data, it is important to distinguish between small-scale experiments that have been individually reported in scientific publications and high-throughput experiments that aim to describe a large number of protein–protein interactions in a single dataset. Both approaches have their strengths and weaknesses. Small-scale experiments are usually more reliable, since they are often based on a combination of experimental techniques, have been subjected to the scrutiny of peer review, and may have been reproduced independently in other labs over time. However, they can be difficult to assess in bulk unless curators have extracted them manually from the scientific literature. Moreover, they usually contain a significant reporting bias: negative results (noninteractors) are often not reported, and even the positive results are sometimes focused on specific areas of interest or have a tendency to confirm previous findings.

With high-throughput experiments, the situation is often the opposite. Reporting bias is usually not much of a problem, since a large set of protein interactions is assessed simultaneously. Similarly, the data are usually accessible quite easily in electronic format, and negative results (noninteractors) are normally part of the reported data. On the downside, the overall quality of high-throughput results often cannot match that of traditional, small-scale laboratory experiments. The rates of false-positive and false-negative interactions are often considerably higher. This is usually inferred not only from observing unusually low reproducibility between independent high-throughput datasets that aim to measure the same interactions, but also from deviations of these data from previously accepted reference data.[5–8]

Numerous different experimental approaches can be used to assemble the data for a protein–protein interaction network. These include *in vivo* techniques, such as yeast two-hybrid screening, and *in vitro* techniques, such as coimmunoprecipitation or protein arrays ("protein chips"). For each of these techniques, it is important to note that the immediate result of a measurement is not just a black-and-white, qualitative statement on whether or not two proteins are interacting. Instead, quantitative readouts are often available, which are then converted into a "yes" or "no" answer using a predetermined procedure and cut-offs; this decision may then be followed by removal of notorious contaminants and/or rescreening. Thus, the final list of protein–protein interactions often hides the complexity of the underlying experiment, and may not contain all of the information from the original measurements. However, when integrating several distinct datasets, it would be better to have this information available: to start with the raw data for each of these datasets, then perform a probabilistic or decision-tree-based data integration, and only afterwards convert the final result into a set of simple qualitative statements on who interacts with whom. While this is not yet done routinely when assembling interaction networks (and not only because the raw data are often not available), it should probably be done more systematically in the future. In addition, many of the experimental datasets contain intrinsic consistency signals, such as repeated measurements of the same interaction or reverse measurements where a given protein is the "bait" in one context and the "prey" in the other. Such intrinsic consistency signals should be used as well, especially when multiple datasets of the same type are to be integrated.

Historically, the yeast two-hybrid technology was the first method to yield genome-wide (or near genome-wide) experimental protein interaction maps.[9,10] This technology is based on artificial fusion proteins ("hybrids"): the two proteins to be tested for an interaction are each fused to one specifically designed protein domain — to a DNA-binding domain in one case, and to a transcriptional transactivation domain in the other case. Normally, these two domains occur in a single protein, forming a so-called transcription factor; in two-hybrid screens, they are fused separately to two test proteins. Only when the two test proteins are capable of forming a specific binding interaction in the nucleus of a cell (usually a *Saccharomyces cerevisiae* cell) does the reconstitution of a complete transcription factor occur, which then drives the expression of a reporter gene. Since its beginnings, the yeast two-hybrid technology has been widely used for screening interactions in a number of organisms (including the large proteomes of higher eukaryotes). A number of improvements and modifications have been made to the original procedure, including the use of shorter gene fragments and multi-purpose versatile cloning vectors such as the Gateway system. In addition, variants of the method have been devised that work in cellular localizations other than the nucleus, for example, in the cytoplasm or at the plasma membrane.[11–14]

While very powerful, yeast two-hybrid screens sometimes suffer from lack of reproducibility between laboratories, and often exhibit limited overlap with previously known protein–protein interactions. This is partly explained by the "foreign" surroundings of the test proteins: these have to fold correctly, and interact correctly, in an organism for which they have not evolved, and often also in a cell compartment where they may normally not be found (the nucleus). In addition, many protein–protein interactions normally require more than two proteins — the formation of protein complexes is thought to often involve cooperative binding and may also require the action of chaperones and other assembly proteins, most of which is presumably not possible inside the nucleus in a two-hybrid screen.

Another widely used experimental technique for protein–protein interaction detection is the biochemical purification of native protein complexes, followed by the identification of their constituent proteins using mass spectrometry (MS).[15,16] This is a very powerful technique, and it has a number of advantages over two-hybrid techniques. First, entire protein complexes that have been purified from their native surroundings are analyzed. This enables the detection of complexes that may need an elaborate series of steps for their assembly, and that may not form easily outside their native surroundings. Second, biochemical purification of the complexes can be performed based on specific and versatile peptide epitopes, which are added to one protein of interest in a complex (the "bait" protein). This means that several entry points (potential baits) exist for each protein complex, so that the same complex can be purified several times independently. The expression of bait proteins can also be kept at endogenous levels, potentially keeping optimal stoichiometry by integrating back the DNA construct encoding the epitope-tagged bait protein into its original genomic context via homologous recombination, thus keeping its normal, endogenous regulation.

Yeast two-hybrid screens and MS analysis of protein complexes have so far clearly been the two major workhorses for the assembly of protein interaction networks. A number of other techniques exist, but these have yet to be applied successfully in proteome-wide screens in high throughput. They include biophysical methods, such as surface plasmon resonance[17] or fluorescence resonance energy transfer,[18] as well as the use of protein microarrays ("protein chips").[19] Compared to nucleic acids, however, proteins are generally much more difficult to work with in a systematic fashion — they need to fold correctly and remain folded, and each protein may require distinct conditions and/or cofactors to do so. Because of this, *in vitro* technologies for detecting protein–protein interactions remain challenging, and may in the end each only work for specific subsets of proteins.

How many distinct and specific protein–protein interaction interfaces exist? This is not yet known (not even for the simplest organisms), but it has been argued that all of the various protein–protein interactions actually realized in nature are ultimately instances of only about 10 000 fundamental "interaction types".[20] Stable protein–protein interaction interfaces are apparently, for the most part, ancient and do not evolve easily — at least those that form interactions of the type that can be analyzed in protein crystals.[21] Because of this limited repertoire of structurally stable protein–protein interfaces, structural genomics initiatives and the increased trend to crystallize entire protein complexes are very useful for filtering and assessing a given protein–protein interaction dataset: a significant fraction of the stable interactions reported by any given experiment should be explainable by structurally known protein interfaces. Of course, the reverse is not true: only a small fraction of the interactions known to be structurally possible is actually realized, especially for highly promiscuous, abundant protein families with many paralogs. Similarly, a large number of the more transient and unstable interactions inside the cell are probably not well represented among crystallized complexes; these include interactions between post-translational modification (PTM) enzymes and their targets, or between transport/sorting proteins and their cargo. Such transient interactions are often mediated by much smaller interaction interfaces, or even just by short peptide segments, and they probably do evolve *de novo* more easily than stable interactions.

3. Networks That Include Indirect Associations

Simple protein–protein interaction networks, while very useful, represent only one type of network out of several types currently in use in biology. Other classes of networks include those that describe genetic interactions between the various loci of an organism, or networks that describe the relations between small organic metabolites and the enzymes that interconvert them (metabolic networks). Yet other networks describe the connections between genes and the transcription factors that control them (transcriptional regulation networks), or they describe specific types of protein–protein interactions such as the interaction between protein kinases and the substrates they phosphorylate. All of these networks are being assembled for much the same reasons as for the protein–protein interaction networks: visualization of complex relationships, data integration, simple browsing, or global evolutionary analysis.

Many of these networks are special instances of a more general type of network: the protein–protein *association* network. In contrast to interaction networks, protein–protein association networks make no assertion as to how exactly two proteins interact — in fact, they may not have to interact at all, in a physical sense. Proteins can show a specific and productive functional interaction without touching each other, for example, by performing subsequent metabolic reactions in the same metabolic pathway, or by regulating each other's abundance through transcription. Even interactions that are thought to be direct physical interactions can in fact be indirect associations, for example, in the case of two proteins that have been isolated together from a protein complex but are in reality located at opposite ends of the complex and may have no specific interaction interface with each other.

In many ways, association networks are a more "natural" way of describing proteins and their mutual interactions. First, many experimental techniques may not reveal whether an interaction is direct or indirect, especially the *in vivo* techniques. Second, many biologically relevant interactions (as, for example, revealed by genetic interaction studies) are not direct either — genes can show a strong functional linkage, but still their products might be located in different parts of the cell or expressed at different times. Third, the distinction between direct physical binding and indirect functional effects can be somewhat artificial: two proteins may interact over only a small segment of their length, or only very briefly (such as when one protein modifies another posttranscriptionally); also, proteins may come into contact simply by chance because they are secluded in the same small intracellular compartment, for example, but it might appear like they are interacting specifically. Last, making the difference between a specific interaction and a nonspecific "infrastructure" interaction is far from obvious: one protein may transport another to a certain place in the cell in a very specific, important, and tightly regulated interaction, or merely as part of a ubiquitous and nonspecific service. Therefore, protein–protein association networks can be said to represent the smallest common denominator of partnership between proteins: whenever two proteins form a specific functional partnership, they can be thought of as being "associated", independently of what the actual mechanism of their association is.

Notably, association is also a very useful concept for another reason: the definition matches the results of certain types of prediction algorithms. Not all computational interaction prediction algorithms predict a direct physical interaction between two proteins; instead, some algorithms merely predict an association. This is particularly true for algorithms that are based on the effects of evolutionary selection. Such predictors work by detecting deviations from a random assortment of genes in multiple genomes. Simply put, a pair of genes — if functionally not related — should evolve more or less independently. Both genes should freely move around within a genome, for example, when chromosomes are broken up and reassembled over time; and the two genes should independently undergo gene duplications, gene losses, gene transfers, and other such events. Any deviation from this random expectation, if detected in a sufficiently large number of genomes, can be taken as evidence that selection has somehow acted similarly on both genes. This in turn means that the two genes somehow share certain aspects of their function — they contribute to the same selectable phenotype and are thus not assorting independently throughout evolution. Interaction prediction algorithms that follow this basic principle are collectively called "genomic context" algorithms,[4,22] and are potentially very powerful as more and more fully sequenced genomes are becoming available.

In summary, association networks are essentially networks of cellular functions that describe how these functions are distributed among proteins. As such, they are most relevant for annotation projects, be it genome annotation or the annotation of protein families or of protein domains. They help to place each protein into a functional context, and support the delineation and description of pathways.

4. Clustering, Modules, and Motifs

Protein networks often contain visually identifiable areas of densely connected proteins, contrasting with other areas where there are hardly any connections between proteins.[23] Starting from this observation, researchers have applied a large variety of clustering algorithms to partition protein networks, hoping to identify natural groupings of proteins.[24] In the case of traditional protein–protein interaction networks, such clustering procedures are thought to reveal protein complexes and larger assemblies. In contrast, in the case of protein–protein association networks, where linkage denotes functional partnership (see above), clustering is seen as a way to uncover "functional modules" or "pathways", i.e. groups of proteins contributing together to a distinct functional process or phenotype. Detecting such modularity in interaction networks is potentially very useful, for example, because it can place uncharacterized proteins into clearly delineated groups of partners, or because it can reduce the complexity of disease association mapping exercises (mapping causal sequence variants to disease phenotypes). In the latter application, instead of considering each gene locus separately, functionally grouped loci can be considered as single units, where it would not matter which of the constituent genes carries a mutation. Such an approach is thought to considerably aid in statistical analysis because fewer independent items need to be considered.[25] Another motivation for clustering protein

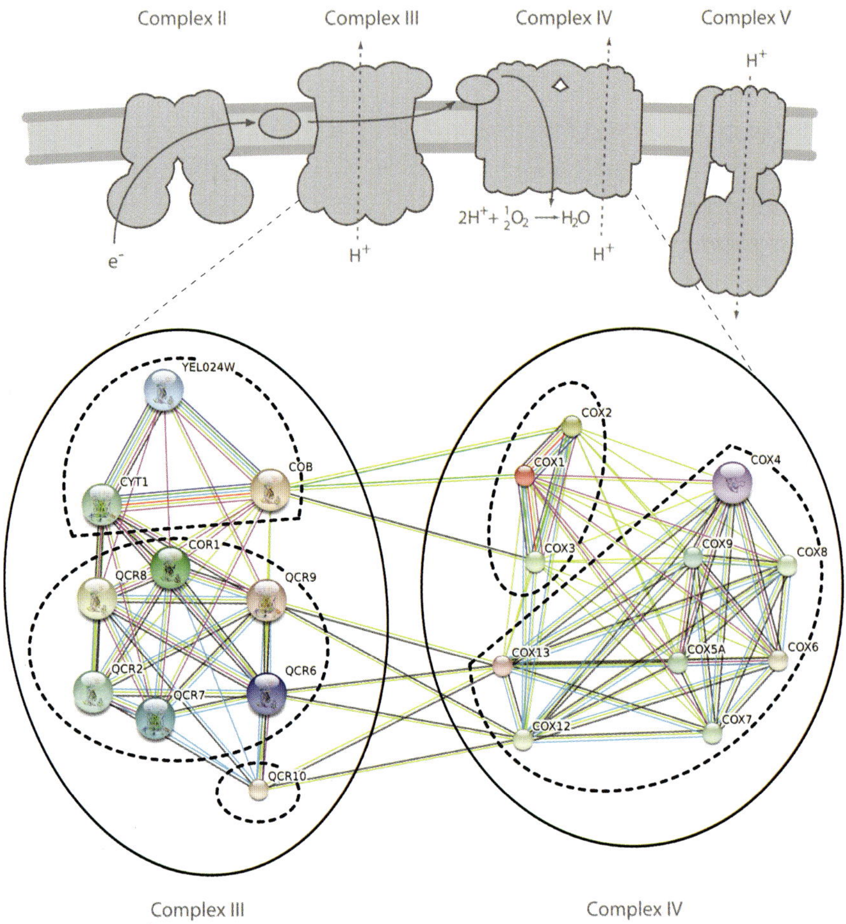

Fig. 1. Unsupervised clustering of a protein–protein association network. A subset of the proteins making up the electron transfer chain of yeast mitochondria is shown in an association network (a schematic overview of the entire transfer chain is shown above; illustration modified from the KEGG database). The association network is from the STRING database (http://string.uzh.ch/, version 7.1). The proteins are seen connected both within and across the two complexes. The line color indicates the type of evidence supporting a functional link: green lines, conserved genomic neighborhood; red lines, gene fusion events; blue lines, co-occurrence across genomes; dark gray lines, similarity in expression regulation; pink lines, experimental interaction evidence; cyan lines, already annotated in common pathway/complex; and bright green lines, mentioned together in the scientific literature. Unsupervised clustering (*K*-means) was applied at two-different cut-offs. With a low-stringency

networks is simply its visual appeal and usefulness — complex protein networks can be difficult to browse and understand, so clustering can help to more easily make out the highly connected "cliques" in such networks (Fig. 1).

For the clustering procedure itself, the choice of algorithm depends not only on the precise question asked, but also on the type of network at hand. Is it a densely connected or a sparse network? Are the edges "weighted", i.e. do they have a strength or confidence value attached to them? Are the edges directed or undirected, and are there various "types" of edges? Apart from standard hierarchical clustering algorithms, such as *K*-means or single-linkage clustering, a number of specialized algorithms have been designed (or at least adapted) specifically for cluster analysis of protein networks. These include Markov clustering (MCL),[26] superparamagnetic clustering (SPC),[27] restricted neighborhood search clustering (RNSC),[28] molecular complex detection (MCODE),[29] and others. When deciding which clustering method to use and — equally important — at which parameter settings to use it, it is essential to evaluate the results in detail and to compare them to a set of trusted functional units that serve as a reference. For any given dataset, the confidence in the results is highest when the results tend to show little dependency on parameter settings, and when they generally give a good overlap with previous expectations and with the reference data.[30,31]

Based on artificial test networks, to which controlled amounts of noise have been added, the four clustering algorithms mentioned above have been tested and compared to each other, and were subsequently also assessed on actual protein interaction data from high-throughput experiments. This has suggested that MCL and RNSC tend to perform better,[24] but tests like these of course depend on the exact type of input data and should be repeated before each application to a new data type.

Fig. 1 (*Continued*) cut-off (solid lines), the two known complexes III and IV are easily recovered. With the high-stringency cut-off (dotted lines), these are further subdivided into functional units. COX1, COX2, and COX3, for example, form a subcomplex together — they constitute the active reaction center of complex IV. Notice how experimental interaction links are seen only within the complexes, whereas functional connections also extend between the two complexes.

Clustering approaches on protein networks do have their limitations, however. They can usually assign a given protein to one cluster only, whereas in reality a protein may function as part of several distinct groups. These distinct groups could be variants of a certain protein complex, or even two completely different functional contexts in which a protein can act, a phenomenon termed "moonlighting".[32] For clustering approaches, such multiple memberships in distinct groups can cause problems and are usually not reproduced in the clustering results. One approach to overcome this is to iteratively execute multiple different clustering runs, using various parameters, thereby generating an ensemble of clusters that can then be filtered and analyzed for proteins present in distinct sets of clusters. This approach has recently been applied to the entire yeast proteome,[33] and has revealed that many protein complexes indeed show considerable variations in their makeup, but that small groups of proteins nevertheless form "cores" or "modules" (subcomplexes) which are quite stable in their composition.

Such small cores of interacting partners (often consisting of two or three proteins only) have also been termed "motifs", especially in studies that focus on their topological arrangement in the network.[34,35] To define a motif, consider that even a small number of nodes can be connected in many different topologies: they can have a minimal number of connections only so that their interaction topology can be stretched out like beads on a string, or they can be fully connected so that each protein is linked with each of the others. Each such configuration is one type of motif. The number of possible connection topologies is particularly high in the case of directed networks such as networks of transcriptional regulatory interactions, where it makes a difference whether protein A regulates protein B or vice versa. In these types of networks, motifs have been described first. Intriguingly, in actual networks derived from experimental data, the various possible types of motifs are not all equally frequent — some are far more abundant than they should be under a random expectation. This is taken as evidence that these modules form distinct regulatory circuits, for example, feedback loops, feedforward loops, switches, delay elements, and so on. Indeed, for many of these motifs, their predicted role and quantitative behavior have been confirmed experimentally (see, for instance, Refs. 36–38).

5. Interpreting Network Topology

Clustering and motif analysis represent two important ways of making use of protein networks, beyond the simple information integration and browsing that networks offer generically. Another, much-studied aspect of networks lies in their higher-level topology. How many interactions exist in the network as a whole? How are these distributed on the nodes? Are there universal structuring principles in biological networks? In a series of very influential papers,[3,39–42] Albert-László Barabási and his colleagues have drawn attention to a number of unique topological features of biological networks. These features have since been found in many types of networks; in fact, they seem to be observable almost universally.

First, biological networks are almost never random networks. A random network would be one where a certain number of edges (interactions) have simply been randomly distributed among a number of nodes (proteins). Such networks usually would have a characteristic average "degree", whereby degree is defined as the number of edges emanating from a given node. For a random network, the average degree simply depends on the number of edges assigned; and for a large fraction of the nodes, the degree will correspond closely to the average degree of the entire network. In contrast, real biological networks have a very different distribution of degree values: most nodes have only a few edges (low degree), and a small number of nodes have very many edges (high degree). In fact, the distribution of degree values is often observed to closely follow a power law, such that the number of nodes with degree k is inversely proportional to k taken to some characteristic power d (degree exponent). Such networks do not have a characteristic average degree, at which most nodes would have their number of connections; while an average degree can indeed be computed, it is rather meaningless. Therefore, such networks have been termed "scale-free". Interestingly, not only biological networks are found to be scale-free — the same is true not only for many technical networks (e.g. the Internet, airline networks), but even for social networks such as friendships or mating networks. In each of these networks, the highly connected nodes serve as "hubs", and only upon their removal does the network as a whole suffer dramatic consequences. Most non-hubs can be removed with relatively little consequences; this could be one of the reasons for the apparent stability of many technical and biological systems.

Second, many biological networks show a natural tendency to form tightly connected subclusters within the network. This can be mathematically expressed by the clustering coefficient *C*, which is defined as being 0 when none of the neighbors of a given node are interconnected ("star" topology) and as being 1 when all neighbors of a node are interconnected among themselves ("clique" topology). Biological networks tend to have large clustering coefficients, suggesting that biologically meaningful groupings can indeed be obtained by automated clustering of these networks. Often, the clustering coefficient of a given node depends again inversely on its degree; if this is the case, the network is said to be hierarchical, with large clusters successively breaking down into smaller ones.

Lastly, many biological networks represent "small worlds". This means that any two nodes in the network can be connected in the network by traversing a small number of other nodes only. This property of networks was first described in an experiment on social networks,[43] which revealed that most people on earth can be connected by as little as six acquaintances.

While most of the above observations are undisputed, there has been some lingering controversy over how exactly to interpret these findings (see, for example, Refs. 44 and 45). Do these topological features represent evidence for deep and fundamental design principles of evolved, stable systems? Or are they mere side effects of much simpler evolutionary processes, most of which we are already familiar with? There is no consensus yet on how to answer this question, but it is clear that scale-free networks are relatively widespread in nature and in technology, and as such their topology is probably not in all cases evidence for a deeper, fundamental design principle. Corroborating this, a number of simple procedures have been proposed that could build scale-free networks from scratch using simple rules — one of these is called "preferential attachment". When a network is grown by attaching new nodes preferentially to already highly connected hubs, then a scale-free network will result. This is a conceivable mechanism that could be at play, for example, in the evolution of protein–protein networks.[46] Nevertheless, whatever the precise cause for the observed topology, it does mean that current networks are far from random and that clustering and motif analysis (and other topology-based approaches) are useful ways of dealing with the huge, experimentally derived sets of protein–protein connections.

6. Online Resources

Creating an organism-wide protein network from heterogeneous data is not a trivial task, as it involves both large-scale automated computation and a number of manual or semiautomatic decisions/judgments. Often, a probabilistic approach is taken for the data integration, whereby individual datasets are first assessed by comparing them against a trusted reference set. They are then integrated in a Bayesian fashion, resulting in weighted interaction networks where the weight of an edge indicates the probability for it to be correct (for examples, see Refs. 47 and 48). Such networks need to be updated frequently, as new data become available, and a separate network needs to be made for each of a number of model organisms. Only a few online resources provide precomputed networks for browsing and downloading (see Table 1).

In general, the quality and coverage of these networks vary widely for different organisms, depending on the complexity of the organism and the quality of the available interaction data. Some of the best networks are those of budding yeast (*S. cerevisiae*) and *Escherichia coli*. Both are well-studied organisms for which many experimental datasets are available, and they are small and simple enough to have many measurements done in a genome-wide fashion. In contrast, most of the higher eukaryotes, such as human, fruit fly, and nematode, are not yet sufficiently covered experimentally to provide for high-quality networks from high-throughput data alone. For these, access to small-scale interaction data as described in the scientific literature is essential, which is usually accessible either through the work of annotation teams at major databases or through unsupervised text mining done by computers. Lastly, an important source of interaction knowledge for complex organisms comes from experiments done on simpler model organisms. The interactions detected for these can be transferred to more complex organisms when clear orthologs can be found for the respective genes. Such an orthologous

Table 1. Online resources for working with protein–protein networks. This is a very incomplete collection, listing some representatives of each category only. For the full list, the reader is referred to the comprehensive Database Summary maintained by the journal *Nucleic Acids Research* (http://www3.oup.co.uk/nar/database/a/). Over 1000 online resources are available there, sorted into categories.

Online Resources		**References**
Interaction databases (primary experimental data)		
IntAct	Maintained at EBI; a driving force of standardization and interconnectivity for all interaction databases.	50
DIP	One of the oldest interaction databases; also contains a tool to align networks (PathBLAST).	51
BioGRID	Focused on eukaryotic model organisms; contains many genetic interactions.	52
HPRD	Limited to human proteins, but conducts extensive curation efforts (also invites curation from the user community).	53
MINT	Actively curates interactions from the literature; includes projections from model organisms to humans (HomoMINT).	54
Pathway databases (interpreted and distilled interaction knowledge)		
KEGG	Main focus is on metabolism; a worldwide reference source for metabolic pathways that covers many organisms.	55
Reactome	State-of-the-art data model and visualizations of pathway knowledge; conducts active curation of knowledge from the literature.	56
PID	Maintained at the National Cancer Institute and at *Nature*; integrates and curates pathway knowledge, focusing on humans.	pid.nci.nih.gov/
Network resources (precomputed interaction networks)		
STRING	Main focus is on association networks; combines experimental data, automated literature searches, and computational predictions; covers many organisms and is regularly updated.	57

(*Continued*)

Table 1. (*Continued*)

Online Resources		References
VisANT	Combines predicted interactions and experimental data; has an interactive network visualization front-end.	58
iHOP	Main focus is on automated literature searches, augmented with imported interaction data; the retrieved relations can be grouped into networks.	59

interaction is termed an "interolog",[49] and its validity depends on the evolutionary distance between the two organisms and on the degree of sequence conservation.

These online resources can serve as a starting point for in-depth analysis of a certain pathway or protein complex. They can also provide proteome-wide background information to better understand a given functional genomics dataset, allowing the following questions, for example, to be raised: Does my set of genetically defined loci cluster on the network? Are certain types of protein modifications found in certain parts of the functional landcape? Can I use the network to reduce complexity in my genotyping data? These and other questions are best addressed by using the powerful information integration provided by organism-wide protein networks.

References

1. Alberts B. (1998) The cell as a collection of protein machines: preparing the next generation of molecular biologists. *Cell* **92**(3): 291–4.
2. Sharan R, Ulitsky I, Shamir R. (2007) Network-based prediction of protein function. *Mol Syst Biol* **3**: 88.
3. Barabási AL, Oltvai ZN. (2004) Network biology: understanding the cell's functional organization. *Nat Rev Genet* **5**(2): 101–13.
4. Skrabanek L, Saini HK, Bader GD, Enright AJ. (2008) Computational prediction of protein–protein interactions. *Mol Biotechnol* **38**(1): 1–17.
5. Bader GD, Hogue CW. (2002) Analyzing yeast protein–protein interaction data obtained from different sources. *Nat Biotechnol* **20**(10): 991–7.

6. Futschik ME, Chaurasia G, Herzel H. (2007) Comparison of human protein–protein interaction maps. *Bioinformatics* **23**(5): 605–11.
7. Reguly T, Breitkreutz A, Boucher L *et al.* (2006) Comprehensive curation and analysis of global interaction networks in *Saccharomyces cerevisiae. J Biol* **5**(4): 11.
8. von Mering C, Krause R, Snel B *et al.* (2002) Comparative assessment of large-scale data sets of protein–protein interactions. *Nature* **417**(6887): 399–403.
9. Ito T, Chiba T, Ozawa R *et al.* (2001) A comprehensive two-hybrid analysis to explore the yeast protein interactome. *Proc Natl Acad Sci USA* **98**(8): 4569–74.
10. Uetz P, Giot L, Cagney G *et al.* (2000) A comprehensive analysis of protein–protein interactions in Saccharomyces cerevisiae. *Nature* **403**(6770): 623–7.
11. Iyer K, Bürkle L, Auerbach D *et al.* (2005) Utilizing the split-ubiquitin membrane yeast two-hybrid system to identify protein–protein interactions of integral membrane proteins. *Sci STKE* **2005**(275): pl3.
12. Möckli N, Deplazes A, Hassa PO *et al.* (2007) Yeast split-ubiquitin-based cytosolic screening system to detect interactions between transcriptionally active proteins. *Biotechniques* **42**(6): 725–30.
13. Obrdlik P, El-Bakkoury M, Hamacher T *et al.* (2004) K^+ channel interactions detected by a genetic system optimized for systematic studies of membrane protein interactions. *Proc Natl Acad Sci USA* **101**(33): 12242–7.
14. Suter B, Fetchko MJ, Imhof R *et al.* (2007) Examining protein–protein interactions using endogenously tagged yeast arrays: the cross-and-capture system. *Genome Res* **17**(12): 1774–82.
15. Gavin AC, Bösche M, Krause R *et al.* (2002) Functional organization of the yeast proteome by systematic analysis of protein complexes. *Nature* **415**(6868): 141–7.
16. Ho Y, Gruhler A, Heilbut A *et al.* (2002) Systematic identification of protein complexes in *Saccharomyces cerevisiae* by mass spectrometry. *Nature* **415**(6868): 180–3.
17. Boozer C, Kim G, Cong S *et al.* (2006) Looking towards label-free biomolecular interaction analysis in a high-throughput format: a review of new surface plasmon resonance technologies. *Curr Opin Biotechnol* **17**(4): 400–5.
18. Piston DW, Kremers GJ. (2007) Fluorescent protein FRET: the good, the bad and the ugly. *Trends Biochem Sci* **32**(9): 407–14.
19. Kung LA, Snyder M. (2006) *Proteome* chips for whole-organism assays. *Nat Rev Mol Cell Biol* 7(8): 617–22.
20. Aloy P, Russell RB. (2004) Ten thousand interactions for the molecular biologist. *Nat Biotechnol* **22**(10): 1317–21.
21. Kiel C, Beltrao P, Serrano L. (2008) Analyzing protein interaction networks using structural information. *Annu Rev Biochem* 77: 415–41.

22. Huynen MA, Snel B, von Mering C, Bork P. (2003) Function prediction and protein networks. *Curr Opin Cell Biol* **15**(2): 191–8.
23. Yook SH, Oltvai ZN, Barábasi AL. (2004) Functional and topological characterization of protein interaction networks. *Proteomics* **4**(4): 928–42.
24. Brohee S, van Helden J. (2006) Evaluation of clustering algorithms for protein–protein interaction networks. *BMC Bioinformatics* 7: 488.
25. Wang K, Li M, Bucan M. (2007) Pathway-based approaches for analysis of genomewide association studies. *Am J Hum Genet* **81**(6). Epub ahead of print.
26. Enright AJ, Van Dongen S, Ouzounis CA. (2002) An efficient algorithm for large-scale detection of protein families. *Nucleic Acids Res* **30**(7): 1575–84.
27. Blatt M, Wiseman S, Domany E. (1996) Superparamagnetic clustering of data. *Phys Rev Lett* **76**(18): 3251–4.
28. King AD, Przulj N, Jurisica I. (2004) Protein complex prediction via cost-based clustering. *Bioinformatics* **20**(17): 3013–20.
29. Bader GD, Hogue CW. (2003) An automated method for finding molecular complexes in large protein interaction networks. *BMC Bioinformatics* **4**: 2.
30. von Mering C, Zdobnov EM, Tsoka S *et al.* (2003) Genome evolution reveals biochemical networks and functional modules. *Proc Natl Acad Sci USA* **100**(26): 15428–33.
31. Pereira-Leal JB, Enright AJ, Ouzounis CA. (2004) Detection of functional modules from protein interaction networks. *Proteins* **54**(1): 49–57.
32. Gancedo C, Flores CL. (2008) Moonlighting proteins in yeasts. *Microbiol Mol Biol Rev* **72**(1): 197–210.
33. Gavin AC, Aloy P, Grandi P *et al.* (2006) Proteome survey reveals modularity of the yeast cell machinery. *Nature* **440**(7084): 631–6.
34. Milo R, Shen-Orr S, Itzkovitz S *et al.* (2002) Network motifs: simple building blocks of complex networks. *Science* **298**(5594): 824–7.
35. Alon U. (2007) Network motifs: theory and experimental approaches. *Nat Rev Genet* **8**(6): 450–61.
36. Basu S, Mehreja R, Thiberge S *et al.* (2004) Spatiotemporal control of gene expression with pulse-generating networks. *Proc Natl Acad Sci USA* **101**(17): 6355–60.
37. Kalir S, Mangan S, Alon U. (2005) A coherent feed-forward loop with a SUM input function prolongs flagella expression in *Escherichia coli*. *Mol Syst Biol* **1**: 2005. 0006. Epub ahead of print.
38. Mangan S, Itzkovitz S, Zaslaver R, Alon U. (2006) The incoherent feed-forward loop accelerates the response-time of the gal system of *Escherichia coli*. *J Mol Biol* **356**(5): 1073–81.

39. Almaas E, Kovács B, Vicsek T *et al.* (2004) Global organization of metabolic fluxes in the bacterium *Escherichia coli. Nature* **427**(6977): 839–43.
40. Jeong H, Mason SP, Barabási AL, Oltvai ZN. (2001) Lethality and centrality in protein networks. *Nature* **411**(6833): 41–2.
41. Jeong H, Tombor B, Albert R *et al.* (2000) The large-scale organization of metabolic networks. *Nature* **407**(6804): 651–4.
42. Ravasz E, Somera AL, Mongru DA *et al.* (2002) Hierarchical organization of modularity in metabolic networks. *Science* **297**(5586): 1551–5.
43. Milgram S. (1967) The small world problem. *Psychol Today* **1**: 61–67.
44. Keller EF. (2005) Revisiting "scale-free" networks. *Bioessays* **27**(10): 1060–8.
45. Norris V, Raine D. (2006) On the utility of scale-free networks. *Bioessays* **28**(5): 563–4.
46. Eisenberg E, Levanon EY. (2003) Preferential attachment in the protein network evolution. *Phys Rev Lett* **91**(13): 138701.
47. Jansen R, Yu H, Greenbaum D *et al.* (2003) A Bayesian networks approach for predicting protein–protein interactions from genomic data. *Science* **302**(5644): 449–53.
48. Lee I, Date SV, Adai AT, Marcotte EM. (2004) A probabilistic functional network of yeast genes. *Science* **306**(5701): 1555–8.
49. Yu H, Luscombe NM, Lu HX *et al.* (2004) Annotation transfer between genomes: protein–protein interologs and protein-DNA regulogs. *Genome Res* **14**(6): 1107–18.
50. Kerrien S, Alam-Farugue Y, Aranda B *et al.* (2007) IntAct — open source resource for molecular interaction data. *Nucleic Acids Res* **35**(Database issue): D561–5.
51. Salwinski L, Miller CS, Smith AJ *et al.* (2004) The database of interacting Proteins: 2004 update. *Nucleic Acids Res* **32**(Database issue): D449–51.
52. Breitkreutz BJ, Stark C, Reguly T *et al.* (2008) The BioGRID interaction database: 2008 update. *Nucleic Acids Res* **36**(Database issue): D637–40.
53. Mishra GR, Suresh M, Kumaran K *et al.* (2006) Human Protein Reference Database — 2006 update. *Nucleic Acids Res* **34**(Database issue): D411–4.
54. Chatr-aryamontri A, Ceol A, Montecchi Palazzi L *et al.* (2007) MINT: the Molecular INTeraction database. *Nucleic Acids Res* **35**(Database issue): D572–4.
55. Kanehisa M, Araki M, Goto S *et al.* (2008) KEGG for linking genomes to life and the environment. *Nucleic Acids Res* **36**(Database issue): D480–4.
56. Vastrik I, D'Eustachio P, Schmidt E *et al.* (2007) Reactome: a knowledge base of biologic pathways and processes. *Genome Biol* **8**(3): R39.
57. von Mering C, Jensen LJ, Kuhn M *et al.* (2007) STRING 7 — recent developments in the integration and prediction of protein interactions. *Nucleic Acids Res* **35**(Database issue): D358–62.

58. Hu Z, Ng DM, Yamada T *et al.* (2007) VisANT 3.0: new modules for pathway visualization, editing, prediction and construction. *Nucleic Acids Res* **35**(Web Server issue): W625–32.
59. Hoffmann R, Valencia A. (2004) A gene network for navigating the literature. *Nat Genet* **36**(7): 664.

Chapter 9

Protein Structure Modeling and Docking at the Swiss Institute of Bioinformatics

Torsten Schwede and Manuel C. Peitsch

1. Introduction

Knowledge of the three-dimensional (3D) structures of proteins and their interactions with other molecules provides invaluable insights into the molecular basis of their functions and a rational basis for empirical functional analysis through site-directed mutagenesis, mapping of disease-related mutations, or the structure-based design of specific inhibitors.[1] While structure determination methods such as X-ray crystallography,[2] high-resolution electron microscopy,[3] and nuclear magnetic resonance (NMR) spectroscopy[4] have greatly progressed, they are still expensive, time-consuming, and not always applicable. Currently, about 51 000 experimental protein structures have been released by the Protein Data Bank (PDB)[5]; these structures correspond to approximately 18 000 different proteins (sharing less than 90% sequence identity among each other). However, the number of structurally characterized proteins is small compared to the 400 000 annotated and curated protein sequences in the Swiss-Prot section of UniProtKB[6] (http://www.expasy.org/sprot/). This number appears even smaller when compared to the 5.9 million known protein sequences in the complete UniProtKB (release June 2008). Even after removing the highly redundant sequences from this database (above), the remaining 4.1 million sequences exceed the number of known 3D structures by two orders of magnitude. Thus, no experimental structure is available for the vast majority of protein sequences. Therefore, the gap in structural knowledge must be bridged by computation. In this context, several computational methods for predicting the 3D structures of proteins have emerged and are the focus of many research and service development efforts. In this chapter, we will describe the approaches and services developed at the Swiss Institute of Bioinformatics (SIB).

2. Protein Structure Prediction with SWISS-MODEL — Methods and Tools

Prediction of the 3D structure of a protein from its amino acid sequence remains a fundamental scientific problem, and it is considered as one of the grand challenges in computational biology. Currently, comparative or homology modeling, which uses experimentally elucidated structures of related protein family members as templates to model the structure of the protein of interest (the "target"), is the most accurate protein structure modeling approach.[7,8] Template-based protein modeling techniques exploit the evolutionary relationship between a target protein and templates with known experimental structures, based on the observation that evolutionarily related sequences generally have similar 3D structures. Most comparative modeling procedures consist of several consecutive steps, which can be repeated iteratively until a satisfactory model is obtained: (a) identification of suitable template structures related to the target protein and the alignment of target and template(s) sequences; (b) modeling of the structurally conserved regions and the prediction of structurally variable regions; (c) refinement of the initial model; and (d) evaluation of the resulting models. See Schwede *et al.*[9] and references therein for details.

Protein modeling requires profound knowledge and understanding of the rules underlying protein structure, suitable hardware and software, as well as expert knowledge in their manipulation. As this combination is not generally available in molecular biology laboratories, protein structure modeling is still not used to its full extent in biomedical research.

The SWISS-MODEL expert system for comparative protein structure modeling and its related resources[10–20] have been developed to facilitate the usage of protein structure information (both experimental and model-based) for the broad biomedical research community.

2.1. *SWISS-MODEL Pipeline*

Fully automated, large-scale protein structure modeling requires a stable and reliable modeling pipeline. Starting from no other input than a protein sequence, the system should (a) identify possible template structures related to the target protein from the template library and select the most suitable ones; (b) generate reliable alignments of the target and template sequences; (c) build a 3D model of the target, including side chains, structurally variable regions, and segments corresponding to gaps in the alignment; and (d) evaluate the resulting models in order to identify errors and inaccuracies, and select the most reliable model. The SWISS-MODEL pipeline combines these individual steps into a workflow,[12,16,21] which powers both the automated modeling steps in the Workspace and the large-scale modeling efforts for the Repository (see below).

2.2. *SWISS-MODEL Template Library*

Comparative protein structure modeling requires high-quality experimental protein structures as templates. The SWISS-MODEL template library (SMTL)[19] is derived from the remediated Protein Data Bank.[22] Each PDB entry is split into individual chains to allow sequence-based template searches. Each template chain is annotated with information about the experimental method, resolution (if applicable), ANOLEA mean force potential,[23] force field energy, and quaternary state assignment to allow for rapid retrieval of the relevant structural information during template selection. Low-quality structures consisting only of $C\alpha$ atoms and short peptide fragments are removed. Searchable sequence indices[24] and a library of hidden Markov models (HMMs)[25] at several levels of sequence redundancy are provided.

2.3. SWISS-MODEL Server and Workspace

Since its outset, the aim of SWISS-MODEL has been to provide a user-friendly interface for fully automated, high-quality protein structure models. With advancing technology, the user interface of SWISS-MODEL has evolved from the first e-mail protein modeling service[10] — over a web-based submission system on ExPASy, the first Internet web server dedicated to molecular biology[26] — to an interactive web-based modeling expert system.[19] Today, SWISS-MODEL Workspace offers for each user a personal web-based integrated working environment where several modeling projects can be carried out in parallel. Protein sequence and structure databases necessary for modeling are accessible; and tools for template selection, model building, and structure quality evaluation can be invoked from within this Workspace.

Depending on the difficulty of the individual modeling task, the Workspace assists the user in building and evaluating protein homology models at different levels of complexity. For models that can be built based on sufficiently similar templates, we provide a fully automated mode where no user input, other than the sequence of the protein to model, is required. This is, of course, the easiest and most user-friendly way to obtain a protein model. On the other hand, for more difficult modeling scenarios with less target–template sequence similarity, the user is given control over individual steps of model building: functional domains in multi-domain proteins can be detected, secondary structure and disordered regions can be predicted for the target sequence, suitable template structures can be identified by searching the SMTL, and target–template alignments can be manually adjusted. The program DeepView[12] is tightly linked to SWISS-MODEL, and allows the visualization and manipulation of modeling projects. As quality evaluation is indispensable for a predictive method like homology modeling, several quality checks are available to assess the expected accuracy of the models. All relevant data of a modeling project are presented in a graphical synopsis.[19]

2.4. SWISS-MODEL Repository

The aim of the SWISS-MODEL Repository is to provide access to an up-to-date collection of high-quality annotated models generated by automated homology modeling, bridging the gap between sequence and structure databases. All models in the Repository are publicly accessible via an interactive website. The current release provides more than 1.3 million annotated 3D models. The SWISS-MODEL Repository is cross-referenced with UniProt and InterPro, complementing the structural information available in these two databases. Each entry in the Repository contains a sequence-based checksum, which uniquely identifies its corresponding target amino acid sequence and guarantees data consistency when exchanging records between different databases to ensure that hyperlinks between websites using database accession codes are referencing identical protein sequences. We have extended this concept to generate a common protein reference for protein models, which allowed us to build a single protein model portal (see below).

Each target sequence in the Repository has a short protein description with links to relevant sequence-based resources. A graphical representation of the InterPro functional and domain-level annotations of the target sequence indicates regions of functional and structural importance in the target sequence. Their location relative to the available models can be visualized directly from the website, or using external visualizing tools such as DeepView.

2.5. *SWISS-MODEL and DeepView — Swiss-PdbViewer*

DeepView[12] — aka Swiss-PdbViewer — is a powerful graphical tool for protein structure visualization, analysis, and manipulation, and can be used as a graphical front-end to SWISS-MODEL Workspace. In cases where automated sequence alignment fails to accurately position insertions and deletions in the target–template alignment, one can often optimize the model by manually altering the alignment in these regions. DeepView allows one to visually inspect these regions of interest, manipulate the alignment (or invoke a search for alternative templates), and submit a new model request to SWISS-MODEL Workspace. In combination, DeepView and SWISS-MODEL are forming a powerful integrated sequence to structure workbench. The program can be downloaded freely from the ExPASy server for Windows™ and Macintosh™ platforms.

3. Large-Scale Protein Structure Prediction and Structural Genomics

Comparative protein structure modeling and experimental protein structure determination complement each other, with the long-term goal of making 3D atomic-level information of most proteins obtainable from their corresponding amino acid sequences. Structural genomics is a worldwide effort, aiming at rapidly determining a large number of protein structures using X-ray crystallography and NMR spectroscopy in a high-throughput mode.[27–29] As a result of concerted efforts in technology and methodology development in recent years, each step of experimental structure determination has become more efficient, less expensive, and more likely to succeed.[30] Structural genomics initiatives are making a significant contribution to both the scope and depth of our structural knowledge about protein families. Although worldwide structural genomics initiatives account for only ~20% of the new structures, these contribute approximately three quarters of new structurally characterized families and over five times as many novel folds as classical structural biology.[31–35]

In light of the ever-growing amount of genome sequencing data, the structure of most proteins, even with structural genomics, will be modeled and not elucidated experimentally. From a modeling-centric perspective, the selection of structural genomics targets should be such that most of the remaining sequences can be modeled with useful accuracy by comparative modeling. The accuracy of comparative models currently declines sharply below 30% sequence identity. Thus, template selection strategies should aim at systematic sampling of protein structures to ensure that most of the remaining sequences are related to at least one experimentally elucidated structure at more than 30% sequence identity; using this cut-off, it has been estimated that a minimum of 16 000 targets must be determined to cover 90% of all protein domain families, including those of membrane proteins.[29] Such estimates show large variations, depending on the level of sequence identity that is assumed to ensure sufficiently accurate model building and on how this coverage is calculated.

4. Protein Model Quality

4.1. *Correctness and Accuracy*

Conceptually, the quality of a model can be determined by two distinct criteria that will determine its applicability to solve biological questions: the correctness of a model, which is largely dictated by the quality of the sequence alignment used to guide the modeling process; and the accuracy of a model, which is essentially limited by the deviation of the used template structure(s) from the target structure.

4.1.1. *Model correctness*

The comparative protein structure modeling methods implemented in SWISS-MODEL rely on the evolutionary relationship between target and template proteins. As sequence similarities reflect these evolutionary relationships, they can be used as a first approximation of structural relatedness. Consequently, the two major limitations of sequence comparison methods are carried over to comparative protein modeling. Firstly, there might simply not be any sequence of known 3D structure with enough similarity to the target sequence; and secondly, sequence comparison algorithms are known to yield erroneous alignments as the level of identity between sequences decreases below 30%. The former limitation simply precludes the building of a comparative model, while the latter is the cause of incorrect models which must be considered with caution or in many cases simply discarded. Indeed, if the sequence alignment is wrong in at least one region of the protein, then the spatial arrangement of the residues in this portion of the model will be incorrect and furthermore adversely impact the conformation of neighboring residues.

Sequence comparison algorithms have become so sensitive that today they can detect very distantly related sequences and allow fold assignments of distantly related proteins.[25] From the perspective of a structural biologist, however, amino acid sequences are little more than a string-based notation attempting to represent a complex structure, analogous to SMILES which represents small molecular entities as strings of characters. By extension, sequence alignments are highly simplified representations of structural similarities that become very limited when the sequence identity levels drop below a certain threshold, usually around 30%. Experience shows that, with decreasing identity levels, alignment errors first appear in loop regions before affecting larger portions of the proteins, and that it is close to impossible to properly align loops of low sequence identity and unequal lengths, leaving it up to the later model-building process to find adequate solutions for loop structures. Furthermore, many protein structures sharing a sequence identity level below 40% contain structurally nonconserved loops, even if they have the same length. Therefore, it becomes apparent that even the alignment of loops with identical length but completely different sequences has little meaning in structural biology.

4.1.2. *Model accuracy*

The accuracy of a protein model is largely limited by the deviation of the used template structure(s) relative to the experimental structure of the target. This limitation is inherent to the method, since comparative models result from a structural extrapolation guided by a sequence alignment. As shown by comparison of the experimentally elucidated structures, there is a direct correlation between the sequence identity level of a protein pair and the deviation of the Cα atoms of their common core.[36] It is therefore generally accepted that the percentage of sequence identity between target and template allows for a reasonable first estimate of the model quality, and that the core Cα atoms of protein models sharing 50% sequence identity with their templates will deviate by approximately 1.0 Å root mean square deviation (RMSD) from their experimentally elucidated structures[36]; this is roughly comparable to the accuracy of a medium-resolution NMR-derived structure or a low-resolution X-ray structure.[37,38] This has led to the definition of three broad classes of model quality based on the level of identity of the core region common to both target and template sequences. Firstly, models based on more than 50% identity will yield high-accuracy models, where inaccuracies are mostly restricted to side-chain packing and loop regions. Secondly, comparative models based on 30% to 50% sequence identity can be considered medium-accuracy models, where the most frequent inaccuracies are found in side-chain packing, slight distortions of the protein core, and inaccurate loop conformations. Thirdly, models based on templates whose cores share less than 30% identity with the template lead to low-accuracy models, where inaccuracies become more important and distortions more severe.

Sequence identity is, however, not the only way templates influence protein model accuracy, and one should not overlook other possible contributions in the course of a modeling process. Indeed, the templates, which are obtained through experimental approaches, are subject to structural variations caused not only by experimental errors and differences in data collection conditions (e.g. temperature[39]), but also by different crystal lattice contacts and the presence or absence of ligands.[40] A direct consequence of the comparative approach is that these influences are carried over to the models derived from these templates, and call for an increased attention to the template selection process and a good understanding of the factors which influenced their experimental structure elucidation.

4.2. *Limitations of Comparative Protein Modeling*

4.2.1. *Template availability and structural diversity*

It is generally accepted that a very small number of different folds account for the majority of known structures,[41] and a recent study has argued that most sequences could already be modeled using known folds (or fragments of known folds) as templates.[42] Thus, for a large proportion of protein domains, a structure with a similar fold would be available in the PDB. However, models based on alignments with low sequence identity generally provide accurate information only about the overall fold of the protein. As the correctness and accuracy of comparative models rapidly decrease when the sequence identity between a target and template drops below 30%–35%, a much denser coverage of the sequence space with experimentally elucidated protein structures is necessary to create adequate protein models for the majority of domains. Ideally, one should find a template in the PDB with 30%–35% (or more) sequence identity for every target. It is, however, important to remember that while the overall fold of proteins is often well conserved even at undetectable levels of sequence similarity, protein function — such as enzyme function and specificity — shows much higher variability,[43,44] even at high levels of sequence identity (above 50%). Functional assignment of protein function thus requires new methods beyond simple homology-based assignments which take into account specific local structural features.

4.2.2. *Unstructured proteins*

Unstructured regions in proteins are implicated in important biological roles such as translation, transcriptional regulation, cell signaling, and molecular recognition. Therefore, they have recently become the focus of much attention. Several studies report examples of disordered proteins implicated in important cellular processes, undergoing transitions to more structured states upon binding to their target ligand, DNA, or other proteins.[45–47] New biological functions linked to native disorder are emerging, such as self-assembly of multi-protein complexes[48] or involvement in RNA and protein chaperones.[49,50] Unstructured proteins pose a serious challenge for experimental structural determination, as they can hinder the crystallization of proteins or interfere with NMR spectroscopy. Consequently, such proteins are also not amenable to comparative modeling techniques. However, computational approaches for detecting regions in protein sequences with high propensity for intrinsic disorder have been successfully developed based on the observation that such protein segments share characteristic sequence properties.[51–54] Furthermore, molecular dynamics (MD)-based methods have been successfully applied to certain proteins of this category, and are used to analyze and understand the path of folding and unfolding in self-assembling multi-protein complexes.[48]

4.2.3. *Membrane proteins*

Membrane proteins are involved in a broad range of central cellular processes, including signaling and intercellular communication, vesicle trafficking, ion transport, and protein translocation. Their particular cellular location and central role in many disease mechanisms make them one of the most frequently studied classes of drug targets. These include protein families and superfamilies such as ion channels, reuptake pumps as targets for antidepressants, and the important group of seven-transmembrane G-protein-coupled receptors (GPCRs). However, membrane proteins pose formidable challenges to experimental structure determination by X-ray crystallography and NMR spectroscopy. Furthermore, human proteins often have no closely related homologs in prokaryotes or archaea, which would facilitate expression and crystallization. As a result, structures of membrane proteins are significantly underrepresented in the PDB. The 3D structures of only ~160 different membrane proteins are currently publicly available (June 2008). Consequently, prediction of membrane protein structures based on physical models that describe intraprotein and protein–solvent interactions in the membrane environment without relying on homologous template structures has been attempted by several groups.[55,56]

An important challenge in the modeling of membrane protein structures is the presumed difference relative to globular proteins. For example, it is believed that membrane proteins are "inside-out" globular proteins, with hydrophobic residues on the outside in contact with the lipid bilayer and polar residues on the inside in the protein core. This design may render the standard scoring functions used for the modeling of globular proteins less suitable for use with membrane proteins. Recently, a new scoring function was developed in Rosetta to account for such differences.[57]

4.3. *Model Quality Evaluation*

Protein structure modeling is maturing and therefore widely used as a scientific research tool today. Consequently, it is increasingly important to evaluate to what extent current prediction methods meet the accuracy requirements of different scientific applications. A good way to assess the reliability of different protein structure modeling methods *a posteriori* is by evaluating the results of blind predictions after the corresponding protein structures have been determined experimentally. One such effort is the biannual Community Wide Experiment on the Critical Assessment of Techniques for Protein Structure Prediction (CASP).[7,58] During a CASP trial, research groups apply their prediction methods to sequences for which the experimental structure is about to be determined; the accuracy of these blind predictions is then assessed independently once the structures are made available.[8,38,59,60] There are also web servers LiveBench[61] and EVA[62] that assess protein structure prediction servers on an automated and continuous basis, using sequences from the PDB, before their structures are released as modeling targets.

Retrospective assessment of the average accuracy of individual modeling methods by projects such as CASP and EVA is valuable for the development of modeling techniques, but unfortunately does not allow drawing any conclusions about the accuracy of a specific model as the correct answer is unknown in a real-life situation. Since the usefulness of predictions crucially depends on their accuracy, a means of reliably predicting the likely accuracy of a protein structure model in the absence of its known 3D structure is an important problem in protein structure prediction. Accurate estimates of the errors in a model are an essential component of any predictive method — protein structure prediction is not an exception.

Different scoring schemes have been developed to determine whether or not a model has the correct fold, to discriminate between native and near-native states, to select the most near-native model in a set of decoys, and to provide quantitative estimates for the coordinate error of the predicted amino acids. A variety of methods have been applied to address these tasks, such as physics-based energies, knowledge-based potentials,[23] combined scoring functions, and clustering approaches. Combined scoring functions integrate several different scores, aiming to extract the most informative features from each of the individual input scores.[63] Clustering approaches use consensus information from an ensemble of protein structure models provided by different methods.[64]

Some structural aspects of a protein model can be verified using methods based on fold recognition methods. These methods rely on empirical pseudo-conformational energy potentials derived from the pairwise interactions observed in well-defined protein structures. These terms are summed over all residues in a model and result in a more (more negative) or less (more positive) favorable energy. These methods can detect a global sequence to structure incompatibility and errors corresponding to topological differences between template and target. They also allow the detection of more localized errors such as β-strands that are "out of register" or buried charged residues. However, none of these methods has allowed the detection of the more subtle structural inconsistencies, which are often localized in nonconserved loops, and thus cannot provide an assessment of the correctness of their geometry.

5. Applications of Protein Models

The suitability of protein models for specific applications depends critically on their correctness and accuracy. There is a wide range of applications for comparative models, such as designing experiments for site-directed mutagenesis or protein engineering, predicting ligand binding sites and docking small molecules in structure-based drug discovery,[65,66] studying the effect of mutations and single nucleotide polymorphisms (SNPs),[67,68] phasing X-ray diffraction data in molecular replacement,[69,70] as well as engineering and designing proteins. Hereafter, we will review a number of applications of models built with SWISS-MODEL. As will be seen, it is important to avoid any dogmatism when considering the applicability of protein models. There are obviously some guidelines that should be considered; for instance, one should use only high-accuracy models (>50% sequence identity between target and template) to run *in silico* docking experiments or molecular replacement attempts. However, these guidelines have their cohort of exceptions, as every modeling project has its own peculiarities.

5.1. *Functional Analysis of Proteins*

The most important aim of genome sequencing projects is to assign a biological function to the many newly discovered genes and their protein transcripts. Insights into the 3D structure of a protein can be of great assistance in assigning its molecular function, while its biological role and localization are much more difficult to relate to its structure. For instance, even if we know that a given protein is a protease and can predict its S1 pocket substrate specificity from its active site configuration, we cannot, from the structure alone, define its biological role and the processes and pathway(s) in which it is involved. In any case, predicting the molecular function of a protein on the sole basis of its 3D structure is in itself a very challenging task. If the active site has been observed previously[71–73] or if the protein has been cocrystallized with a substrate analog, we have a better chance of succeeding; however, correctly predicting specificity remains a relatively rare event.[74–76] As protein models inherit many of the features of their templates, but with reduced accuracy, it is not surprising that the detection and functional assignment of a known active or binding site requires models of good quality.[77–79]

The relationships between sequence identity, model correctness, and model accuracy have been described above. One should not overlook the fact that these sequence identity levels are an average value computed over the whole of the protein or, at least, of the core elements common to the target and modeling template. In most cases, however, enzyme active sites or ligand-binding regions are more conserved than the rest of the structure. Therefore, one can often build high-quality models for active sites, while other regions of the protein model are less accurate or even display irresolvable correctness issues in distal loops.

Nonetheless, low-accuracy models can provide additional hints to protein function and can help confirm sequence similarity searches and the results of fold recognition. Indeed, it is an advantage to build models[75] in such cases, as structural insights can confirm hypotheses derived from homology detection. In our hands, we were able to verify and confirm the assignment of several *C. elegans* insulin-like genes using low-accuracy models.[80] Similarly, the trimeric nature of the CD40 ligand was first proposed based on a low-accuracy model in which the target and the TNFα template share less than 26% sequence identity.[81]

5.1.1. *Studying the impact of mutations and SNPs on protein function*

Functional analysis of proteins greatly benefits from the observation of naturally occurring mutations. Indeed, diseases, or less severe phenotypic variations, which can be univocally assigned to single-point mutations provide a good framework for understanding the molecular function and biological role of a protein. Therefore, protein models can be readily applied to interpret the impact mutations can have on the overall structure and, thus, function of a protein.[67,68] While objective scoring functions to assess this impact are still not reliable, visual inspection associated with a good knowledge and understanding of the rules underlying protein structure has proven useful in defining the broad reasons for mutant malfunction (for concrete examples, see Refs. 67, 68 and 82). There is an increasingly large body of data on naturally occurring mutations (over 43 000 human sequence variants are reported in SWISS-PROT) and SNPs, of which a sizable proportion will alter the translated protein sequences. Interpreting the potential functional effects of these mutants will be crucial to elucidate the molecular basis of human diseases.

5.1.2. *Planning site-directed mutagenesis experiments*

One definite advantage of 3D structure and models in functional protein analysis is that they provide a solid base for site-directed mutatgenesis experiments aimed at the elucidation of the molecular function of proteins. Even medium- and low-accuracy models can be used as solid frameworks for experiment planning, and guide the selection of key mutants designed to test functional hypotheses[83,84] (Fig. 1) or to modulate biophysical properties.[85] These experimentally generated mutants complement the naturally occurring ones mentioned in the previous subsection and, together with the mapping of other facts such as glycosylation sites, greatly contribute to the elucidation of protein function.[1] For instance, the comparative models generated for the Fas ligand, its protein family members,[86] and its receptor illustrate how models can be applied to (a) the understanding of the impact of naturally occurring mutations[87–89]; (b) experimental mutagenesis; and (c) to the interpretation and mapping of other known features, such as glycosylations, to understand the finer molecular function of a protein.

5.2. *Molecular Replacement*

Solving the phase problem, i.e. knowing the phase of the diffracted waves, in crystallography experiments is a crucial step towards reconstructing

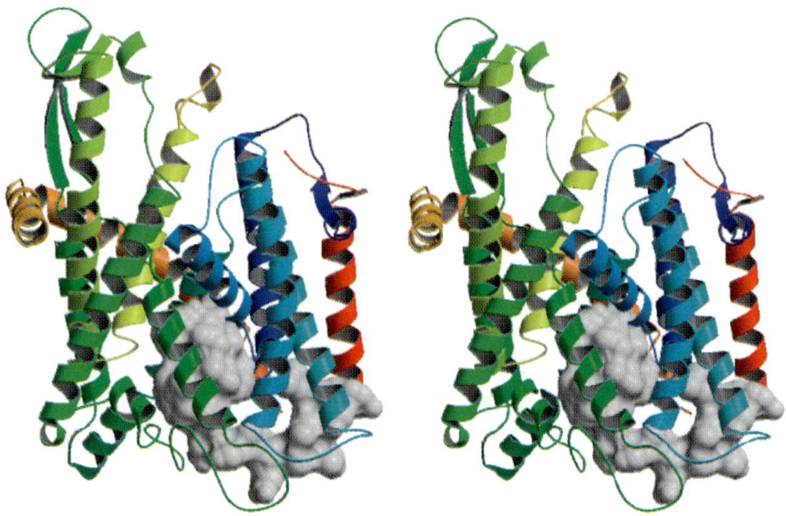

Fig. 1. Comparative structure model of the yeast Sec61 complex modeled on the structure of the *M. jannaschii* SecYEβ translocon.[90] The model, shown as a ribbon representation in stereo, allowed designing a deletion mutant of the plug domain (residues 52–74, displayed as a space-filling contour) for studying the functional role of the plug domain *in vivo*.[83]

atomic structures that optimally fit the experimental data. As these cannot be measured directly, they have to be obtained using experimental methods such as heavy-atom isomorphous replacement, anomalous scattering, or molecular replacement.[2,69,70,91] The latter approach generally requires a good-quality atomic structure of a related protein that is then rotated and translated into the new crystal system until there is a good match with the experimental data. The first application of a model built with SWISS-MODEL in molecular replacement was performed by Karpusas *et al.*[92] to obtain a 2 Å resolution X-ray structure of the human CD40 ligand (PDB entry 1aly). The authors used our published murine homology model[14] (PDB entry 1cda) to build a human model of CD40L, and then applied the latter "model of a model" to the molecular replacement approach. A more recent example can be seen in Hou *et al.*[93]

5.3. *Protein Design*

Biotechnology often requires that proteins be redesigned to improve their thermal stability[94] or solubility,[95] graft an epitope onto an immunogenic protein,[96,97] alter and optimize the binding properties of an antibody,[98–102] or change the substrate specificity of enzymes.[103] It is certainly reasonable to consider that high-accuracy models can be used effectively in protein redesign. However, one will have to carefully consider each case, as every redesign project will have its own challenges.

5.4. *Docking*

Drug discovery and design are certainly one of the more desirable, but demanding, applications of protein models. Early in the drug discovery process, once a suitable drug target has been identified for a given disease, it is crucial to identify small molecules that bind to this target and alter its activity (i.e. inhibition or activation). This is generally performed by screening large collections of compounds in dedicated bioassays. The resulting "hits" can, if confirmed by one or more other relevant biological assays, become the starting point of a compound optimization process aimed at identifying highly efficacious analogs with low toxicity. While compound screening is predominantly an experimental approach, structure-based computational approaches provide an alternative and complementary way to identify such hits. In cases where high-quality experimental 3D structures or models of the target protein alone or in complex with a ligand are available, molecular docking approaches can be used to simulate the nonbonded chemical interactions between the target protein and individual compounds stored in large libraries.[104]

Notably, this approach is of particular interest when none or only very few active compounds are available for a protein target.[105] There are, however, many challenges in docking and drug design. For instance, the ligand binding site on a protein structure can be very flexible and adopt very different shapes, depending on ligand binding[106,107]; conversely, the conformation of the ligand itself can change upon binding.[108,109]

Nevertheless, recent publications from both academia and industry convincingly demonstrate that careful application of structure-based virtual screening in combination with follow-up experimental verification can indeed lead to the discovery of new compounds active against diverse enzymes, receptors, and clinically relevant drug targets, such as NF-κB,[110] the nuclear receptor PPARγ,[111] the histone arginine methyltransferase,[112] a cytochrome P450,[113] a fish estrogen receptor,[114] or the CK2 protein kinase[66]; and are able to complement assay-based high-throughput screening approaches.[115] Further examples of successful structure-based virtual screening have recently been summarized in Cavasotto and Orry.[116]

6. Protein Model Portal

While the PDB currently holds approximately 51 000 experimentally derived coordinate entries representing 18 000 different proteins, several millions of comparative protein models have been generated for the protein sequences contained in the UniProtKB database using these experimentally elucidated structures as templates.[13,17,18,20,117] Databases of annotated comparative models such as SWISS-MODEL Repository[17,20] and others[117–119] allow cross-referencing with other nonstructure-centric resources, and make comparative models accessible to nonexperts.

We have developed the Protein Model Portal (http://www.proteinmodelportal.org) as part of the PSI Structural Genomics Knowledgebase to provide a single portal to all structure information available for a given protein within various databases, thereby implementing the first step of the community workshop recommendation[120] on archiving structural models of biological macromolecules. The Protein Model Portal currently allows querying models from six structural genomics centers, MODBASE, and SWISS-MODEL Repository, as well as PDB experimental structures, through a single search interface. All models are mapped to a common unique sequence reference system based on UniProt/UniParc,[121] which allows one to cross-query all resources simultaneously and to dynamically annotate the model target sequences with functional[121] and domain annotation.[122] The current release

(June 2008) consists of 5.8 million comparative protein models for 1.97 million distinct UniProt entries.

7. Future Outlook

With the deluge of new DNA sequence information being generated by large-scale genomics and metagenomics projects, protein structure modeling will be the only feasible approach for inferring structural information for the vast majority of proteins in the foreseeable future. The development of reliable, accurate, and easy-to-use resources for protein modeling is mandatory to make best use of the investments in experimental structure determination, and to provide structure information for biomedical research projects which are not directly supported by experimental structure determination efforts. However, automated protein structure models still fall short of experimental structures, and significant research will be required to ensure the reliability and applicability of models.

One major task will be to develop suitable methods to assess the applicability of models for specific applications, and to derive appropriate model quality estimates if the accuracy of a given protein model (or a part thereof) is expected to be sufficient for a given application. Several tools for estimating the overall relative accuracy of protein models have been developed recently.[64,123–125] However, significantly more work is required to correctly estimate the expected absolute local errors within protein models.

As errors and inaccuracies from template structures are propagated into the models, further development of the SMTL will be crucial with respect to the quality of HMM profiles or the correction of experimental errors and inaccuracies.[126,127] A well-curated and annotated template library is also a prerequisite to extend our automated modeling pipeline to automatically model oligomeric proteins in their correct biologically active state, and to include essential cofactors and ligands in the models.

In parallel, we will further develop the user interface of the SWISS-MODEL expert system. Recent advances in web technology will allow us to provide richer functionality directly within the web page with respect to visualization and structure comparisons, annotations, and cross-references with other resources and tools; and provide direct seamless workflows for the most relevant structure-related questions in biomedical research. The following fictive example may illustrate the possibilities of such an integrated web-based workflow at the SIB: Starting from a protein sequence in the UniProt[121] database on ExPASy, a 3D model can be constructed using SWISS-MODEL Workspace.[19] Annotated SNPs and known mutations can be mapped on the model and visualized. If the model is of sufficiently high quality, ionization states and partial charges could be assigned to prepare the model as a target molecule for docking of a specific ligand[128] or for virtual screening.[104,108]

Although the gap between the number of predicted protein sequences and experimentally determined protein structures will remain large for the foreseeable future, we will continue our efforts to bridge this gap by developing integrated and easy-to-use protein modeling resources such as SWISS-MODEL and the Protein Model Portal.

References

1. Peitsch MC. (2002) About the use of protein models. *Bioinformatics* **18**(7): 934–8.
2. Stirnimann CU, Grütter MG. (2008) New frontiers in X-ray crystallography. In: Schwede T, Peitsch MC (eds.). *Computational Structural Biology*. Singapore: World Scientific Publishing, pp. 601–622.
3. Engel A. (2008) New frontiers in high-resolution electron microscopy. In: Schwede T, Peitsch MC (eds.). *Computational Structural Biology*. Singapore: World Scientific Publishing, pp. 623–654.
4. Nilges M, Markwich P, Malliavin T *et al.* (2008) New frontiers in characterizing structure and dynamics by NMR. In: Schwede T, Peitsch MC (eds.). *Computational Structural Biology*. Singapore: World Scientific Publishing, pp. 655–680.
5. Berman H, Henrick K, Nakamura H, Markley JL. (2007) The worldwide Protein Data Bank (wwPDB): ensuring a single, uniform archive of PDB data. *Nucleic Acids Res* **35**(Database issue): D301–3.
6. Bairoch A, Apweiler R, Wu CH *et al.* (2005) The Universal Protein Resource (UniProt). *Nucleic Acids Res* **33** Database issue: D154–9.
7. Moult J, Fidelis K, Kryshtafovych A *et al.* (2007) Critical assessment of methods of protein structure prediction — round VII. *Proteins* **69**(S8): 3–9.

8. Kopp J, Bordoli L, Battey JND *et al.* (2007) Assessment of CASP7 predictions for template-based modeling targets. *Proteins* **69**(S8): 38–56.
9. Schwede T, Sali A, Eswar N, Peitsch MC. (2008) Protein structure modeling. In: Schwede T, Peitsch MC (eds.). *Computational Structural Biology.* Singapore: World Scientific Publishing, pp. 3–36.
10. Peitsch MC. (1995) Protein modelling by e-mail. *Biotechnology* **13**: 658–60.
11. Peitsch MC, Herzyk P, Wells TN, Hubbard RE. (1996) Automated modelling of the transmembrane region of G-protein coupled receptor by SWISS-MODEL. *Receptors Channels* **4**(3): 161–4.
12. Guex N, Peitsch MC. (1997) SWISS-MODEL and the Swiss-PdbViewer: an environment for comparative protein modeling. *Electrophoresis* **18**(15): 2714–23.
13. Peitsch MC. (1997) Large scale protein modelling and model repository. *Proc Int Conf Intell Syst Mol Biol* **5**: 234–6.
14. Peitsch MC, Wilkins MR, Tonella L *et al.* (1997) Large-scale protein modelling and integration with the SWISS-PROT and SWISS-2DPAGE databases: the example of *Escherichia coli. Electrophoresis* **18**(3–4): 498–501.
15. Peitsch MC, Schwede T, Guex N. (2000) Automated protein modeling — the proteome in 3D. *Pharmacogenomics* **1**(3): 257–66.
16. Schwede T, Kopp J, Guex N, Peitsch MC. (2003) SWISS-MODEL: an automated protein homology-modeling server. *Nucleic Acids Res* **31**(13): 3381–5.
17. Kopp J, Schwede T. (2004) The SWISS-MODEL Repository of annotated three-dimensional protein structure homology models. *Nucleic Acids Res* **32**(Database issue): D230–4.
18. Kopp J, Schwede T. (2004) Automated protein structure homology modeling: a progress report. *Pharmacogenomics* **5**(4): 405–16.
19. Arnold K, Bordoli L, Kopp J, Schwede T. (2006) The SWISS-MODEL Workspace: a web-based environment for protein structure homology modelling. *Bioinformatics* **22**(2): 195–201.
20. Kopp J, Schwede T. (2006) The SWISS-MODEL Repository: new features and functionalities. *Nucleic Acids Res* **34**(Database issue): D315–8.
21. Peitsch MC. (1996) ProMod and SWISS-MODEL: Internet-based tools for automated comparative protein modelling. *Biochem Soc Trans* **24**(1): 274–9.
22. Henrick K, Feng Z, Bluhm WF *et al.* (2008) Remediation of the Protein Data Bank archive. *Nucleic Acids Res* **36**(Database issue): D426–33.
23. Melo F, Feytmans E. (2008) *Scoring Functions for Protein Structure Prediction.* Singapore: World Scientific Publishing.
24. Altschul SF, Madden TL, Schaffer AA *et al.* (1997) Gapped BLAST and PSI-BLAST: a new generation of protein database search programs. *Nucleic Acids Res* **25**(17): 3389–402.
25. Soding J. (2005) Protein homology detection by HMM-HMM comparison. *Bioinformatics* **21**(7): 951–60.

26. Appel RD, Bairoch A, Hochstrasser DF. (1994) A new generation of information retrieval tools for biologists: the example of the ExPASy WWW server. *Trends Biochem Sci* **19**(6): 258–60.
27. Burley SK. (2000) An overview of structural genomics. *Nat Struct Biol* 7 (Suppl): 932–4.
28. Thornton J. (2001) Structural genomics takes off. *Trends Biochem Sci* **26**(2): 88–9.
29. Vitkup D, Melamud E, Moult J, Sander C. (2001) Completeness in structural genomics. *Nat Struct Biol* **8**(6): 559–66.
30. Slabinski L, Jaroszewski L, Rodrigues AP *et al.* (2007) The challenge of protein structure determination — lessons from structural genomics. *Protein Sci* **16**(11): 2472–82.
31. Chandonia JM, Brenner SE. (2006) The impact of structural genomics: expectations and outcomes. *Science* **311**(5759): 347–51.
32. Marsden RL, Lewis TA, Orengo CA. (2007) Towards a comprehensive structural coverage of completed genomes: a structural genomics viewpoint. *BMC Bioinformatics* **8**: 86.
33. Todd AE, Marsden RL, Thornton JM, Orengo CA. (2005) Progress of structural genomics initiatives: an analysis of solved target structures. *J Mol Biol* **348**(5): 1235–60.
34. Liu J, Montelione GT, Rost B. (2007) Novel leverage of structural genomics. *Nat Biotechnol* **25**(8): 849–51.
35. Levitt M. (2007) Growth of novel protein structural data. *Proc Natl Acad Sci USA* **104**(9): 3183–8.
36. Chothia C, Lesk AM. (1986) The relation between the divergence of sequence and structure in proteins. *EMBO J* **5**(4): 823–6.
37. Baker D, Sali A. (2001) Protein structure prediction and structural genomics. *Science* **294**(5540): 93–6.
38. Read RJ, Chavali G. (2007) Assessment of CASP7 predictions in the high accuracy template-based modeling category. *Proteins* **69**(S8): 27–37.
39. Tilton RF Jr, Dewan JC, Petsko GA. (1992) Effects of temperature on protein structure and dynamics: X-ray crystallographic studies of the protein ribonuclease-A at nine different temperatures from 98 to 320 K. *Biochemistry* **31**(9): 2469–81.
40. Muller CW, Schlauderer GJ, Reinstein J, Schulz GE. (1996) Adenylate kinase motions during catalysis: an energetic counterweight balancing substrate binding. *Structure* **4**(2): 147–56.
41. Orengo CA, Thornton JM. (2005) Protein families and their evolution — a structural perspective. *Annu Rev Biochem* **74**: 867–900.
42. Zhang Y, Skolnick J. (2005) The protein structure prediction problem could be solved using the current PDB library. *Proc Natl Acad Sci USA* **102**(4): 1029–34.

43. Rost B. (2002) Enzyme function less conserved than anticipated. *J Mol Biol* **318**(2): 595–608.
44. Tian W, Skolnick J. (2003) How well is enzyme function conserved as a function of pairwise sequence identity? *J Mol Biol* **333**(4): 863–82.
45. Dyson HJ, Wright PE. (2005) Intrinsically unstructured proteins and their functions. *Nat Rev Mol Cell Biol* **6**(3): 197–208.
46. Radivojac P, Iakoucheva LM, Oldfield CJ *et al.* (2007) Intrinsic disorder and functional proteomics. *Biophys J* **92**(5): 1439–56.
47. Fink AL. (2005) Natively unfolded proteins. *Curr Opin Struct Biol* **15**(1): 35–41.
48. Dima RI. (2008) Protein–protein interactions and aggregation processes. In: Schwede T, Peitsch MC (eds.). *Computational Structural Biology*. Singapore: World Scientific Publishing, pp. 299–324.
49. Namba K. (2001) Roles of partly unfolded conformations in macromolecular self-assembly. *Genes Cells* **6**(1): 1–12.
50. Tompa P, Csermely P. (2004) The role of structural disorder in the function of RNA and protein chaperones. *FASEB J* **18**(11): 1169–75.
51. Bordoli L, Kiefer F, Schwede T. (2007) Assessment of disorder predictions in CASP7. *Proteins* **69**(S8): 129–36.
52. Obradovic Z, Peng K, Vucetic S *et al.* (2005) Exploiting heterogeneous sequence properties improves prediction of protein disorder. *Proteins* **61** (Suppl) 7: 176–82.
53. Ward JJ, McGuffin LJ, Bryson K *et al.* (2004) The DISOPRED server for the prediction of protein disorder. *Bioinformatics* **20**(13): 2138–9.
54. Schlessinger A, Liu J, Rost B. (2007) Natively unstructured loops differ from other loops. *PLoS Comput Biol* **3**(7): e140.
55. Barth P, Schonbrun J, Baker D. (2007) Toward high-resolution prediction and design of transmembrane helical protein structures. *Proc Natl Acad Sci USA* **104**(40): 15682–7.
56. Zhang Y, Devries ME, Skolnick J. (2006) Structure modeling of all identified G protein-coupled receptors in the human genome. *PLoS Comput Biol* **2**(2): e13.
57. Yarov-Yarovoy V, Schonbrun J, Baker D. (2006) Multipass membrane protein structure prediction using Rosetta. *Proteins* **62**(4): 1010–25.
58. Moult J. (2005) A decade of CASP: progress, bottlenecks and prognosis in protein structure prediction. *Curr Opin Struct Biol* **15**(3): 285–9.
59. Battey JN, Kopp J, Bordoli L *et al.* (2007) Automated server predictions in CASP7. *Proteins* **69**(S8): 68–82.
60. Jauch R, Yeo HC, Kolatkar PR, Clarke ND. (2007) Assessment of CASP7 structure predictions for template free targets. *Proteins* **69**(S8): 57–67.
61. Rychlewski L, Fischer D. (2005) LiveBench-8: the large-scale, continuous assessment of automated protein structure prediction. *Protein Sci* **14**(1): 240–5.

62. Koh IY, Eyrich VA, Marti-Renom MA *et al.* (2003) EVA: evaluation of protein structure prediction servers. *Nucleic Acids Res* **31**(13): 3311–5.
63. Capriotti E, Marti-Renom MA. (2008) *Asessment of Protein Structure Predictions.* Singapore: World Scientific Publishing.
64. Wallner B, Elofsson A. (2007) Prediction of global and local model quality in CASP7 using Pcons and ProQ. *Proteins* **69** (Suppl 8): 184–93.
65. Hillisch A, Pineda LF, Hilgenfeld R. (2004) Utility of homology models in the drug discovery process. *Drug Discov Today* **9**(15): 659–69.
66. Vangrevelinghe E, Zimmermann K, Schoepfer J *et al.* (2003) Discovery of a potent and selective protein kinase CK2 inhibitor by high-throughput docking. *J Med Chem* **46**(13): 2656–62.
67. Feyfant E, Sali A, Fiser A. (2007) Modeling mutations in protein structures. *Protein Sci* **16**(9): 2030–41.
68. Wattenhofer M, Di Iorio MV, Rabionet R *et al.* (2002) Mutations in the TMPRSS3 gene are a rare cause of childhood nonsyndromic deafness in Caucasian patients. *J Mol Med* **80**(2): 124–31.
69. Qian B, Raman S, Das R *et al.* (2007) High-resolution structure prediction and the crystallographic phase problem. *Nature* **450**: 259–264.
70. Raimondo D, Giorgetti A, Giorgetti A *et al.* (2007) Automatic procedure for using models of proteins in molecular replacement. *Proteins* **66**(3): 689–96.
71. Barker JA, Thornton JM. (2003) An algorithm for constraint-based structural template matching: application to 3D templates with statistical analysis. *Bioinformatics* **19**(13): 1644–9.
72. Bartlett GJ, Porter CT, Borkakoti N, Thornton JM. (2002) Analysis of catalytic residues in enzyme active sites. *J Mol Biol* **324**(1): 105–21.
73. Laskowski RA, Thornton JM, Humblet C, Singh J. (1996) X-SITE: use of empirically derived atomic packing preferences to identify favourable interaction regions in the binding sites of proteins. *J Mol Biol* **259**(1): 175–201.
74. Rost B, Liu J, Nair R *et al.* (2003) Automatic prediction of protein function. *Cell Mol Life Sci* **60**(12): 2637–50.
75. Devos D, Valencia A. (2000) Practical limits of function prediction. *Proteins* **41**(1): 98–107.
76. Pizzi E, Tramontano A, Tomei L *et al.* (1994) Molecular model of the specificity pocket of the hepatitis C virus protease: implications for substrate recognition. *Proc Natl Acad Sci USA* **91**(3): 888–92.
77. Ausiello G, Via A, Helmer-Citterich M. (2005) Query3d: a new method for high-throughput analysis of functional residues in protein structures. *BMC Bioinformatics* **6** (Suppl 4): S5.
78. Ausiello G, Zanzoni A, Peluso D *et al.* (2005) pdbFun: mass selection and fast comparison of annotated PDB residues. *Nucleic Acids Res* **33**(Web Server issue): W133–7.

79. Via A, Peluso D, Gherardini PF *et al.* (2007) 3dLOGO: a web server for the identification, analysis and use of conserved protein substructures. *Nucleic Acids Res* **35**(Web Server issue): W416–9.
80. Duret L, Guex N, Peitsch MC, Bairoch A. (1998) New insulin-like proteins with atypical disulfide bond pattern characterized in *Caenorhabditis elegans* by comparative sequence analysis and homology modeling. *Genome Res* **8**(4): 348–53.
81. Peitsch MC, Jongeneel CV. (1993) A 3D model for the CD40 ligand predicts that it is a compact trimer similar to the tumor necrosis factors. *Int Immunol* **5**(2): 233–8.
82. Jimenez JL, Bashir R. (2007) *In silico* functional and structural characterisation of ferlin proteins by mapping disease-causing mutations and evolutionary information onto three-dimensional models of their C2 domains. *J Neurol Sci* **260**(1–2): 114–3.
83. Junne T, Schwede T, Goder V, Spiess M. (2006) The plug domain of yeast Sec61p is important for efficient protein translocation, but is not essential for cell viability. *Mol Biol Cell* **17**(9): 4063–8.
84. Junne T, Schwede T, Goder V, Spiess M. (2007) Mutations in the Sec61p channel affecting signal sequence recognition and membrane protein topology. *J Biol Chem* **282**(45): 33201–9.
85. Schwede TF, Badeker M, Langer M *et al.* (1999) Homogenization and crystallization of histidine ammonia-lyase by exchange of a surface cysteine residue. *Protein Eng* **12**(2): 151–3.
86. Peitsch MC, Tschopp J. (1995) Comparative molecular modelling of the Fas-ligand and other members of the TNF family. *Mol Immunol* **32**(10): 761–72.
87. Schneider P, Bodmer JL, Holler N *et al.* (1997) Characterization of Fas (Apo-1, CD95)–Fas ligand interaction. *J Biol Chem* **272**(30): 18827–33.
88. Hahne M, Peitsch MC, Irmler M *et al.* (1995) Characterization of the non-functional Fas ligand of gld mice. *Int Immunol* 7(9): 1381–6.
89. Notarangelo LD, Peitsch MC. (1996) CD40Lbase: a database of CD40L gene mutations causing X-linked hyper-IgM syndrome. *Immunol Today* **17**(11): 511–6.
90. Van den Berg B, Clemons WM Jr, Collinson I, Modis Y, Hartmann E, Harrison SC, Rapoport TA. (2004) X-ray structure of a protein-conducting channel. *Nature* **427**: 36–44.
91. Tramontano A. (2008) The biological applications of protein models. In: Schwede T, Peitsch MC (eds.). *Computational Structural Biology*. Singapore: World Scientific Publishing, pp. 111–128.
92. Karpusas M, Hsu YM, Wang JH *et al.* (1995) 2 Å crystal structure of an extracellular fragment of human CD40 ligand. *Structure* **3**(10): 1031–9.
93. Hou X, Chen M, Chen L *et al.* (2007) X-ray sequence and crystal structure of luffaculin 1, a novel type 1 ribosome-inactivating protein. *BMC Struct Biol* 7: 29.

94. Xiong AS, Peng RH, Cheng ZM *et al.* (2007) Concurrent mutations in six amino acids in beta-glucuronidase improve its thermostability. *Protein Eng Des Sel* **20**(7): 319–25.
95. Li H, Cocco MJ, Steitz TA, Engelman DM. (2001) Conversion of phospholamban into a soluble pentameric helical bundle. *Biochemistry* **40**(22): 6636–45.
96. Corthesy B, Kaufmann M, Phalipon A *et al.* (1996) A pathogen-specific epitope inserted into recombinant secretory immunoglobulin A is immunogenic by the oral route. *J Biol Chem* **271**(52): 33670–7.
97. Crottet P, Peitsch MC, Servis C, Corthesy B. (1999) Covalent homodimers of murine secretory component induced by epitope substitution unravel the capacity of the polymeric Ig receptor to dimerize noncovalently in the absence of IgA ligand. *J Biol Chem* **274**(44): 31445–55.
98. Morea V, Lesk AM, Tramontano A. (2000) Antibody modeling: implications for engineering and design. *Methods* **20**(3): 267–79.
99. Kim SJ, Park Y, Hong HJ. (2005) Antibody engineering for the development of therapeutic antibodies. *Mol Cells* **20**(1): 17–29.
100. Teillaud JL. (2005) Engineering of monoclonal antibodies and antibody-based fusion proteins: successes and challenges. *Expert Opin Biol Ther* **5**(Suppl 1): S15–27.
101. Kusharyoto W, Pleiss J, Bachmann TT, Schmid RD. (2002) Mapping of a hapten-binding site: molecular modeling and site-directed mutagenesis study of an anti-atrazine antibody. *Protein Eng* **15**(3): 233–41.
102. Nevanen TK, Hellman ML, Munck N *et al.* (2003) Model-based mutagenesis to improve the enantioselective fractionation properties of an antibody. *Protein Eng* **16**(12): 1089–97.
103. Ziegelmann-Fjeld KI, Musa MM, Phillips RS *et al.* (2007) A *Thermoanaerobacter ethanolicus* secondary alcohol dehydrogenase mutant derivative highly active and stereoselective on phenylacetone and benzylacetone. *Protein Eng Des Sel* **20**(2): 47–55.
104. Friesner RA, Repasky M, Farid R. (2008) Small molecule docking. In: Schwede T, Peitsch MC (eds.). *Computational Structural Biology*. Singapore: World Scientific Publishing, pp. 469–500.
105. Podvinec M, Schwede T, Peitsch MC. (2008) Docking for neglected diseases as community efforts. In: Schwede T, Peitsch MC (eds.). *Computational Structural Biology*. Singapore: World Scientific Publishing, pp. 683–704.
106. Rosenfeld R, Vajda S, DeLisi C. (1995) Flexible docking and design. *Annu Rev Biophys Biomol Struct* **24**: 677–700.
107. Waszkowycz B. (2002) Structure-based approaches to drug design and virtual screening. *Curr Opin Drug Discov Devel* **5**(3): 407–13.
108. Goodsell DS, Morris GM, Olson AJ. (1996) Automated docking of flexible ligands: applications of AutoDock. *J Mol Recognit* **9**(1): 1–5.

109. Ma B, Shatsky M, Wolfson HJ, Nussinov R. (2002) Multiple diverse ligands binding at a single protein site: a matter of pre-existing populations. *Protein Sci* **11**(2): 184–97.
110. Leban J, Baierl M, Mies J *et al.* (2007) A novel class of potent NF-kappaB signaling inhibitors. *Bioorg Med Chem Lett* **17**(21): 5858–62.
111. Scarsi M, Podvinec M, Roth A *et al.* (2007) Sulfonylureas and glinides exhibit peroxisome proliferator-activated receptor gamma activity: a combined virtual screening and biological assay approach. *Mol Pharmacol* **71**(2): 398–406.
112. Ragno R, Simeoni S, Castellano S *et al.* (2007) Small molecule inhibitors of histone arginine methyltransferases: homology modeling, molecular docking, binding mode analysis, and biological evaluations. *J Med Chem* **50**(6): 1241–53.
113. Lafite P, Andre F, Zeldin DC *et al.* (2007) Unusual regioselectivity and active site topology of human cytochrome P450 2J2. *Biochemistry* **46**(36): 10237–47.
114. Marchand-Geneste N, Cazaunau M, Carpy AJ *et al.* (2006) Homology model of the rainbow trout estrogen receptor (rtERalpha) and docking of endocrine disrupting chemicals (EDCs). *SAR QSAR Environ Res* **17**(1): 93–105.
115. Doman TN, McGovern SL, Witherbee BJ *et al.* (2002) Molecular docking and high-throughput screening for novel inhibitors of protein tyrosine phosphatase-1B. *J Med Chem* **45**(11): 2213–21.
116. Cavasotto CN, Orry AJ. (2007) Ligand docking and structure-based virtual screening in drug discovery. *Curr Top Med Chem* 7(10): 1006–14.
117. Pieper U, Eswar N, Davis FP *et al.* (2006) MODBASE: a database of annotated comparative protein structure models and associated resources. *Nucleic Acids Res* **34**(Database issue): D291–5.
118. Castrignano T, De Meo PD, Cozzetto D *et al.* (2006) The PMDB Protein Model Database. *Nucleic Acids Res* **34**(Database issue): D306–9.
119. Migliavacca E, Adzhubei AA, Peitsch MC. (2001) MDB: a database system utilizing automatic construction of modules and STAR-derived universal language. *Bioinformatics* **17**(11): 1047–52.
120. Berman HM, Burley SK, Chiu W *et al.* (2006) Outcome of a workshop on archiving structural models of biological macromolecules. *Structure* **14**(8): 1211–7.
121. Wu CH, Apweiler R, Bairoch A *et al.* (2006) The Universal Protein Resource (UniProt): an expanding universe of protein information. *Nucleic Acids Res* **34**(Database issue): D187–91.
122. Mulder NJ, Apweiler R, Attwood TK *et al.* (2007) New developments in the InterPro database. *Nucleic Acids Res* **35**(Database issue): D224–8.
123. Bhattacharya A, Wunderlich Z, Monleon D *et al.* (2008) Assessing model accuracy using the Homology Modeling Automatically software. *Proteins* **70**(1): 105–18.

124. Benkert P, Tosatto SC, Schomburg D. (2008) QMEAN: a comprehensive scoring function for model quality assessment. *Proteins* **71**(1): 261–77.
125. Cozzetto D, Kryshtafovych A, Ceriani M, Tramontano A. (2007) Assessment of predictions in the model quality assessment category. *Proteins* **69**(Suppl 8): 175–83.
126. Davis IW, Leaver-Fay A, Chen VB *et al.* (2007) MolProbity: all-atom contacts and structure validation for proteins and nucleic acids. *Nucleic Acids Res* **35**(Web Server issue): W375–83.
127. Weichenberger CX, Sippl MJ. (2007) NQ-Flipper: recognition and correction of erroneous asparagine and glutamine side-chain rotamers in protein structures. *Nucleic Acids Res* **35**(Web Server issue): W403–6.
128. Grosdidier A, Zoete V, Michielin O. (2007) EADock: docking of small molecules into protein active sites with a multiobjective evolutionary optimization. *Proteins* **67**(4): 1010–25.

Chapter 10

Molecular Modeling of Proteins: From Simulations to Drug Design Applications

Vincent Zoete, Michel Cuendet, Ute F. Röhrig, Aurélien Grosdidier and Olivier Michielin

1. Introduction

Over the last 30 years, molecular modeling has become an essential tool to study the properties of large molecular systems in various fields of chemistry, biology, and medicine. Nowadays, molecular modeling is routinely used to understand the molecular function of complex protein systems, or to perform rational protein and drug design. The range of examples includes the detailed understanding of ion conduction through the potassium channel,[1] the mechanism of action of the GroEL chaperonin,[2] electron transfer in DNA molecules,[3] the estimation of protein pKa's,[4] the design of improved immunoglobulin,[5] and the design of a small-molecule inhibitor of the Bcr-Abl aberrant tyrosine kinase.[6] As such, molecular modeling aims at bridging the gap between the static structure provided by X-ray or nuclear magnetic resonance (NMR) experiments and the biological function.

The molecular dynamics (MD) method was first introduced by Alder and Wainwright in the late 1950s[7,8] to simulate the dynamics of 150 argon atoms described as interacting hard spheres. Many important insights concerning the behavior of simple liquids emerged from their studies. The next major advance was in 1964, when Rahman carried out the first simulation using a realistic potential for liquid argon.[9] The first MD simulation of a realistic system was done by Rahman and Stillinger in their simulation of liquid water in 1974.[10] The first protein simulations appeared in 1977 with the simulation of the bovine pancreatic trypsin inhibitor (BPTI).[11] The delay between the first simulations of simple liquids and those of proteins reflects the difficulty in obtaining a reliable set of force field parameters for complex molecules like proteins.

In Sec. 2, we will present the CHARMM force field as an example of one of the most widely used semiempirical force fields. The different types of MD simulations that can be performed using such a force field are the object of Sec. 3. Free energy simulations, one of the most important applications of MD simulations, are presented in Sec. 4. Finally, Sec. 5 will cover some of the applications that are currently being pursued at the Molecular Modeling Group of the Swiss Institute of Bioinformatics (SIB).

2. Molecular Force Fields

In this section, we present the statistical mechanics foundations that provide the theoretical background behind MD simulations and, more generally, behind all molecular modeling techniques. Molecular modeling provides detailed knowledge of the *micro*scopic states of a system that can be converted into *macro*scopic values using statistical mechanics. For example, the macroscopic pressure of a gas is just the average force per unit area exerted by its particles as they collide with the container walls. Statistical mechanics provides, therefore, the link between the microscopic world of *in silico* simulations and the macroscopic quantities that can be directly used to compare or predict results from chemical or biological experiments. In what follows, we will put a major emphasis on the *affinity* between two molecules, one of the most important macroscopic quantities for biological and medical applications, and illustrate how this quantity can be computed using its statistical mechanics definition.

2.1. *Statistical Mechanics Connection*

It was shown by Boltzmann that all thermodynamic properties of a system at temperature T and confined in a volume V can be deduced from a central function, called the partition function, defined as

$$Z = \sum_i e^{-\beta Ei}, \tag{1}$$

where the summation runs over all microstates of the system $\beta = 1/K_B T$, in which K_B is the Boltzmann constant, T is the temperature, and E_i is the energy of microstate i. Note that this definition holds for discrete states and that a corresponding definition exists for continuous states, as will be developed below. One can show from first principles that most thermodynamic properties can be expressed in terms of the partition function. For example, pressure is expressed as

$$p = K_B T \left(\frac{\delta \ln Z}{\delta V} \right) \Bigg|_{N,T} \tag{2}$$

and the free energy G is expressed as

$$G = K_B T \ln(Z). \tag{3}$$

Most computations in molecular modeling aim at obtaining an approximation of the partition function. According to Eq. (1); this entails computing the energy of each microstate i, which will be the object of the following subsection on the CHARMM force field, as well as being able to sample all of the relevant microstates for the process under investigation, as hidden in the summation sign. The latter will be addressed in Sec. 3.

2.2. *The CHARMM Force Field*

The CHARMM force field was one of the first semiempirical atomic force fields to be tested successfully on proteins.[11] Nowadays, several general-purpose force fields are available, such as the AMBER, GROMOS, and OPLS force fields. Since most of the key features are fairly similar between these different implementations, we will describe the CHARMM force field in more detail below.

The all-atom PARAM 22 CHARMM force field[10] was parametrized using quantum calculations of small-molecule properties as well as experimental results. The bonded and nonbonded energy terms are illustrated in Fig. 1 and the full functional form of the potential is given in Eq. (4).

$$
\begin{aligned}
V(r) = {} & \frac{1}{2}\sum_{\text{Bonds}} K_b(b-b_0)^2 + \frac{1}{2}\sum_{\text{Angles}} K_\theta(\theta-\theta_0)^2 \\
& + \frac{1}{2}\sum_{\text{Impropers}} K_\omega(\omega-\omega_0)^2 \\
& + \frac{1}{2}\sum_{\text{Dihedrals}} K_\phi(1+\cos(n\phi-\delta)) \\
& + \sum_{ij} 4\varepsilon_{ij}\left[\left(\frac{\sigma_{ij}}{r_{ij}}\right)-\left(\frac{\sigma_{y}}{r_{ij}}\right)\right] + \sum_{i,j}\frac{q_i q_j}{\varepsilon r_{ij}}.
\end{aligned} \tag{4}
$$

The first four terms of Eq. (4) are collectively referred to as bonded energy terms; and the last two, the van der Waals and electrostatic terms, are called nonbonded energy terms. The bonded energy terms describe the energy associated with the covalent structure of the molecules. The nonbonded terms have to be computed for all possible atom pairs in the system, resulting in a numerically expensive N^2 computation, where N is the number of atoms. To limit the computation cost, cut-offs are frequently introduced, where atom pairs distant of more than a certain value are not considered interacting. Compared to nonbonded interactions, the cost of bonded terms is negligible.

The van der Waals energy is described by the Lennard-Jones term, which is the fifth term in Eq. (4); and accounts for the unspecific interaction between all atoms. The attractive term results from fluctuations in the electronic cloud that induces an instantaneous dipole momentum, which polarizes the electronic cloud of neighboring atoms or molecules and vice versa. The net effect of the dipole–dipole interaction is attractive

$$K_b(b-b_0)^2$$

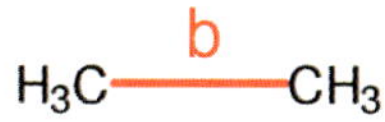

$$K_\theta(\theta-\theta_0)^2$$

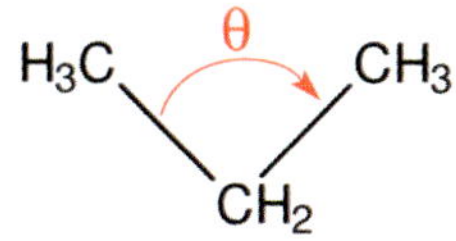

$$K_\phi\left[1+\cos(n\phi-\delta)\right]$$

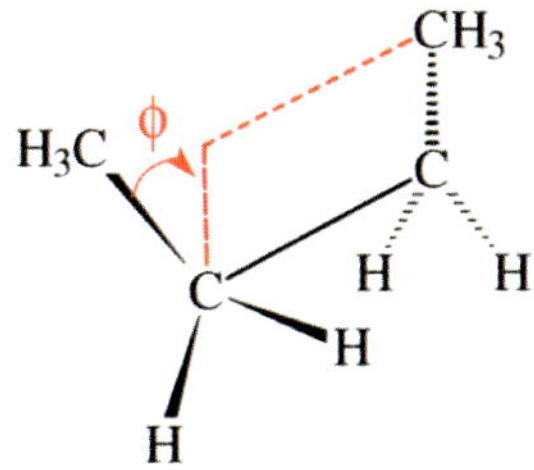

$$\frac{q_i\, q_j}{\varepsilon r_{ij}}$$

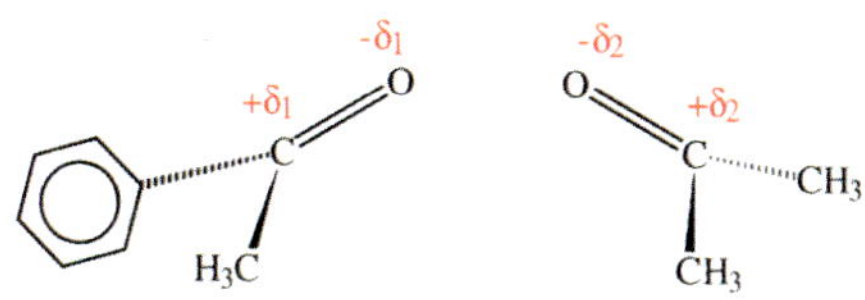

$$4\varepsilon_{ij}\left[\left(\frac{\sigma_{ij}}{r_{ij}}\right)^{12}-\left(\frac{\sigma_{ij}}{r_{ij}}\right)^{6}\right]$$

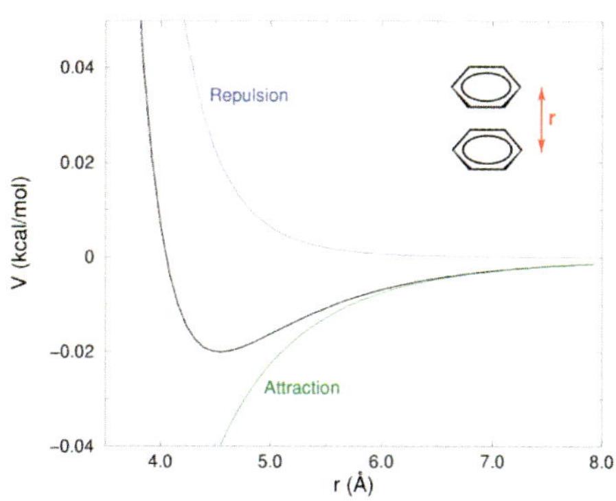

Fig. 1. The CHARMM force field. Each energy term listed on the left enters in the computation of the total energy at each point of an MD simulation.

and is called the van der Waals interaction. The short-range repulsive term results from the Pauli exclusion principle.

The electrostatic term, which is the sixth term in Eq. (4); accounts for the interaction between point charges located at the atomic nuclei. This functional form is an approximation, since the molecular electronic density is sometimes far from being spherical, as in covalent bonds. The electrostatic potential is the longest-range interaction with a $1/r$ dependence; and its treatment requires specific techniques to lower the computational cost, like the Ewald summation.[12]

Several limitations are inherent to the functional form of $V(r)$ given in Eq. (4). First of all, the partial charges of all atoms are fixed at the beginning of the simulation. This approach clearly neglects the electronic polarization, though the conformational polarization is still present, notably in the water molecules that solvate the system. Polarizable force fields are now emerging[13] and might become the standard in a few years. Second, the quadratic nature of the bond-stretching energy terms does not allow bond braking or bond creation. This strongly limits the use of this approach for the study of enzymatic reactions with covalent bond modifications. To address this limitation, the most common approach is to use Morse potentials[14] or a hybrid quantum mechanics/molecular mechanics (QM/MM) approach, where the system is divided into a quantum part in which bond creation takes place[15] and a classical part surrounding the quantum one.

Despite their inherent simplicity and evident shortcomings, force fields have been used successfully over the last few decades and will provide the basis of molecular modeling for several decades to come. The improvement of semiempirical methods still represents a very active domain of research.

3. Molecular Dynamics Simulations

3.1. *Integration of the Equation of Motion*

Given the potential energy function V of Eq. (4); one can compute the force acting on atom i:

$$F_i = -\frac{\delta V}{\delta r_i}. \tag{5}$$

Using Newton's equation, one gets the corresponding acceleration a_i:

$$-\frac{\delta V}{\delta r_i} = m_i a_i. \tag{6}$$

Starting from given initial positions and after assigning random velocities, one can propagate the system by numerical integration of Eq. (6). Such an integrator can be obtained by considering a Taylor expansion of the position with respect to small time increments δt:

$$r_i(t+\delta t) = r_i(t) + v_i(t)\delta t + \frac{1}{2}a_i(t)\delta t^2 \tag{7}$$

$$r_i(t-\delta t) = r_i(t) - v_i(t)\delta t + \frac{1}{2}a_i(t)\delta t^2. \tag{8}$$

Summing the two equations, we obtain the so-called Verlet algorithm:

$$r_i(t+\delta t) = 2r_i(t) - r_i(t-\delta t) + a_i(t)\delta t^2. \tag{9}$$

Thus, knowledge of the system at times $t - \delta t$ and t allows us to obtain the positions at time $t + \delta t$. Velocities can be subsequently computed based on the positions:

$$v_i(t) = \frac{r_i(t+\delta t) - r_i(t-\delta t)}{2\delta t}. \tag{10}$$

Note that these average velocities are computed for time t, whereas the positions in Eq. (9) were obtained for time $t + \delta t$. This inaccuracy can be accounted for in other implementations.

Repeating this procedure, one can simulate the complete time evolution of the system, as represented in Fig. 2. More sophisticated algorithms have been developed following similar principles, like the velocity Verlet, leap-frog, and Beeman algorithms.

A general protocol to run an MD simulation starting from a given molecular structure is presented in Fig. 3. The initial coordinates

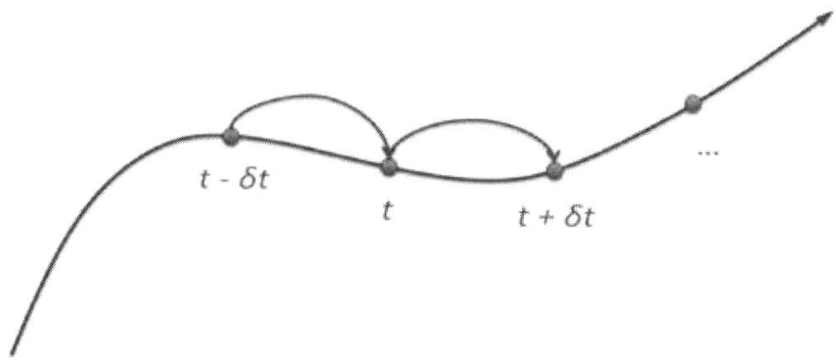

Fig. 2. Graphical representation of the stepwise time solution obtained by the Verlet algorithm.

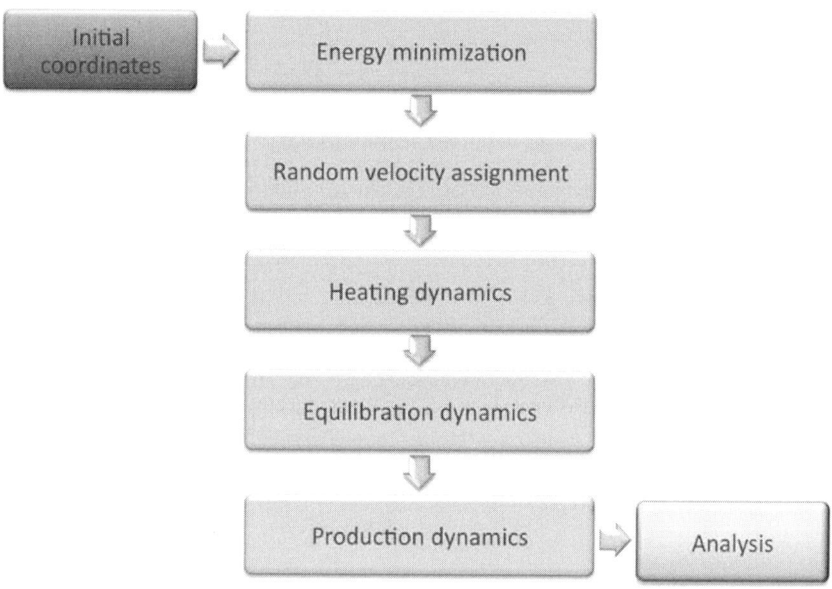

Fig. 3. Main steps involved in the MD simulation of a given structure.

obtained from an experimental structure (X-ray or NMR) or a theoretical model (homology modeling or *ab initio* folding) are first minimized to prevent large unphysical motions at the start of the dynamics resulting from potential steric clashes. Initial velocities are assigned randomly using Gaussian distribution at a low temperature, usually around 100K. The system is then heated up to the desired temperature using various techniques (e.g. scaling of the velocities, equilibration with a heat bath) and equilibrated to allow energy redistribution. After all of these steps, the equilibrated system can be studied and the production dynamics started.

3.2. *Thermodynamic Ensembles*

Newton's equation (6), where F_i represent interatomic forces, describes the evolution of an isolated system. The associated integrator [Eq. (9)] generates trajectories at a fixed energy E determined by the initial conditions. Let the Hamiltonian function

$$H(r,p) = \sum_{1} \frac{p_i^2}{2m_i} + V(r)$$

represent the instantaneous energy of the system. We say that the Newtonian dynamics (or the equivalent Hamiltonian dynamics) generates the NVE or microcanonical ensemble. This means that allowed system conformations (r, p) in phase space are restricted to a hypersurface of constant energy E, described by the microcanonical density of states

$$\rho_{\mathrm{NVE}}(r,p) \propto \delta[H(r-p-E)].$$

However, many laboratory experiments are conducted in solution, where the system is thermally coupled to its environment. In this case, the temperature of the system not its energy, is fixed. This corresponds to the NVT or canonical ensemble. A fundamental reason to study systems coupled to heat baths is that isolated systems cannot display any dissipative behavior. For the system to be able to relax to a state of maximum entropy, as the second law of thermodynamics describes for all macroscopic systems, there has to be a way to exchange energy with the environment. Indeed, the canonical distribution function can be defined as the distribution maximizing the entropy[16] and is expressed as

$$\rho_{\mathrm{NVT}}(r,p) \propto e^{-\beta H(r,p)}.$$

The Helmholtz free energy F corresponding to the canonical ensemble is

$$F = U - TS. \tag{11}$$

where U represents the average energy of the system, and S the entropy.

A number of factors impede a precise computer simulation of the NVT statistical mechanical ensemble. First, one can obviously not simulate a realistic heat bath with a large number of degrees of freedom. Second, the nature of the coupling between the system and the bath, which is deliberately disregarded in standard statistical mechanical theory, needs to enter explicitly the MD equations of motion. The modified dynamics that serve this purpose are called thermostats.

A thermostat maintains the average temperature of the system at T by absorbing any excess energy that might appear in the simulation. Spurious energy sources are due to various inaccuracies in the MD algorithm used. These include the use of cut-offs for long-range interactions, the skipping of time steps for determination of atom pairs within the cut-off, the use of a special algorithm to constrain bond length, and numerical drift if the equations of motion are integrated with large time steps. All of these effects tend to heat up the system. If, in addition, work is performed by an external agent to drive a given process, dissipation will tend to increase the temperature. A thermodynamic picture of an MD system coupled to a thermostat is shown in Fig. 4. There, the energy variation ΔU of the system expected from Eq. (11) is mediated by its different couplings to the outside. These couplings are expressed as different kinds of work: the external work W_{ext}, the work W_{thermo} provided by the thermostat, or W_{MD} resulting from the side effects of the algorithms used.

An example of thermostating dynamics is the Nosé–Hoover (NH) thermostat.[17] In addition to the physical degrees of freedom (r, p), an

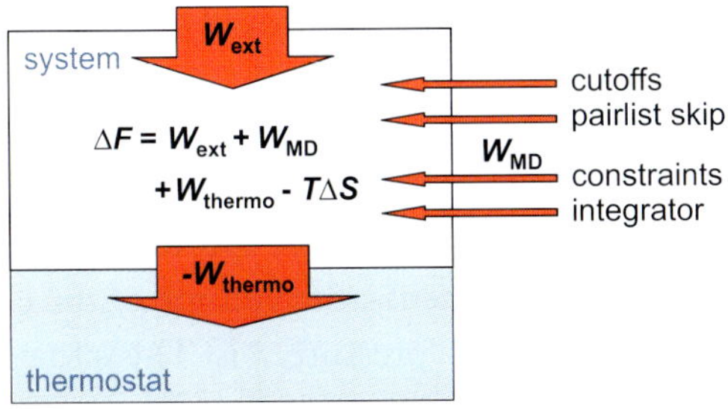

Fig. 4. Schematic representation of a thermostated MD system. Red arrows represent heat or work flows.

auxiliary variable ζ is introduced that plays the role of a time-dependent friction coefficient. The resulting dynamics is no longer Newtonian:

$$r_i = \frac{p_i}{m_i}$$

$$p_i = -\frac{\delta V}{\delta r_i} - \frac{\zeta}{Q} p_i$$

$$\zeta = \sum_i \frac{p_i^2}{m_i} - N_{dj} k_B T.$$

Here, Q is a pseudomass determining the time scale of the thermostat and N_{df} is the number of degrees of freedom in the system. We see that the time derivative of ζ is essentially determined by the difference between the instantaneous temperature (or kinetic energy) of the system and the target temperature T. Note that the NH equations of motion are not Hamiltonian. The main property of the NH dynamics, however, is that it is formally proven to produce a canonical distribution for the physical degrees of freedom (r, p).[18] The NH thermostat can be improved in various ways, including chains of thermostats for better ergodicity or control of higher moments of the velocity distribution.

Besides the NH thermostat, different types of thermostats have been proposed,[19] one of the first being the weak coupling or Berendsen thermostat.[20] The Berendsen thermostat is based on a first-order (exponential) relaxation of the system temperature, as opposed to the second-order (oscillatory) relaxation of the NH thermostat. Although the Berendsen thermostat was never formally shown to reproduce any known thermodynamical ensemble,[21] it is still widely used in MD applications because of its fast convergence and the absence of spurious oscillations.

Often in laboratory experiments, the volume of the experimental system is not fixed, but the external pressure P is. This situation corresponds to the isobaric-isothermic or NPT ensemble. The corresponding distribution function is

$$\rho_{\mathrm{NPT}}(r, p) \propto e^{-\beta(H(r,p)+pV)}.$$

The NPT ensemble can be generated for explicit solvent simulations, in which the volume of the periodic box containing the system is allowed to fluctuate. The size of the system becomes a dynamical variable controlled by a feedback mechanism adjusting the instantaneous pressure to the reference pressure P, in a fashion similar to the NH temperature control.

4. Free Energy Calculations

Free energy represents the most important quantity to describe the behavior of a molecular system. The probabilities of the different states of a system are, indeed, directly related to the value of their free energy. In the case of proteins, for example, the conformational change between two states, the folding process, the association between two monomers, or the affinity of a small molecule for its receptor are all described by the free energy. For this reason, much effort has been devoted to the development of computational methods that allow reliable estimates of this quantity for a given molecular system and a given process under investigation.

The theoretical foundations of free energy simulations can, to a large extent, be attributed to John Kirkwood for his pioneering developments in the 1930s on the computation of free energy differences using thermodynamic integration (TI),[22] and to Robert Zwanzig for the free energy perturbation (FEP) method.[23] The first applications of this formalism to biological problems came in the early 1980s with the work of Tembe and McCammon on protein–ligand binding,[24] followed by Peter Kollman and coworkers with the first alchemical simulations (see section below) to estimate the binding free energy difference between a wild-type and a point-mutated protein.[25] Since these early days, free energy simulation techniques have been the subject of intense research efforts. Only recently have these methods become reliable due, on the one hand, to the better sampling provided by the more powerful computers available today, but, more importantly, to improved theoretical approaches with better convergence properties. In this section, we will review some of the basic methods used in the field as well as the most recent theoretical developments. This survey is not exhaustive, but is centered on the techniques that are most often used by the Molecular Modeling Group to address the biological problems encountered in the development of new cancer therapeutic agents.

Consider a well-defined state A described by the potential energy function $V_A(r)$ and the corresponding Hamiltonian $H_A(r, p)$. For a given number N of particles at constant volume and temperature T, state A is described by the partition function

$$Z_A = \frac{1}{h^{3N} N!} \int e^{-\beta H_A(r,p)} \mathrm{d}r \mathrm{d}p,$$

where $\beta = K_B T$. The normalization constant contains Plank's constant h, a measure of the elementary volume in phase space, and the factor $N!$, which should be present only when the particles are indistinguishable. The statistical mechanics definition of the free energy of a system in a given state A is

$$G_A = -k_B T \ln Z_A.$$

In complex systems, however, such absolute free energies are intrinsically impossible to compute because the partition function is essentially a measure of the full configuration space accessible to the system. In experiments as well as in simulation, free energies are always computed relatively to a reference state. Let state B be described by H_B and characterized by Z_B. The free energy difference between two states A and B is given by a ratio of partition functions:

$$\Delta G_{AB} = -k_B T \ln \frac{Z_B}{ZA}.$$

The main idea behind the methods presented below is to avoid direct computation of the individual partition functions Z_A and Z_B by using the fact that the variations between states A and B of interest are often localized in relevant regions of the configuration space; elsewhere, the corresponding partition functions Z_A and Z_B have a high degree of similarity. Most approaches correspond, therefore, to reformulating Eq. (12) such that common parts of Z_A and Z_B not directly relevant to the process under investigation cancel out. A fundamental aspect of these approaches is that they express, as we will see, the free energy difference in terms of an ensemble average, which can be directly measured or calculated in a simulation, unlike an absolute free energy.

In Sec. 4.1, we present methods derived from first principles that are exact at the statistical mechanics level. For these methods, the quality of the results (for a given model or force field) depends mainly on the quality of the sampling and on convergence properties. In Secs. 4.2 and 4.3, approximate methods will be described. These methods are not exact at the statistical mechanics level, but do show interesting convergence properties that make them very useful in some applications.

4.1. *Exact Statistical Mechanics Methods for Free Energy Differences*

Here, we briefly derive the free energy perturbation (FEP) and thermodynamic integration (TI) expressions, which can be applied in MD or in Monte Carlo simulations to calculate free energy differences.[26] In the following, λ is an external parameter changing the functional

As an illustration of endpoint methods, we will review the molecular mechanics–Poisson-Boltzmann surface area (MM-PBSA) models.[37,38]

MM-PBSA

In MM-PBSA, ΔG_{bind} is written as the sum of the gas phase contribution, $\Delta H^{\text{gas}}_{\text{bind}}$; the energy difference due to translational and rotational degrees of freedom, $\Delta H_{\text{trans/rot}}$; the desolvation free energy of the system upon binding, ΔG_{desolv}; and an entropic contribution, $-T\Delta S$[37,38]:

$$\Delta G_{\text{bind}} = \Delta H^{\text{gas}}_{\text{bind}} + \Delta H_{\text{trans/rot}} + \Delta G_{\text{desolv}} - T\Delta S.$$

The term $\Delta H^{\text{gas}}_{\text{bind}}$ contains the van der Waals and electrostatic interaction energies between the two partners in the complex, as well as the internal energy variation (including bond, angle, and torsional angle energies) between the complex and the isolated molecules, ΔH_{intra}. In the classical limit, $\Delta H_{\text{trans/rot}}$ is equal to 3RT; this constant term is generally omitted in MM-PBSA calculations. ΔG_{desolv} is the difference between the solvation free energy, ΔG_{solv}, of the complex and that of the isolated parts. ΔG_{solv} is divided into the electrostatic, $\Delta G_{\text{elec,solv}}$, and the nonpolar, $\Delta G_{\text{np,solv}}$, contributions:

$$\Delta G_{\text{solv}} = \Delta G_{\text{elec,solv}} + \Delta G_{\text{np,solv}}.$$

In MM-PBSA, $\Delta G_{\text{elec,solv}}$ is calculated by solving the Poisson or the Poisson–Boltzmann equation,[39,40] depending on whether the salt concentration is zero or nonzero. Recently, an approach related to MM-PBSA, where $\Delta G_{\text{elec,solv}}$ is determined using a generalized Born (GB) model,[41] has been introduced as molecular mechanics–generalized Born surface area (MM-GBSA).[41,42] Despite its approximations, the GB model variant is attractive since it is much faster than the PB model variant. Recent advances of GB models[43,44] in reproducing the PB solvation energies of macromolecules as well as desolvation energies upon binding further support the use of GB models in this context.[45] The term $\Delta G_{\text{np,solv}}$, which can be considered as the sum of a cavity term and a solute–solvent van der Waals term, is assumed to be proportional to the solvent-accessible surface area (SASA):

$$\Delta G_{\text{np,solv}} = \gamma SASA + b.$$

This well-known and often-used approximation comes from the fact that the ΔG_{solv} of saturated nonpolar hydrocarbons is linearly related to the SASA.[46,47] Several linear models exist. The surface tension γ and the constant b can be set to 0.00542 kcal mol^{-1} $Å^{-2}$ and 0.92 kcal mol^{-1}, respectively, if $\Delta G_{\text{elec,solv}}$ is calculated from PB models.[48] Values of 0.0072 kcal mol^{-1} $Å^{-2}$ and 0 kcal mol^{-1},[49] or 0.005 kcal mol^{-1} $Å^{-2}$ and 0 kcal mol^{-1}, can be used together with GB models.[50] Recently, an alternative model for $\Delta G_{\text{np,solv}}$ using a cavity solvation free energy term plus an explicit solute–solvent van der Waals interactions energy term has been tested; this model led to better results in estimating ΔG_{bind} for the Ras–Raf association, although the transferability of the results was questioned.[50]

The entropy term, due to the loss of degrees of freedom upon association, is decomposed into translational (S_{trans}); rotational, (S_{rot}); and vibrational (S_{vib}); contributions. These terms are calculated using standard equations of statistical mechanics.[51] S_{rot} is a function of the moments of inertia of the molecule, whereas S_{trans} is a function of the mass and the solute concentration. S_{trans} is the only term in the free energy of an ideal solution that depends on solute concentration, leading to the concentration dependence of the binding reactions. S_{vib} is calculated with the quantum formula from a normal mode analysis (NMA).[51] A quasi-harmonic analysis of the MD simulations is also possible; however, it has been found that it does not always yield convergent values, even using very long MD simulation trajectories, and also led to large deviations from the results obtained with NMA, giving an overall unreasonable entropic contribution.[50]

In the standard MM-PB (GB)SA protocol, the energy terms are averaged over 200–500 frames extracted from MD simulation trajectories, typically performed in explicit solvent, using periodic or stochastic boundary conditions. Explicit water molecules are removed prior to energy calculations, since the solvent effect is described according to a PBSA or GBSA implicit solvent model. More recently, some studies also performed the MD simulations using implicit solvent models.[52,53] The normal modes are usually calculated on a smaller number of frames, due to the central processing unit (CPU) requirement of such calculations. Short 0.5–1-ns trajectories are generally performed, yielding to converged energy terms. Longer simulations have been tested, up to 10 ns in length,[50] but they were not found to provide better results, most probably because long simulations emphasize force field errors and limitations. Indeed, it has been found that MM-PBSA yields better results with MD simulations restrained around the X-ray structure, compared to unrestrained simulations.[54]

Two possibilities arise concerning the number of MD simulations to perform. In principle, one should make three trajectories, one for the complex and each of the isolated partners, and calculate the energy terms using the adequate simulation. However, a popular alternative consists of performing only one MD simulation for the complex. In this variant, the terms relative to one isolated partner are calculated after removing the atoms of the other partner in the frames extracted from the MD simulation of the complex. As a consequence, the reorganization energy of the molecules upon association is neglected ($\Delta H_{\text{intra}} = 0$); however, this variant is less CPU-demanding and leads to increased convergence due to cancellation of errors, reduction of noise arising from flexible remote regions relative to the binding site, and conformational restraints imposed by the complex geometry. Thus, this one-simulation variant is attractive when ΔH_{intra} may be reasonably neglected. Comparisons between one- and three-trajectory results can be found in the literature.[36,45,54]

MM-PB(GB)SA is expected to estimate absolute ΔG_{bind} without adjustable parameters. Although several studies were able to reproduce experimental ΔG_{bind} for protein–protein association with an error lower than 2 kcal mol^{-1},[45,50] these results are open to discussion. Indeed, the approach contains several "hidden" parameters, such as the force field used, the choice of the PB or GB variant and that of the nonpolar solvation model, the use of one or three trajectories, and the different terms that can be included or neglected. As a consequence, it is sometimes possible to find a combination of such hidden parameters apparently allowing a fine estimation of ΔG_{bind} for a given system. Of course, the transferability of such results to other systems is questionable. Nevertheless, MM-PB(GB)SA proved to be useful for several applications less sensitive to the choice of hidden parameters, such as the determination of relative affinities for different small ligands in DD applications, comparison of relative stabilities of macromolecular conformations, and estimation of the effect of mutations on association processes and fold stability.

Although some studies aimed at determining absolute ΔG_{bind} for ligand–protein association, MM-PB(GB)SA is generally used to estimate relative affinities for different ligands targeting the same protein. This allows additional approximations, like the neglect of the entropy terms for ligands of similar masses binding to the same site. Also, despite the fact that this approach is expected to tackle chemically diverse ligands, it is often applied to a series of chemically related ligands. This clearly simplifies the problem thanks to the additional cancellation of errors, but it also reflects the usual DD processes that generally focus on families of similar ligands. A recent and detailed review of the numerous studies using MM-PB(GB)SA in the context of DD can be found elsewhere.[55] MM-PB(GB)SA has given variable results, ranging from poor correlations between experimental and calculated ΔG_{bind} to very good ones, with correlation coefficients up to 0.96.[55] The performance seems to be a function of the nature of the targeted protein and of the range of activities encompassed by the ligands. Not surprisingly, the ranking is better for a broader range of affinities.[56]

MM-PB(GB)SA has been found to perform well at determining the effect of mutations on association processes, and at identifying the "hot spots" of protein–protein complexes.[42,50,57–60] Two main approaches exist. First, it is possible to perform a so-called computational alanine scanning (CAS).[57,58] The alanine mutation is introduced by modifying the frames extracted from the MD simulation of the wild-type system. The difference in ΔG_{bind} between the wild-type system and the mutants may be compared directly to the results of an experimental alanine scanning (AS).[57,58] The second possibility is to perform a binding free energy decomposition (BFED)[42] for the wild-type system. This process aims at calculating the contributions to ΔG_{bind} arising from each atom or groups of atoms (typically side chains). Like CAS, BFED also identifies the nature of the energy change in terms of interaction and solvation energies, or entropic contributions. A detailed description of the BFED process can be found elsewhere.[42,59]

The MM-GBSA variant is attractive for BFED, not only because it is much faster than MM-PBSA, but also because the pairwise nature of the GB equation allows the decomposition of $\Delta G_{elec,slov}$ into atomic contributions in a straightforward manner.[42] It is, however, interesting to note that the decomposition of a PB-calculated $\Delta G_{elec,slov}$ can also be performed,[60] though it is more computationally demanding. Although its results cannot be compared directly to an experimental AS, BFED offers a faster alternative to CAS, since it only requires one binding free energy calculation. Also, it allows studying the contributions from nonmutable groups of atoms, such as backbone atoms. In addition, contrary to CAS, BFED is a nonperturbing approach that does not require introducing a mutation into the system. A comparison between CAS and BFED results can be found in Zoete and Michielin.[59] Obviously, these methods cannot be expected to provide results exactly comparable to values obtained from an experimental AS, since they both neglect the effect of the mutations on the protein conformation. However, fair agreements between the experimental and theoretical results have been found in several studies and open the way to rational protein engineering.[42,58,60,61]

It has been found that the side chain contributions to S_{vib} play an important role, and increase the quality of the correlation between experimental and calculated energy changes.[42] A theoretically exact way to calculate the contribution of a given group of atoms to S_{vib} is to zero their mass and recalculate the normal modes and the corresponding total entropy.[42] The difference between the wild-type system S_{vib} and that of the system with some zeroed masses gives the contribution of the corresponding atoms. This approach is very time-consuming, since it requires an NMA for each group of atoms. Consequently, the entropic contribution is often neglected in such studies or calculated for the most important residues only. However, a new vibrational entropy decomposition scheme has been introduced recently to circumvent this problem: the linear decomposition of the vibrational entropy (LDVE) approach,[59] which necessitates only one NMA for the wild-type system and is thus much faster. It is based on the idea that the most important contributions to S_{vib} originate from side chains which contribute most to the vibrational amplitude.

Recently, the CMEPS (computational mutations to estimate protein stability) approach[61] has been introduced. It uses MM-GBSA calculations to study the impact of mutations on protein structural stability and determine the most important residues for the protein fold. It is based on the notion that the ΔG_{bind} corresponding to the alchemical complexation of a given side chain (considered as a pseudo-ligand) into the rest of the protein (considered as a pseudo-receptor) reflects the importance of this side chain to the thermodynamic stability of the protein. This method has been applied successfully to the study of insulin,[61] p53,[62] and PPAR[63] structural stability.

5. Examples of Applications

To illustrate the application of some of these techniques to biological and medical questions, we will review some projects recently developed at the Molecular Modeling Group of the SIB.

5.1. *Protein Design*

The specific cellular immune response is based on the recognition by cytotoxic T lymphocytes (CTLs) of an immunogenic peptide (p) presented at the surface of the target cell by a class I major histocompatibility complex (MHC; see Fig. 6). Binding of the T cell receptor (TCR) to the p–MHC complex results in activation of the CTLs and target cell destruction. Because of the central role of the T lymphocyte in the immune response, the molecular basis of the interaction between the TCR and the p–MHC is of general interest in immunology and medicine, and more specifically in cancer immunotherapy. The major determinant of T cell activation is the affinity of the TCR for the p–MHC complex, though kinetic parameters are also important. Therefore, methods aimed at understanding the role played by each residue in the recognition between the different components of the TCR–p-MHC complex, not

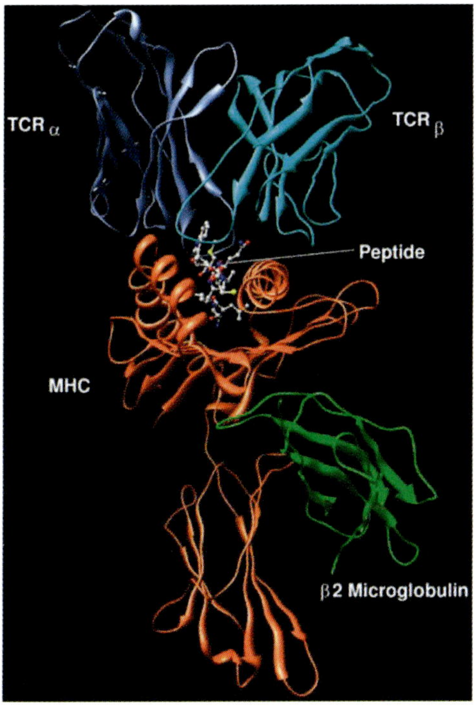

Fig. 6. Structure of a complex between a TCR and a p–MHC. The peptide, shown in the context of the MHC molecule, is in ball-and-stick representation.

only at the structural level but also from a thermodynamic point of view, are of major interest. Such molecular modeling approaches can be used to guide new experimental investigations, and to rationally optimize tumor-specific TCRs.

Recently, we performed a study of the 2C TCR/SIYR/H-2Kb system using a binding free energy decomposition based on the MM-GBSA approach (see above). The results show that the TCR–p-MHC binding free energy decomposition including entropic terms provides a detailed and reliable description of the interactions between the TCR and p–MHC molecules at an atomistic level.[59] Comparison of the calculated binding free energy changes upon mutation with experimentally determined activity differences for alanine mutants yields a correlation of 0.67 when the entropy is neglected, and 0.72 when the entropy is taken into account.

We are now applying the method to study the interactions between TCRs and cancer-related p–MHC systems, like, for instance, the histocompatibility leukocyte antigen (HLA)-A2 tumor epitope NY-ESO-1. NY-ESO-1 is a cancer testis antigen peptide expressed not only in melanoma, but also in several other types of cancers. It has been observed at high frequencies in melanoma patients with unusually positive clinical outcome and, therefore, represents an interesting target for cancer immunotherapy using adoptive transfer with modified TCRs. As explained in the previous section, the MM-GBSA approach not only allows decomposition of the calculated binding free energy between two proteins into contributions coming from the different residues, but also provides information about which energy terms are responsible for a residue's contribution, in terms of van der Waals or electrostatic interactions, for instance. Therefore, the results of the MM-GBSA approach identify those TCR residues that are more or less important for the association process with p–MHC, and those that could be mutated. It also helps in designing sequence modifications that are expected to increase the affinity and selectivity of the protein–protein binding.

Sequence mutations of TCRs potentially increasing the affinity for this epitope have been proposed based on this theoretical approach and tested experimentally. A large number was found experimentally to increase the affinity of the TCR for this p–MHC, showing a qualitative agreement between the theory and experiments, and opening the way to adoptive transfer immunotherapy. Successfully predicted mutations include drastic changes like, for instance, mutations from alanine to glutamate, illustrating the efficiency of the approach. In addition, a significant quantitative agreement was also found between the binding free energy change upon mutation calculated using the theoretical approach, and the change in affinity between TCR and p–MHC measured using an ELISA experiment for the different mutations, with a correlation of about 0.80 (see Fig. 7). The predictive ability of our approach shows that a rational fine-tuning of the TCR sequence can be obtained.

5.2. *Drug Design*

The heme-containing enzyme indoleamine 2,3-dioxygenase (IDO; EC 1.13.11.52) has been implicated in the establishment of pathological

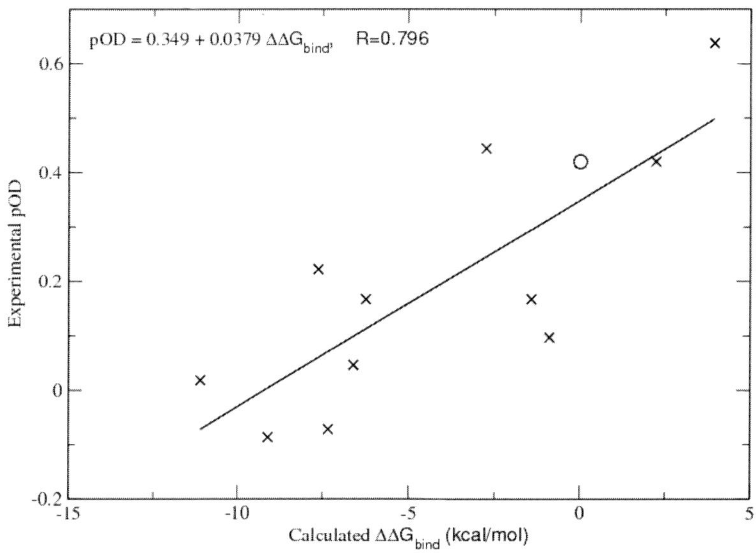

Fig. 7. Correlation between the cologarithm of the optical density (pOD) determined experimentally by ELISA for a series of rationally designed TCR mutants (Melita B. Irving, private communication) and the binding free energy change upon the corresponding mutation ($\Delta\Delta G_{bind}$) calculated using the MM-GBSA approach. Higher-affinity mutants are characterized by low pOD and negative $\Delta\Delta G_{bind}$ values. The circle corresponds to the wild-type TCR.

immune tolerance by tumors. IDO catalyzes the initial and rate-limiting step in the catabolism of tryptophan (Trp) along the kynurenine pathway.[64] By depleting Trp locally, IDO blocks the proliferation of T lymphocytes, which are extremely sensitive to Trp shortage. The observation that many human tumors constitutively express IDO introduced the hypothesis that its inhibition could enhance the effectiveness of cancer immunotherapy. Results from *in vitro* and *in vivo* studies suggest that the efficacy of therapeutic vaccination of cancer patients may indeed be improved by concomitant administration of an IDO inhibitor.[65,66]

Most known IDO inhibitors display affinities in the micromolar range, but recently some submicromolar inhibitors have been discovered.[67–69] The crystal structures of human IDO[70] complexed with 4-phenylimidazole (PIM) can serve as a scaffold for the *in silico* design of new and more potent IDO inhibitors.

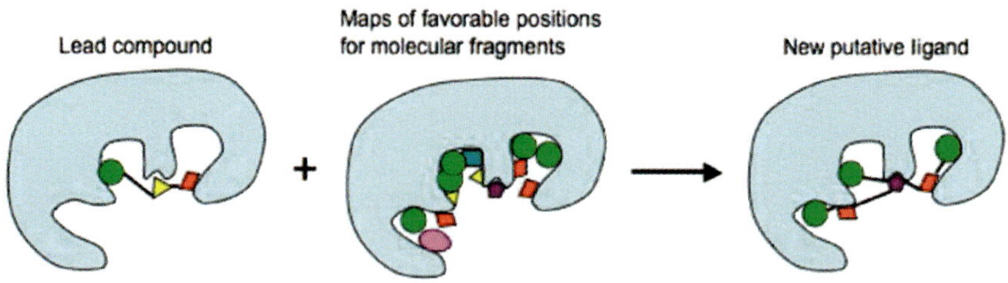

Fig. 8. The principle of fragment-based drug design (DD): chemical building blocks are docked separately into an active site. New putative ligands are constructed from these maps of favorable poses by linking several building blocks together or eventually to a known lead compound according to geometric constraints.

We are using a fragment-based approach to design small organic IDO ligands. This approach is based on the work of Bemis and Murcko,[71,72] who analyzed commercially available drugs and came up with a limited set of a few tens of molecular frameworks and side chains that are able to describe the chemical space of many drugs. In an *in silico* drug design approach, this concept can be used by separately docking these fragments into the protein active site (Fig. 8). For each fragment, a map with the 50 most favorable poses in the IDO active site is produced using EADock.[73] Geometric constraints, which are derived from molecular bond lengths, bond angles, and dihedral angles, are then used to determine whether several fragments can be linked together to create a new virtual compound. If so, the putative ligand is subjected to the docking procedure to check if it adapts the intended binding mode and makes favorable interactions. This approach can be adopted for both virtual lead design and optimization.

A strong hint for the validity of this approach is given by the rediscovery of the IDO ligand PIM from the X-ray structure[70] (Fig. 9). When the most favorable pose of the imidazole and the second most favorable pose of the phenyl rings are linked, the structure of the PIM-bound X-ray structure is reproduced with a root mean square deviation (RMSD) lower than 1 Å.

So far, we have identified 21 promising candidate ligands that we have tested experimentally. Thirteen out of these 21 molecules have shown some IDO inhibitory activity in an *in vitro* enzymatic assay. This high success

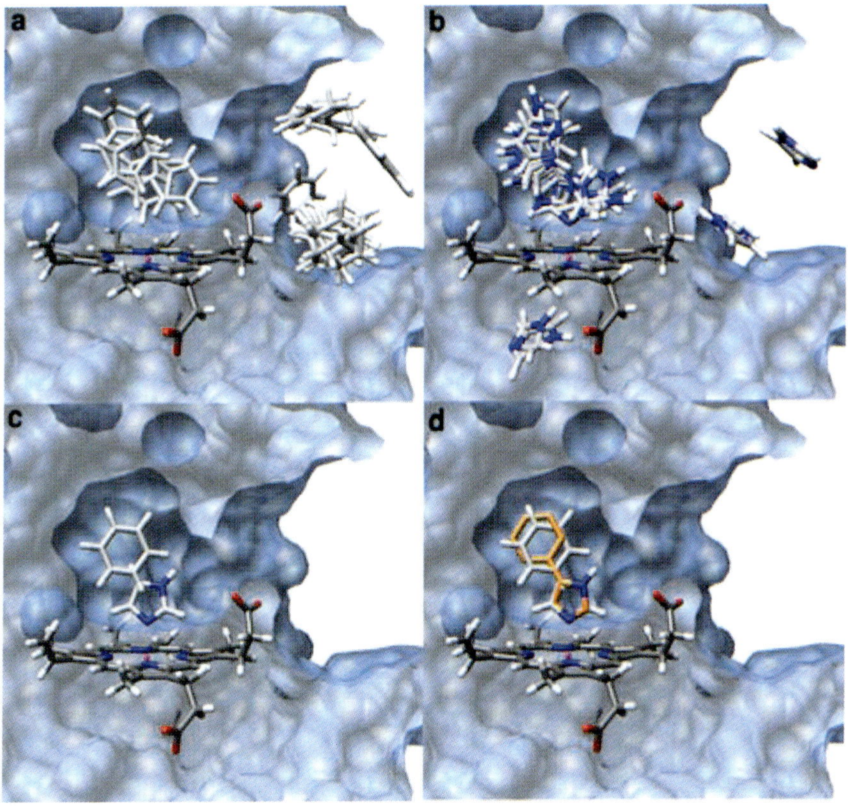

Fig. 9. Retrieving PIM by fragment-based virtual lead design. (a) Map of most favorable phenyl poses in IDO active site. (b) Map of most favorable imidazole poses. (c) Map of second most favorable phenyl pose and most favorable imidazole pose. (d) Superposition of (c) with the PIM X-ray structure (in orange).

rate of more than 60% clearly demonstrates the efficiency of the fragment-based virtual lead design approach. All active substances represent straightforward organic frameworks of low molecular weight with many purchasable or easily synthesizable derivatives.

An example of a successful ligand design by this fragment-based approach is given in Fig. 10. The quinoline framework has been linked with a hydroxy and an amine side chain, thereby yielding the commercially available compound 5-amino-8-hydroxyquinoline. Subsequent testing in the enzymatic assay confirmed its activity as an IDO inhibitor

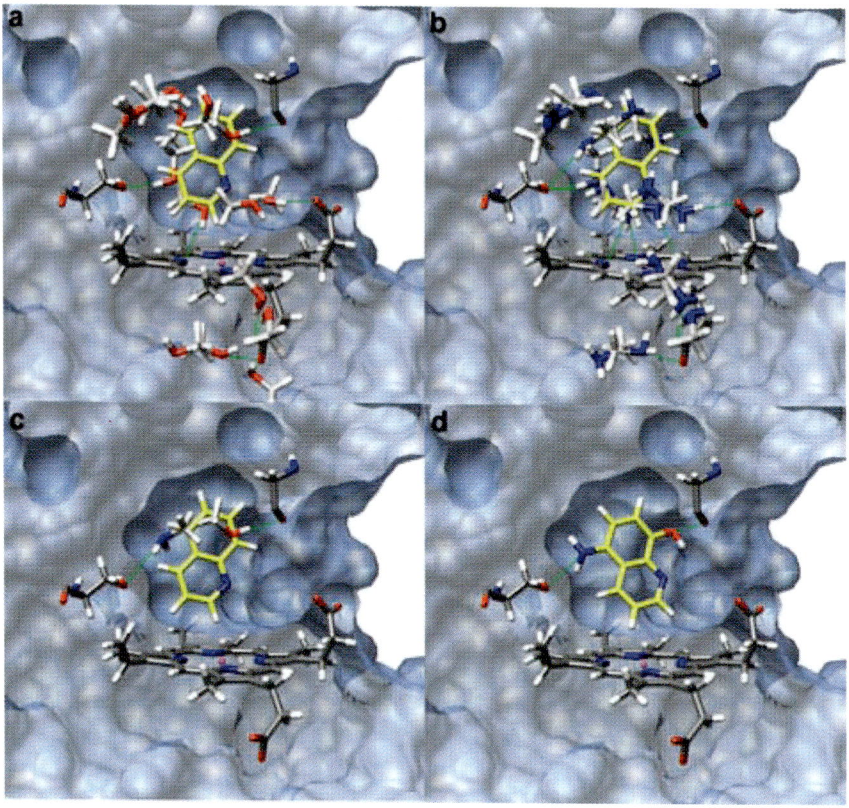

Fig. 10. Example of successful fragment-based virtual lead design. (a) Docked quinoline molecule with a map of most favorable methanol poses in IDO active site. (b) Quinoline with a map of most favorable methylamine poses. (c) Favorable poses for linking between quinoline and side chains. (d) Docking pose of new ligand (experimental K_i 300 nM).

with a K_i of 300 nM. Based on these results, another cycle of docking/linking experiments has been carried out in order to create further interactions with other parts of the active site and to enhance the specificity of the ligand.

References

1. Berneche S, Roux B. (2001) Energetics of ion conduction through the K^+ channel. *Nature* **414**: 73–7.

2. Ma J, Karplus M. (1998). The allosteric mechanism of the chaperonin GroEL: a dynamic analysis. *Proc Natl Acad Sci USA* **95**: 8502–7.
3. Dal Peraro M, Ruggerone P, Raugei S *et al.* (2007) Investigating biological systems using first principles Car–Parrinello molecular dynamics simulations. *Curr Opin Struct Biol* **17**: 149–56.
4. van Vlijmen HW, Schaefer M, Karplus M. (1998) Improving the accuracy of protein pKa calculations: conformational averaging versus the average structure. *Proteins* **33**: 145–58.
5. Morea V, Lesk AM, Tramontano A. (2000) Antibody modeling: implications for engineering and design. *Methods* **20**: 267–79.
6. Liu Y, Gray NS. (2006). Rational design of inhibitors that bind to inactive kinase conformations. *Nat Chem Biol* **2**: 358–64.
7. Alder BJ, Wainwright TE. (1957) Phase transition for hard sphere systems. *J Chem Phys* **27**: 1208.
8. Alder BJ, Wainwright TE. (1959) Studies in molecular dynamics. I. General method. *J Chem Phys* **31**: 459.
9. Rahman A. (1964) Correlations in the motion of atoms in liquid argon. *Phys Rev A* **136**: 405–11.
10. Stillinger FH, Rahman A. (1974) Improved simulation of liquid water by molecular dynamics. *J Chem Phys* **60**: 1545–57.
11. McCammon JA, Gelin BR, Karplus M. (1977) Dynamics of folded proteins. *Nature* **267**: 585–90.
12. Darden T, Perera L, Li L, Pedersen L. (1999) New tricks for modelers from the crystallography toolkit: the particle mesh Ewald algorithm and its use in nucleic acid simulations. *Structure* **7**: R55–60.
13. Lamoureux G, Roux B. (2006) Absolute hydration free energy scale for alkali and halide ions established from simulations with a polarizable force field. *J Phys Chem B* **110**: 3308–22.
14. Olsen L, Rydberg P, Rod TH, Ryde U. (2006) Prediction of activation energies for hydrogen abstraction by cytochrome p450. *J Med Chem* **49**: 6489–99.
15. Dinner AR, Blackburn GM, Karplus M. (2001) Uracil-DNA glycosylase acts by substrate autocatalysis. *Nature* **413**: 752–5.
16. Chandler D. (1987). *Introduction to Modern Statistical Mechanics.* New York, NY: Oxford University Press.
17. Hoover WG. (1985) Canonical dynamics: equilibrium phase-space distributions. *Phys Rev A* **31**: 1695.
18. Tuckerman ME, Liu Y, Ciccotti G, Martyna GJ. (2001) Non-Hamiltonian molecular dynamics: generalizing Hamiltonian phase space principles to non-Hamiltonian systems. *J Chem Phys* **115**: 1678.
19. Huenenberger PH. (2005) Thermostat algorithms for molecular dynamics simulations. *Adv Polym Sci* **173**: 105.

20. Berendsen HJC, Postma JPM, van Gunsteren WF *et al.* (1984) Molecular dynamics with coupling to an external bath. *J Chem Phys* **81**: 3684.
21. Morishita T. (2000) Fluctuation formulas in molecular-dynamics simulations with the weak coupling heat bath. *J Chem Phys* **113**: 2976.
22. Kirkwood JG. (1935) Statistical mechanics of fluid mixture. *J Chem Phys* **3**: 300–13.
23. Zwanzig RW. (1954) High-temperature equation of state by a perturbation method. I. Nonpolar gases. *J Chem Phys* **22**: 1420–6.
24. Tembe BL, McCammon JA. (1984) Ligand–receptor interactions. *Comput Chem* **8**: 281–3.
25. Bash PA, Singh UC, Brown FK *et al.* (1987) Calculation of the relative change in binding free energy of a protein-inhibitor complex. *Science* **235**: 574–6.
26. Frenkel D, Smit B. (2002) *Understanding Molecular Simulation: From Algorithms to Applications.* San Diego, CA: Academic Press.
27. Bennett CH. (1976) Efficient estimation of free energy differences from Monte Carlo data. *J Comput Phys* **22**: 245–68.
28. Pitera JW, van Gunsteren WF. (2002) A comparison of non-bonded scaling approaches for free energy calculations. *Mol Simul* **28**: 45–65.
29. Beutler TC, Mark AE, van Schaik RC *et al.* (1994) Avoiding singularities and numerical instabilities in free energy calculations based on molecular simulations. *Chem Phys Lett* **222**: 529–39.
30. Straatsma TP, McCammon JA. (1991) Multiconfiguration thermodynamic integration. *J Chem Phys* **95**: 1175–88.
31. van Gunsteren WF, Beutler TC, Fraternali F *et al.* (1993) Computation of free energy in practice: choice of approximations and accuracy limiting factors. In: van Gunsteren WF, Weiner PK, Wilkinson AJ (eds.). *Computer Simulation of Biomolecular Systems: Theoretical and Experimental Applications*, Vol. 2. Leiden, The Netherlands: Escom Science Publishers, pp. 315–48.
32. Fasnacht M, Swendsen RH, Rosenberg JM. (2004) Adaptive integration method for Monte Carlo simulations. *Phys Rev E Stat Nonlin Soft Matter Phys* **69**: 056704.
33. Ytreberg FM, Swendsen RH, Zuckerman DM. (2006) Comparison of free energy methods for molecular systems. *J Chem Phys* **125**: 184114.
34. Michielin O, Karplus M. (2002) Binding free energy differences in a TCR–peptide-MHC complex induced by a peptide mutation: a simulation analysis. *J Mol Biol* **324**: 547–69.
35. Oostenbrink C, van Gunsteren WF. (2005) Free energies of ligand binding for structurally diverse compounds. *Proc Natl Acad Sci USA* **102**: 6750–4.
36. Swanson JM, Henchman RH, McCammon JA. (2004) Revisiting free energy calculations: a theoretical connection to MM/PBSA and direct calculation of the association free energy. *Biophys J* **86**: 67–74.

37. Srinivasan J, Cheatham TE, Cieplak P *et al.* (1998) Continuum solvent studies of the stability of DNA, RNA, and phosphoramidate-DNA helices. *J Am Chem Soc* **120**: 9401–9.
38. Kollman PA, Massova I, Reyes C *et al.* (2000) Calculating structures and free energies of complex molecules: combining molecular mechanics and continuum models. *Acc Chem Res* **33**: 889–97.
39. Gilson MK, Honig BH. (1988) Calculation of the total electrostatic energy of a macromolecular system: solvation energies, binding energies, and conformational analysis. *Proteins* **4**: 7–18.
40. Gilson MK, Honig BH. (1988) Energetics of charge–charge interactions in proteins. *Proteins* **3**: 32–52.
41. Still WC, Tempczyk A, Hawley RC, Hendrickson T. (1990) Semianalytical treatment of solvation for molecular mechanics and dynamics. *J Am Chem Soc* **112**: 6127–9.
42. Gohlke H, Kiel C, Case DA. (2003) Insights into protein–protein binding by binding free energy calculation and free energy decomposition for the Ras–Raf and Ras–RalGDS complexes. *J Mol Biol* **330**: 891–913.
43. Lee MS, Feig M, Salsbury FR Jr, Brooks CL III. (2003) New analytic approximation to the standard molecular volume definition and its application to generalized Born calculations. *J Comput Chem* **24**: 1348–56.
44. Lee MS, Salsbury FR Jr, Brooks CL III. (2002) Novel generalized Born methods. *J Chem Phys* **116**: 10606–14.
45. Zoete V, Meuwly M, Karplus M. (2005) Study of the insulin dimerization: binding free energy calculations and per-residue free energy decomposition. *Proteins* **61**: 79–93.
46. Amidon GL, Yalkowsky SH, Anik ST, Valvani SC. (1975). Solubility of nonelectrolytes in polar solvents. V. Estimation of the solubility of aliphatic monofunctional compounds in water using a molecular surface area approach. *J Phys Chem* **79**: 2239–46.
47. Hermann RB. (1972) Theory of hydrophobic bonding. II. Correlation of hydrocarbon solubility in water with solvent cavity surface area. *J Phys Chem* **76**: 2754–9.
48. Sitkoff D, Sharp KA, Honig B. (1994). Accurate calculation of hydration free energies using macroscopic solvent models. *J Phys Chem* **98**: 1978–88.
49. Jayaram B, Sprous D, Beveridge DL. (1998) Solvation free energy of biomacromolecules: parameters for a modified generalized Born model consistent with the AMBER force field. *J Phys Chem B* **102**: 9571–6.
50. Gohlke H, Kuhn LA, Case DA. (2004) Change in protein flexibility upon complex formation: analysis of Ras–Raf using molecular dynamics and a molecular framework approach. *Proteins* **56**: 322–37.
51. Tidor B, Karplus M. (1994) The contribution of vibrational entropy to molecular association. *J Mol Biol* **238**: 405–14.

52. Moreira IS, Fernandes PA, Ramos MJ. (2007). Computational alanine scanning mutagenesis — an improved methodological approach. *J Comput Chem* **28**: 644–54.
53. Rizzo RC, Toba S, Kuntz ID. (2004) A molecular basis for the selectivity of thiadiazole urea inhibitors with stromelysin-1 and gelatinase-A from generalized Born molecular dynamics simulations. *J Med Chem* **47**: 3065–74.
54. Pearlman DA. (2005) Evaluating the molecular mechanics Poisson–Boltzmann surface area free energy method using a congeneric series of ligands to p38 MAP kinase. *J Med Chem* **48**: 7796–807.
55. Foloppe N, Hubbard R. (2006) Towards predictive ligand design with free-energy based computational methods? *Curr Med Chem* **13**: 3583–608.
56. Kuhn B, Gerber P, Schulz-Gasch T, Stahl M. (2005) Validation and use of the MM-PBSA approach for drug discovery. *J Med Chem* **48**: 4040–8.
57. Huo S, Massova I, Kollman PA. (2002) Computational alanine scanning of the 1: 1 human growth hormone-receptor complex. *J Comput Chem* **23**: 15–27.
58. Massova I, Kollman PA. (1999) Computational alanine scanning to probe protein–protein interactions: a novel approach to evaluate binding free energies. *J Am Chem Soc* **121**: 8133–43.
59. Zoete V, Michielin O. (2007) Comparison between computational alanine scanning and per-residue binding free energy decomposition for protein–protein association using MM-GBSA: application to the TCR–p-MHC complex. *Proteins* **67**: 1026–47.
60. Lafont V, Schaefer M, Stote RH *et al.* (2007) Protein–protein recognition and interaction hot spots in an antigen–antibody complex: free energy decomposition identifies "efficient amino acids". *Proteins* **67**: 418–34.
61. Zoete V, Meuwly M. (2006) Importance of individual side chains for the stability of a protein fold: computational alanine scanning of the insulin monomer. *J Comput Chem* **27**: 1843–57.
62. Yip YL, Zoete V, Scheib H, Michielin O. (2006) Structural assessment of single amino acid mutations: application to TP53 function. *Hum Mutat* **27**: 926–37.
63. Michalik L, Zoete V, Krey G *et al.* (2007) Combined simulation and mutagenesis analyses reveal the involvement of key residues for peroxisome proliferator-activated receptor alpha helix 12 dynamic behavior. *J Biol Chem* **282**: 9666–77.
64. Sono M, Roach MP, Coulter ED, Dawson JH. (1996) Heme-containing oxygenases. *Chem Rev* **96**: 2841–88.
65. Uyttenhove C, Pilotte L, Theate I *et al.* (2003) Evidence for a tumoral immune resistance mechanism based on tryptophan degradation by indoleamine 2,3-dioxygenase. *Nat Med* **9**: 1269–74.
66. Muller AJ, DuHadaway JB, Donover PS *et al.* (2005). Inhibition of indoleamine 2,3-dioxygenase, an immunoregulatory target of the cancer suppression gene Bin1, potentiates cancer chemotherapy. *Nat Med* **11**: 312–9.

67. Brastianos HC, Vottero E, Patrick BO *et al.* (2006) Exiguamine A, an indoleamine-2,3-dioxygenase (IDO) inhibitor isolated from the marine sponge *Neopetrosia exigua*. *J Am Chem Soc* **128**: 16046–7.
68. Pereira A, Vottero E, Roberge M *et al.* (2006) Indoleamine 2,3-dioxygenase inhibitors from the Northeastern Pacific marine hydroid *Garveia annulata*. *J Nat Prod* **69**: 1496–9.
69. Kumar S, Malachowski WP, DuHadaway JB *et al.* (2008) Indoleamine 2,3-dioxygenase is the anticancer target for a novel series of potent naphthoquinone-based inhibitors. *J Med Chem* **51**: 1706–18.
70. Sugimoto H, Oda S, Otsuki T *et al.* (2006). Crystal structure of human indoleamine 2,3-dioxygenase: catalytic mechanism of O_2 incorporation by a heme-containing dioxygenase. *Proc Natl Acad Sci USA* **103**: 2611–6.
71. Bemis GW, Murcko MA. (1996) The properties of known drugs. 1. Molecular frameworks. *J Med Chem* **39**: 2887–93.
72. Bemis GW, Murcko MA. (1999) Properties of known drugs. 2. Side chains. *J Med Chem* **42**: 5095–9.
73. Grosdidier A, Zoete V, Michielin O. (2007) EADock: docking of small molecules into protein active sites with a multiobjective evolutionary optimization. *Proteins* **67**: 1010–25.

Section III

PHYLOGENETICS AND EVOLUTIONARY BIOINFORMATICS

Chapter 11

An Introduction to Phylogenetics and Its Molecular Aspects

Gabriel Jîvasattha Bittar and Bernhard Pascal Sonderegger

1. Introduction

In general terms, phylogenetics is the study of the relationships between evolving objects. The term derives from the Greek *phylon* ("race, tribe, (old) family") and *genos* ("birth, origin"), *gennêtikos* ("related to generation, to the genesis (of something)").

Phylogenetics is part biology and part information science. It is a discipline that helps to shed light on the ways in which biological objects (organisms, genes) have appeared and evolved. Evolutionary science has demonstrated that all living things have a common ancestry, but the root of the tree of life on this planet lies nearly 4 billion years in the past. Since then, evolution and phylogenesis have shaped a huge variety of biological objects. On the whole, the tree of life is becoming more complex over time. The process in which evolution tends to form more and more diverse forms of life, in spite of (or maybe because of) catastrophic extinctions, is called cladogenesis (from the Greek *kladon* ("branch")). In addition to this process, the life forms themselves generally tend to become more and more complex — a process called anagenesis (from the Greek *ana-* ("up, to the top")). Phylogenetics incorporates both cladogenetics and anagenetics.

The essence of the cladogenetic part of phylogenetics can be illustrated by means of a simple graph. In Fig. 1, points C, D, and E — also

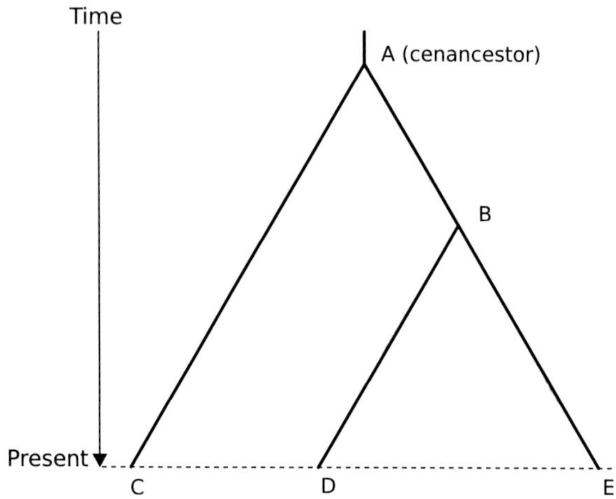

Fig. 1. Dendrogram illustrating the evolution of present-day forms C, D, and E, from ancestral forms B and A.

called terminal nodes — are the present-time forms of an object capable of duplication and evolution. Some time ago, an ancestral form B bifurcated into two diverging forms, ending in present-day forms D and E. An earlier ancestral form A had previously bifurcated into two diverging forms, resulting in the ancestral form B and the present-day form C. In addition, one must keep in mind that an evolving object can be subject to extinction (e.g. if there were no points D and E, with B an end node representing an ancient form which is now extinct).

Because of its branching structure, this kind of graph is called a tree or dendrogram. Formally, it can be characterized as a graph presenting a single path between any two nodes. A branch is normally the segment connecting two adjacent nodes (e.g. branch [AB]).[a] Branching occurs at internal nodes, where branches bifurcate (in terms of organisms, it denotes a speciation event). A series of successive branches that form a lineal descent is called a lineage.

[a] The term "branch" can also be used more loosely to identify a terminal subtree, e.g. the branch formed by the set of nodes A, B, D, E and their interconnecting branches (with D and E as terminal nodes).

To keep with the tree analogy, the terminal nodes (C, D, and E here) are often termed leaves; and the node of origin, the root. The most recent common ancestor of a set is called their cenancestor, or direct ancestor (their last ancestor before the speciation event from which the two or more lineages were born). Thus, in Fig. 1, node B is the cenancestor of D and E; while node A is the cenancestor of C, D, and E.

Trees can also be represented in a parenthesis format: each internal node is represented as a pair of parentheses with the descendant nodes between them separated by commas, only the terminal nodes being labeled. Generally, a semicolon is used to indicate the end of the tree. The dendrogram in Fig. 1 can therefore be represented as

$$(C, (D, E)).$$

As we are beginning to see, an entire family of terms and concepts has evolved around phylogenetics, and we will look at some of these in more detail as we progress through the chapter. Phylogenetics first developed as a science for classifying organisms into phylons and phyla[b] according to their evolutionary history. A phylon is a set of organisms that normally includes all of the descendants from a given ancestral node and some or all of the ancestral forms they share. In phylogenetics, an organism or a group of organisms, irrespective of its position in the taxonomic hierarchy, is often referred to as a taxon.[c] It must be stressed that a taxon can represent a single organism (e.g. subspecies *Homo sapiens sapiens*) or a group of organisms (e.g. hominoids).

The term "phyletic" refers to the nodes of lineage, without addressing the manner in which phylons were formed — a phyletic tree is a simple dendrogram, while a phylogenetic tree is a graphical construction which displays anagenetic information in addition to the nodes of the tree (the cladogenetic part). This supplementary information may be an estimate of

[b] The term "phylum" has come to mean, particularly in zoology, a higher-order rank of living organisms whose cenancestor is situated relatively early in the tree of life (again, the phylum usually includes the cenancestor and any other ancestor descending from it). Phyla are thus particular kinds of phylons.

[c] Both "taxons" and "taxa" are used as the plural form.

the number of mutations along a branch or an estimate of the time the evolutionary process took; as we shall see, these estimates are often represented as proportional to the length of the branches along one graph axis.

This introductory section has presented the simple tree structure, which is the basis of phylogenetics, and introduced some basic terminology. In the sections that follow, let us go to the heart of the matter.

2. Homology and Homoplasy: Look-alikes are Not Necessarily Closely Related

2.1. *Characters and Their States*

The first step in any phylogenetic analysis is to identify and observe comparable characters displayed by the taxa or genes to be analyzed. In nature, these characters can be morphological (e.g. color of the tail, presence or absence of a bony endoskeleton, number of legs, etc.) or molecular (e.g. the nucleotidic base or the amino acid at a specific position of a gene or a protein). The phylogenetic model states that two comparable characters should present states which, on the one hand, tend to be similar due to a common evolutionary origin (their cenancestor) and, on the other hand, tend to be dissimilar due to evolutionary divergence from this common origin.

It is rarely possible to make a meaningful statement regarding the phyletic history of a number of taxons just by analyzing the evolution of a single character. Generally, the larger the number of homologous characters for a given set of taxons, the better the phylogenetic analysis (We will explain further the meaning of "homologous" in Sec. 2.2). This can translate into a fairly large dataset which is best displayed in the form of a matrix of the states of characters, where a row corresponds to a taxon, and a column to a character. If the characters within any taxon have a natural linear or sequential order (as in DNA or a protein), each row within the matrix is called a sequence of characters.[d] In Fig. 2, a number of

[d] Often, characters in a sequence are called sites.

```
rank of character in sequence         10        20        30        40
Euglena_gracilis             MSRYRGPRLRIVRRIGK-LPSL--TNKTSKKRKSPGQPATS
Marchantia_polymorpha        MSRYRGPRVKIIRRLGA-LPGL--TNKTLKLKSGYI----N
Polypodium_vulgare           ???????RLKLIRRLKN-LPGFVGKSKMPNLGEFRV--ATD
Cycas_revoluta               ???????RLKIIRRLRT-LPGL--TNKRPKSRNVPT----N
Taxus_baccata                ???????RLKIINRLKT-LPGL--SKKKPNRIEEPR--TKK
Pinus_thunbergii             MSRYRGPRLKIIRRLKT-LPGL--TSKRPKNRKDSM----N
Spinacia_oleracea            MSRYRGPRFKKIRRLGA-LPGL--TNKRPRAGSDLR----N
Dipsacus_fullonum            ???????RFKKIRRLGD-LPGL--TNKKPRAGSDPR----N
Epifagus_virginiana          MSRYRGPSLKKIRRLGA-LPGL--TNKRSKAENDFI----K
Nicotiana_tabacum            MSRYRGPRFKKIRRLGA-LPGL--TNKKPRNGSDLR----N
Nelumbo_nucifera             ???????RFKKIRRLGT-LPGL--TSKRPKAGSDFR----N
Ranunculus_lingua            ???????RFKKIRRLGV-LPGL--TSKRPTPGSDLR----N
Laurus_nobilis               ???????RFKKIRCLGA-LPGL--TSKRPRSGSDLR----N
Acorus_calamus               ???????RLKKIRRLGA-LPGL--TSKGTRPGSDLR----N
Arum_italicum                ???????RFKKIRRLGA-LAGL--TSKRPRSGSDLK----N
```

Fig. 2. A character matrix of peptide sequences, as observed for 15 plant species. The displayed sequence is for the first amino acids of a polypeptide, coded for by the 5′ end of the chloroplastic gene *rps4*. For nine of these sequences, the beginning of the gene has not been sequenced; accordingly, the unknown states are symbolised by "?". Some characters, such as Char. 18 (amino acid leucine, "L") and Char. 0 (the initiating amino acid methionine, "M"), display the same state in all 15 sequences. But most characters are multi-state in this matrix, e.g. Char. 10 which presents three states: isoleucine ("I", 5 times), leucine ("L", once), and lysine ("K", 9 times).

peptide sequences are displayed. In the case of molecular sequences, the alignment matrix is often called a multiple alignment in reference to the way it is constructed.

Technically, the total length of this alignment could be reduced to 37 characters. However, the shortest alignment is not necessarily the best because, in the course of evolution, characters can be deleted and new ones can be inserted. For example, we have good reason to believe that the two characters "VG" in the *Polypodium* sequence are both specific inserts: they have no equivalent in the other sequences, where a double gap ("—") is shown. Also the gap for Char. 17 is kept because this is an extract from a larger data matrix with many more sequences, some of them displaying an amino acid insert on site 17. We shall further discuss the alignment procedure which produced this matrix in Sec. 3.2.

2.2. Homology — A Phylogenetic Hypothesis

Homology is a hypothesis, made when the state of one or more characters is phylogenetically compared in a number of taxa or sequences. The basis for a comparison is that there is reason to think that the characters are derived from a common ancestor. By comparing the amino acids in column 24 of Fig. 2, we are assuming that they are descendants from an amino acid present in the cenancestor of all the sequences. The state of the character is the same (threonine, "T") in all displayed taxa except two, *Polypodium vulgare* (where it is a lysine, "K") and *Taxus baccata* (where it is a serine, "S"). The placement in the same character column means it is hypothesized that all of these threonines as well as the lysine and the serine are descendants from the same ancestral amino acid (most probably a "T").

More formally, homology is the hypothesis made while defining, within two different sequences s and t, the two states K_s and K_t which are thought to describe the same character K. This is done because there is reason to assume that states K_s and K_t derive from a common origin K_0 and are therefore phylogenetically comparable. When a given character K can be found in two different taxons and their degree of similarity is due to a common origin, this character is said to be homologous in the two taxons.[c]

Homology, being a hypothesis, can be either true or false. Accordingly, it does not make sense to state that two quantifiable characters are, say, 79% homologous. They are 79% similar (21% dissimilar); and if this similarity is indeed due to a shared origin, then they are homologous. The same applies to a set of characters, and by extension also to taxa. Two men who look strongly similar are not necessarily closely related — similarity is not homology.

The principle of parsimony, also known as Ockham's razor, states that, all things being equal, the simplest explanation is best. In phylogenetics, the principle of parsimony translates into the following rule: *apart*

[c] Obviously, there is a difference between the notion of character and of its state; however, a less rigorous use of terminology, where it is said that so-called "characters" K_s and K_t are homologous, is so widespread as to be acceptable.

from pure chance, the simplest hypothesis for explaining similar characteristics in a set of evolving objects is that the similarity is due to common ancestry. It is according to this rule that one can infer that, considering Char. 24 in Fig. 2 is a "T" not only in most land plants, but also in the alga *Euglena gracilis*, it was most probably a "T" in the cenancestor of all these organisms.

It must be stressed that this is a practical, operational hypothesis; however, natural processes, and in particular evolution, are not necessarily simple or parsimonious. Most often, they are chaotic and recurring, and their results are largely indeterminable, unpredictable, and redundant — there are complex, random forces at work, and the life factory often appears as a giant apparatus where organisms are assembled by blind workers in a somewhat haphazard manner, fine-tuning being then left to the forces of natural selection. Before addressing the difficulties that this reality may entail for an analysis of a large number of characters, it is necessary to clarify the basic phylogenetic concepts as applied to a single character.

2.3. *Ancestral or Derived — Qualifying the State of a Character*

In a phylogenetic analysis, the state of a character at a given node can be considered, relatively to an ancestral node, either as unchanged (thus, still ancestral), or as changed (thus, derived).[f] In most cases, it is through a comparison with the state of the character in an outgroup (a group phyletically external to the group being studied) that the analyst can establish whether the state is ancestral or derived.

A character in an ancestral state is said to be plesiomorphic, whereas one in a derived state is said to be apomorphic (Greek *plêsios* ("neighbor"), *apo-* ("far from"), *morphê* ("form")).[g]

---

[f] The terms "primitive" and "evolved" are also used, though they can be confusing since they are often associated with a subjective notion of progress; furthermore, a descendant can regress to a quite "primitive" state.

[g] While, strictly speaking, the terms "ancestral" and "derived" qualify the state of a character, they are often used to qualify a character as such; in this case, they are synonyms for "plesiomorphic" and "apomorphic", respectively.

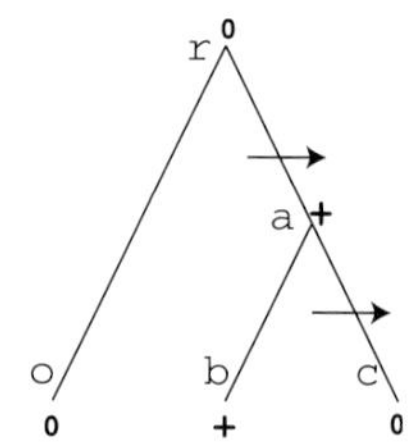

Fig. 3. Apomorphy is relative (an arrow indicates the transformation of a character). The character of interest is absent ("0") at the root *r* and in present-day forms *o* and *c*. It is present ("+") in ancestor *a* and present-day form *b*. The character is plesiomorphic in *b* relative to *a*, but apomorphic in *b* relative to *r*.

One must realise that plesiomorphy and apomorphy are not absolute qualifiers. They depend on the set of taxons being investigated. In Fig. 3, for instance, at node "b" the character is plesiomorphic with respect to node "a" but apomorphic with respect to node "r". Clearly, a character cannot be apomorphic or plesiomorphic in and by itself.

2.4. *Homoplasy — Pitfall in Phylogenetics*

Let us now address the consequences of making an incorrect assumption of homology. As mentioned above (Sec. 2.2), natural processes are not necessarily simple or parsimonious. In particular, evolutionary processes are highly correlated with environmental variables, and identical conditions often produce somewhat identical results. Therefore, similar characteristics or characters might appear in very distant lineages because of adaptation to similar conditions, not because of any memory of a distantly shared ancestor.

In evolutionary biology, homoplasy is the occurrence of similar states of character that are not a result of shared lineage. This can occur either through chance or through adaptation to similar selective pressures. At a morphological scale, identical conditions may lead to quite similar results. On the other hand, at the DNA level, especially with the limited four-letter alphabet of nucleotide sequences, pure chance easily leads to similarity between different strings of nucleotides (homologous or not).

Convergence or parallelism is the phyletically independent occurrence of the same states of character within different lineages.[h] Reversion is the return to an ancestral state of character.

These notions apply to any category of objects evolving within some kind of environment, whether natural or artificial, through duplication and extinction events. Consider the design process of automobiles: two cars may look very alike, although they are not of the same model. There are three possible explanations. The cars may be produced by the same manufacturer and therefore have certain design features in common (this is straight homology). Alternatively, two manufacturers may design similar cars based on a comparable perception of consumer taste and car functionality (this is convergence homoplasy). Finally, a newly conceived car may resemble an earlier model when a certain look comes back into fashion (this is reversion homoplasy).

In evolutionary biology, two textbook examples illustrate the notion of homoplasy due to shared environmental constraints (homoplasy due to chance being trivial enough):

(1) *Active flight in tetrapods — thrice upon a time...* Active flight developed at least three times among the Tetrapoda — a number of times in ancient reptiles such as pterosaurs, once in avians (birds) — and once or twice in chiropterans — pteropods (flying foxes) and bats. Although the forelimbs of all tetrapods are homologous organs, laws of aerodynamics and not common ancestry are responsible for the similarities of their wings, which have roughly converged in their aerodynamic morphology because the function of the forelimbs of these taxa as true wings is identical.

(2) *Return to the sea.* Cetaceans (dolphins and whales) and sirenians (manatees and dugongs) are mammals that have fully adapted to an aquatic habitat. The necessary transformations for this adaptation occurred long ago twice: starting from two different land-dwelling mammalian ancestors, the two lineages evolved their capacity for life

[h] The term "convergence" is used for distant lineages, while "parallelism" is reserved for closer ones.

> in water independently of each other. The two groups are morphologically similar; however, this is not because of common ancestry, but because they were both subject to the same physical constraints of an aquatic environment. Cetaceans and sirenians thus show convergence homoplasy with respect to their roughly similar hydrodynamic morphology. This being said, when compared to fish, a vertebrate group with which both sirenians and cetaceans share a very ancient ancestor, these two mammalian lineages each show reversion homoplasy in relation to their hydrodynamic morphology.

A phylogeneticist who inappropriately compares true wings in birds and in chiropterans and furthermore bases his analysis on homoiothermy (warm-bloodedness) might well create an evolutionary tree where birds are descendants of some sort of ancestral bat: his analysis would indicate that some ancient reptile evolved into mammals, that certain mammals specialized for active fight, and that some of these developed feathers and other apomorphies to become birds. This analysis has been trapped by the pitfall of homoplasy. In reality, an analysis based on a wider selection of characters, e.g. egg laying,[i] establishes that homoiothermy and real wings are two characters which have independently evolved in the same direction in mammals and birds (convergence homoplasy).

Not all cases of homoplasy are as obvious as the examples given above. It is especially difficult to detect homoplasy in molecular data. The relatively high mutation rate and the small alphabet of nucleotides both make the likelihood of random homoplasies, through convergence or reversion, far greater than with morphological characters. To further complicate matters, proteins are highly modular: there are domains common to a wide variety of proteins. It is not simple to resolve the branches of an evolutionary tree of slowly evolving proteins when the structural constraints on a domain are such that only a few amino acids are truly free to mutate. This example leads us again to a notion that needs to be

[i] Egg laying outside water is common to most reptiles, all birds, and only two genera of mammals (the echidna and the platypus).

developed further: one can reconstruct a phylogenetic tree not only of organisms, but also of biological molecules such as proteins or genes.

3. Molecular Phylogenetics

Until the mid-20th century, the classification of organisms was based exclusively on anatomy and morphology. In the language of genetics, only phenotypic characters (i.e. visible ones) could be compared. Today, with the large amount of genetic sequence data available in public databases and the relatively low cost of sequencing and computing technology, numerical phylogenetics has shifted strongly towards molecular characters.

Such a shift has its advantages and its disadvantages. While it is very tricky to study molecular characters of fossils (extracting intact DNA from fossils without contaminating it is quite challenging), their morphological and anatomical characters can readily be compared to those of modern-day organisms. Paleontology may give a phylogeneticist direct information on whether a phenotypic character's state is ancestral or derived. On the other hand, the information content of molecular data and the sheer number of characters available for comparison make the molecular approach very attractive.

Since 1966, it is known that genes are basically sequences of molecular characters: four nucleic bases constitute the four possible states for each nucleotide in the sequence (a gap in an alignment may be considered as a special kind of fifth state). Notwithstanding the technical difficulties of a multiple alignment, comparisons are straightforward with molecular sequences: when comparing two bases, these are identical or not; quantification problems typical of morphological data, such as where to draw the line between long legs and short legs, are avoided.

3.1. *Gene Duplication vs. Speciation, Paralogy vs. Orthology*

As we have seen in Sec. 1, evolutionary processes tend to increase the complexity of the tree of life. Not only does the number of living branches tend to increase over the long term (there are generally more and more different living organisms), but, furthermore, the organisms themselves tend to become more complex. With the advent of molecular biology and genetics, the study of these parallel build-ups of biological complexity has become an integral part of phylogenetics.

A complex organism, endowed with tens of thousands of genes, does not evolve overnight. Its intrinsic genetic complexity is the result of millions of years of successive gene duplications and more complex (e.g. chromosomal) rearrangements. After every one of these duplicative events, separate copies of genes had the possibility to evolve and differentiate, each in its own direction. The most common outcome is the complete degeneration of all but one copy of the gene. On occasion, however, some selective advantage is gained by having more than one copy and they are maintained — it can also be pure chance, of course. Generally, in this scenario, each copy ends up fulfilling a slightly different function and continues to evolve. In the end, each copy becomes an individual gene in its own right, sometimes with a radically different function from the cenancestor gene.

Therefore, at the molecular level, evolution leads not only to differences between the genes of one organism and the homologous genes of another organism, but also to the divergence of multiple copies of a gene within a single lineage (a gene family). With the present abundance of molecular data, genes themselves can be studied as evolving objects capable of duplication, transformation, and disappearance. There is thus a need to reconstruct not only the evolutionary trees of the families of living and past organisms, but also the evolutionary trees of present (and possibly extinct) families of genes.

Since 1970, two terms have been in use to clearly distinguish between the two types of homology that phylogenetics now addresses. Related genes that diverged after a speciation event and a gene duplication event are called orthologs and paralogs, respectively.

The classic example of paralogous genes is the family of oxygen-transporting globins (Fig. 4). Hemoglobins form a complex family of proteins, all of them descendants by gene duplication from a single ancestor gene which existed some 500 million years ago. Further back, some

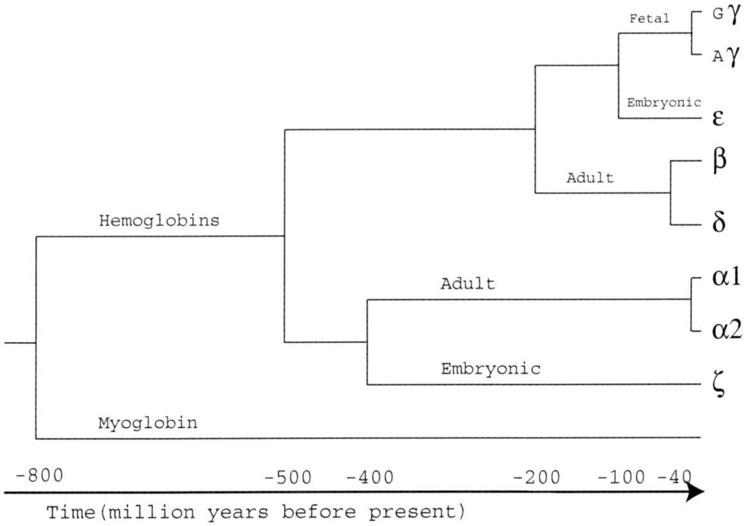

Fig. 4. Oxygen-transporting globins. Successive gene duplications have created a diverse gene family. Each gene in the family fulfills a different function. Hemoglobins transport oxygen in the blood, with various specializations, while myoglobin transports oxygen in muscles.

800 million years in the past, another gene duplication event created the separate lineages of myoglobin and hemoglobins.

Both the pig and the human genomes contain genes for various hemoglobins. Given a data set consisting of all the available α-hemoglobin sequences from tetrapods, in which human β-hemoglobin rather than human α-hemoglobin would inadvertently have been included, an erroneous phyletic tree would be produced in which the human species would be shown to have diverged from many other tetrapods 500 million years ago, instead of a few tens of millions of years ago from a smaller number of tetrapods, all mammalian (this is readily seen by following in Fig. 5 the path from human β-hemoglobin to the first internal node, then to pig α-hemoglobin). Hence, another pitfall in phylogenetic reconstruction.

Such a result would evidently appear as wrong to the knowledgeable, leading the data to be viewed with suspicion. However, not all cases of incorrect trees produced through an inappropriate mixture of orthologs

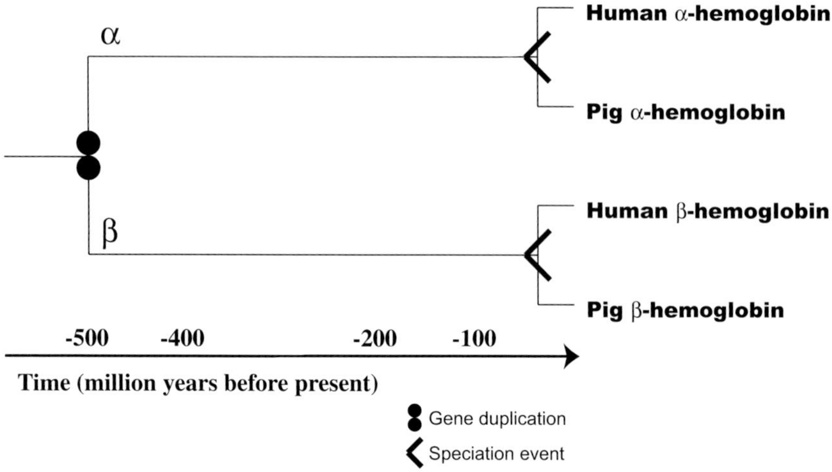

Fig. 5. Orthology versus paralogy. A gene duplication event 500 million years ago led to the coexistence of two hemoglobin genes: α- and β-hemoglobin. Much more recently, a speciation event allowed the primates to diverge from the ungulates, this divergence applying to all genes within both taxa and thus showing up in the lineages of both α- and β-hemoglobins of both pig and human.

and paralogs are so easy to identify. Nowadays, with the possibility of downloading huge amounts of molecular data from the sequence databases, including those from unclassified environmental samples, and with automated tools claiming to easily identify orthologs (see Sec. 5.1), the danger of underestimating the problem of paralogy is larger than ever. An analyst should always be very careful when establishing a set of homologous sequences, and never blindly trust the results of blackbox-type number crunching.

3.2. *Sequence Alignment — A Homology Hypothesis*

As with morphological characters, the first step of a phylogenetic analysis based on molecular data is the selection of comparable characters. Therefore, at the outset, a hypothesis of homology must be made.

Multiple sequence alignments are used to decide which characters will be compared. Producing these multiple alignments is far from trivial, and a great deal of research has gone into the development of automated methods to do so. A description of some of the problems encountered follows.

- As we have seen, there is always the possibility that in the course of evolution, nucleotides have been inserted into a gene or deleted from it (similarly, that amino acids have been inserted or deleted in a peptide sequence). Since it is often impossible to say beforehand which of the two events took place, the term "indel" (insertion/deletion) is commonly used. Phylogenetically, a column where there is a nucleotide or an amino acid in one sequence and an indel in another is analogous to a morphological character which is present in one species but not in another. As we have seen in Fig. 2, an indel in a multiple alignment is represented by a gap ("-") in some of the sequences and by a nucleotide (or an amino acid) in others. As a further complication, an insertion or deletion event is not necessarily limited to a single site: several nucleotides or amino acids may be inserted or deleted at once.
- Of course, a multiple alignment is not meant to produce indels *ad libitum*, as this would be contrary to biological common sense and to the principle of parsimony. Accordingly, the clever analyst and automated algorithms allocate a "penalty" for gaps. Once a scoring system for mismatched nucleotides and a penalty for indels have been defined, finding the optimal alignment is an apparently straightforward mathematical problem. Solving it for two sequences is fairly simple and can be done at tremendous speeds by today's computers. However, the computation time and memory needed for the calculations grow exponentially with the number of sequences, thus posing a major challenge to computers. Several shortcuts or heuristics (Greek *heuriskein* ("to find")) have been devised to speed up the process, but they all work at the cost of alignment quality. Consequently, many scientists in phylogenetics and in other domains of biology choose to manually improve computer-generated alignments, drawing on their biochemical and taxonomic knowledge base.

- Even if the mathematically optimal result has been calculated, it may not be correct. Unsatisfactory results are often obtained if large portions of a gene are missing in some of the sequences (the "?" in Fig. 2). The insertion or deletion of an entire protein domain can lead to prohibitively large penalties which would not allow the correct alignment to be found. This is another reason for manually reviewing and correcting alignments.
- Insertions and deletions are not the only evolutionary events that take place. Complex rearrangements, more akin to macromutational events, also happen, but these are rarely taken into account by multiple-alignment software.
- The sheer complexity and variability of genetic information and biological mechanisms can lead to further problems. These include synonymous codons and underlying protein sequence in coding regions; pseudogenes; mRNA editing; splicing boundaries in eukaryotic genes; regulatory regions such as promoters; secondary and tertiary structures in proteins, rRNA, and tRNA, etc. Each dataset has its own supplementary constraints that can be leveraged to improve alignment, but one tool cannot do it all.

Once a multiple alignment has been created, a second step is performed to select the columns of the alignment that are to be included in the phylogenetic analysis. One must be certain that the characters are rich in information and that the alignment is reliable. Since there will always be a mathematically optimal alignment, even for purely random sequences, it is necessary to take some preliminary precautions. Generally, only molecular sequences that are clearly identifiable and have roughly the same length are used. Regions coding for proteins fit the bill perfectly, especially if a secondary, tertiary, or even quaternary structure is known and can be used to improve homology assumptions. This allows for a functional verification of alignment results.

Several approaches can be used to approximate the information content of the selected characters. A simple one is to perform an exhaustive cladistic maximum parsimony (CMP) search (see Sec. 4.3) of all possible trees that can be generated from a small number of sequences. If there are 12 sequences, then there will be 654 729 075 possible perfectly bifurcating unrooted trees. A histogram of the lengths of all these trees is drawn. If the distribution obtained is symmetric and bell-shaped, we are dealing with virtually random sequences with little or no phylogenetic signal. If, on the other hand, the distribution curve is skewed to the right (Fig. 6: longer tail towards the short trees, steep slope on the side of the longer trees, mode > median > mean), then there is a better likelihood of having a phylogenetic signal.

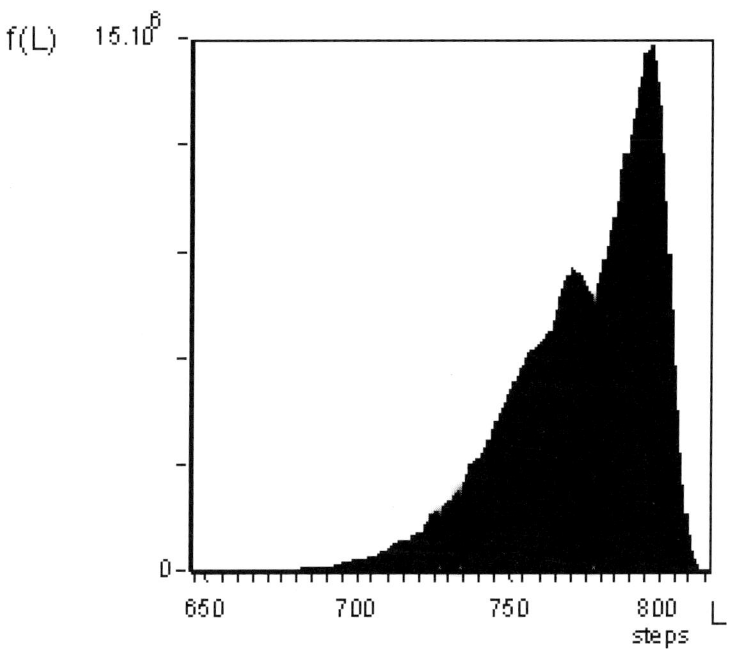

Fig. 6. Distribution of tree lengths from a CMP search of all possible trees for 12 aligned sequences. The distribution is shifted to the right (to the longer trees, those with a higher number of "steps"), indicating a possibly strong phylogenetic signal (good signal-to-noise ratio). The x-axis shows the tree length L in CMP steps. The y-axis shows the frequency count of trees with length L.

3.3. *Evolutionary Time*

3.3.1. *Evolutionary distance and the course of time*

As said earlier, phylogenetics is a matter not only of cladogenesis, but also of anagenesis, which involves *inter alia* the fundamental problem of estimating some sort of evolutionary distance between nodes in a tree.

Generally, this represents either a straight pairwise measurement of differences between the homologous characters of an ancestor and those of a descendant (a dissimilarity), or the accumulated number of mutations which have occurred along the lineage from this ancestor to a present-day form. When applied to present-day forms, it is either their pairwise dissimilarity or the sum of the accumulated number of mutations from the cenancestor to each of the present-day forms.

These two different estimators of evolutionary distances have very different properties. A pairwise dissimilarity between two sequences, which is incremented by one unit when two homologous nucleotides are different, cannot be additive.[j] Statistically, it is a monotonically increasing function of time, but definitely not a linear function of time, if only because a character can mutate more than once in a lineage and of course return to an ancestral state while so doing.[k]

On the other hand, an estimate of a counter which is incremented by one unit every time a nucleotide is transformed to any one of the three others can be additive, and should be. This kind of mutational distance between two nodes can more appropriately be treated as proportional to the time distance between them.

A dendrogram in which each branch length is proportional to an estimate of evolutionary distance between the two interconnected nodes, whether additive or not, is often called a phylogram.

---

[j] If distance d is additive, C being the cenancestor of terminal nodes A and B, then $d[A, B] = d[A, C] + d[B, C]$.

[k] With nonadditive metric dissimilarity D, $D(A, B)$ can only be said to be smaller or equal to $D(A, C) + D(B, C)$.

3.3.2. *Time and time again: paleontology and molecular evolution*

There are two basic approaches to measuring evolutionary time. The first and most straightforward belongs mainly to the colorful realm of paleontology and involves using fossil records to date ancestral forms of life, and thus reconstruct both cladogenesis and anagenesis. Generally speaking, this approach involves macromutations, which are rare, extraordinary events at the macromolecular level, such as a change in the number of chromosomes. Often, a macromutation can incur an identifiable change of phenotype, such as a morphological change, allowing correlation to paleontological time. This is a tremendous advantage leading to an absolute calibration of the phylogenetic timescale. On the other hand, macromutations occur at highly irregular intervals, and in the absence of fossils this renders a quantitative evaluation of the passing of time very problematic. In addition, this approach is normally not applicable to paralogous analysis.

The second approach belongs to the realm of molecular evolution and involves an estimation of the total number of molecular mutations which took place since a given event of speciation or gene duplication. It favors the quantification of micromutations, which are small ("point") mutations at the DNA level, possibly translating into a change of amino acid. This approach has gained favor thanks to the massive, quickly growing amount of genetic sequence data available and to the amenability of this data to quantitative methods.

3.3.3. *Micromutations and the molecular clock*

Mutation rates in living cells, at their most basic level, are fairly constant throughout time and organisms. However, after DNA repair processes and organism reproduction constraints, the number of mutations that are actually conserved in a population, and therefore seen in the molecular data available, can over time present a picture of highly heterogeneous rates of transformation across the different sections of a phylogram.

Nevertheless, because micromutations are relatively easy to quantify, their use and modelization in phylogenetics have been developed. There are data sets for which the measured mutation rate can be considered more or less stable and uniform. For such molecular data, one can reasonably make a "molecular clock" hypothesis and estimate durations from mutational distances. Even if, in contrast to the radioactive-decay methods used to date fossils, no simple formula exists to convert evolutionary distances into an exact timescale, this approach has been very productive for the evolutionary analysis of gene families (paralogs) and even of taxons (orthologs).

As explained above, rates of evolution can be heterogeneous. To make matters more complicated, this applies not only to lineages, but also to characters within a single lineage. Some parts of a protein are more conserved than others, i.e. an amino acid change within a conserved part is often lethal to the organism and is rarely transmitted to future generations. This is another source of caution for phylogeneticists, yet they can use this to their advantage: they use quickly evolving genes or parts of genes to assess phyletic relationships between organisms having diverged a short time ago, and slowly evolving genes or parts of genes for those having diverged a long time ago.

The rationale for this differentiation is that the high frequency of micromutations, combined with the small number (four) of possible nucleotide states, can rapidly lead to signal saturation at large evolutionary distances. After a certain number of mutations, it becomes impossible to estimate how many mutations took place and therefore how large the phylogenetic distance between two taxa or two genes really is. To simplify, it is impossible to estimate how many mutations occurred since two genes diverged if their dissimilarity gets too close to the statistical upper bound of $(4-1)/4 = 0.75$.[1] Conversely, dissimilarities that are too small are not appropriate for a quantitative approach: if all pairwise dissimilarities have values ranging from, say, 0 to 2, for nucleotide sequence lengths in the hundreds, any phylogenetic signal would be buried under random noise.

In summary, while there is a molecular clock that runs fairly precisely, the observation of the clock is hampered by biological constraints such

[1] Actually, the "twilight zone" for nucleotide sequences, i.e. the upper threshold beyond which dissimilarity values start to be too large for multiple alignments to make phylogenetic sense, is often estimated to be as low as 0.5.

as fixation frequencies in a population, and by nonlinear statistical effects of signal saturation when the accumulated number of mutations becomes too high.

4. Tree Reconstitution

After preparation of the input data, a reliable method is needed in order to create a tree from the selected characters. This section discusses these methods.

Many different algorithms exist, and explaining them all in detail is beyond the scope of this chapter. However, most of the algorithms can be grouped together into three main categories. For alternative methods, see Semple and Steel.[1]

Before we get down to the details of the three categories, we will describe the theoretical framework common to them all.

4.1. *The Tree Graph Model — Transmission of Phylogenetic Information*

As we have seen, a tree graph, as opposed to a reticular (network-like) graph, has certain inherent properties. For instance, there can only be a single path from one node to another. This holds true for all nodes, whether internal or terminal. Furthermore, when time is taken into account in a phylogenetic tree, information always flows in the same direction. Branches are unequivocally oriented from the root to the terminal nodes (in Fig. 7, from node "A" along the respective lineages to "C", "D", and "E").

In genealogy (the study of family trees), it is possible for two representatives of two separate lineages to breed and thereby reconnect their lineages. By contrast, in the phylogenetic model, which applies to a higher taxonomic level than the individual, in principle lineages cannot cross each other. Horizontal gene transfer[m] is sufficiently rare to be

[m] Horizontal gene transfer is the exchange of genetic information through means other than simple heredity. This usually involves transposable elements, viruses, or bacteria. The result is a reticulate phylogenetic network that is no longer a tree.

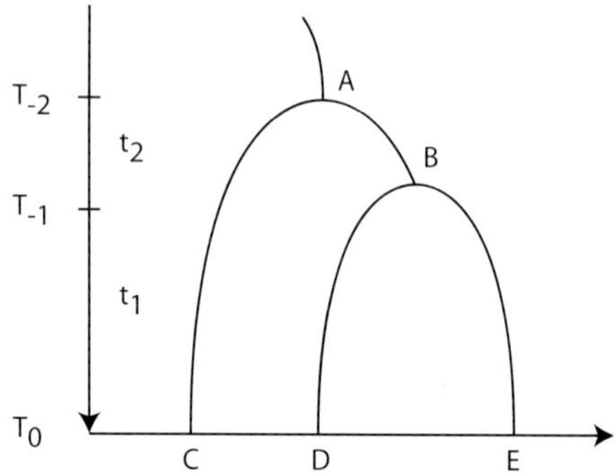

Fig. 7. A tree graph with the direction of time indicated. C, D, and E are present-day taxa, while A and B are ancestral forms. T_{-1} and T_{-2} are the points in time where branching took place, while T_0 is the present time. t_1 and t_2 represent the length of time between the branching event at B and present day and the length of time between the branching event at A and the branching event at B, respectively.

insignificant in most cases (notable exceptions include viruses and bacteria). Transfer of genetic information is thus practically always "vertical". This verticality is what allows the reconstruction of an entire tree despite only the terminal taxa being known — if it were not for this property, there would not be enough structural constraints on the available data for tree reconstitution.

4.2. *Numerical Taxonomic Phenetics (NTP)*

Numerical taxonomic phenetics (NTP) methods, also called taxometry methods, start with the calculation of pairwise dissimilarities between taxons. The matrix of character states is converted into a semimatrix of dissimilarities D_{ij} .[n]

Rudimentary clustering methods such as UPGMA (Unweighted Pair Group Method using Arithmetic averages) simply cluster taxons on the basis of their similarity to each other. No phylogenetic hypotheses — such as the possibility of homoplasy — enter into these methods. Even so, some of them, for example WPGMA (Weighted Pair Group Method

[n] Obviously, $D_{ij} = D_{ji}$ and $D_{jj} = 0$.

using Arithmetic averages), are more satisfying from a phylogenetic point of view than others (WPGMA is not as highly sensitive to sampling bias as UPGMA is).

4.2.1. *The neighbor joining algorithm*

An example of a more sophisticated and more phylogenetically valid method is neighbour joining (NJ), which operates on a numerically derived evolutionary distance. In contrast to raw dissimilarity, this evolutionary distance tries to satisfy the property of additivity ($[AB] + [BC] = [AC]$).

The NJ algorithm is based on the following four-point property: given a tree with additive distances and four leaves A, B, C, D, of the three possible sum of distance pairs $D_{AB} + D_{CD}$, $D_{AC} + D_{BD}$, and $D_{AD} + D_{BC}$, two must be equal and the third must be smaller than the other two (see Fig. 8).

NJ is a much-studied and widely accepted method that takes the heterogeneity of evolutionary rates in the different lineages into account, but cannot detect homoplasy.

4.2.2. *A common NTP artefact*

Not only do NTP methods fail to allow for homoplasy, but most of them also tend to draw together slowly evolving lineages, while moving quickly evolving lineages to a basal (external or peripheral) position in the tree. Figure 9 shows a hypothetical evolutionary tree (the true tree) and an

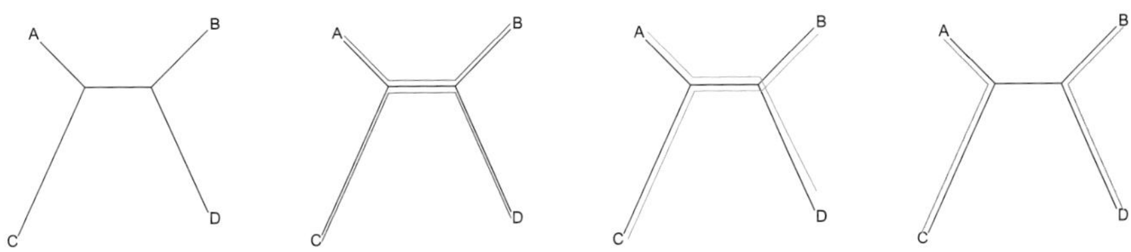

Fig. 8. The NJ algorithm is based on the four-point property, which can be stated as follows; for any four-leaf tree with additive distances, of the three possible sum of distance pairs, two must be equal and the third must be smaller than the other two.

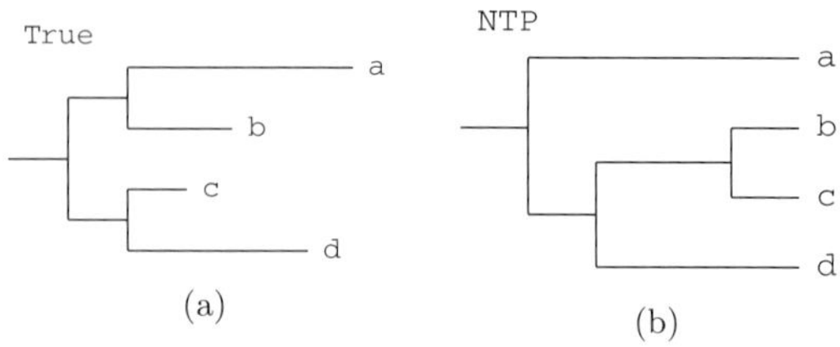

Fig. 9. (a) A true evolutionary tree with heterogeneous rates of evolution between branches and (b) an attempt at reconstruction by a simple clustering NTP method.

attempt to reconstruct it by a simple clustering NTP method. *b* is phyletically more closely related to *a* than it is to *c*, even though the dissimilarity between *b* and *a* is greater than the one between *b* and *c*. The dissimilarity between *b* and *c* is small because both have evolved slowly from their common ancestor at the root of the tree: they look uncannily similar, despite being on lineages having diverged long ago, because they are "living fossils". NTP methods, with the notable exception of NJ, fail to detect this pitfall and place *b* and *c* together as though they were very closely related.

4.3. *Cladistic Maximum Parsimony (CMP) Methods*

The cladistic maximum parsimony (CMP) method and the related character compatibility (CC) method explicitly reconstruct ancestral states of characters.

In order to understand how CMP methods work, we need to expand the terminology concerning ancestral and derived states presented in Sec. 2.3.

4.3.1. *Symplesiomorphy, synapomorphy, and autapomorphy*

A symplesiomorphy is a plesiomorphy shared by a number of taxa relative to the same ancestor (Greek *syn* ("(shared) with")). A synapomorphy is an apomorphy shared by a number of taxa relative to the same ancestor

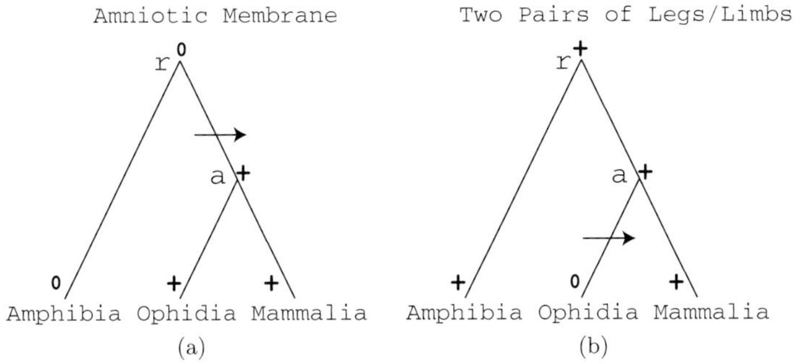

Fig. 10. (a) Synapomorphy of the amnion character in Ophidia and Mammalia relative to *r*, but symplesiomorphy of the amnion in Ophidia and Mammalia relative to *a*. (b) Symplesiomorphy of the character "two pairs of legs/limbs" in Amphibia and Mammalia relative to *r*, but autapomorphy of the character in Ophidia (snakes have no apparent limbs).

(Fig. 10). An autapomorphy is an apomorphy proper to a single taxon (Greek *autos* ("self")).

The two examples in Fig. 10 illustrate these concepts. In both dendrograms, we are looking at present-day taxa Amphibia (frogs and newts), Mammalia (humans, etc.), and Ophidia (snakes).

The left dendrogram is a graphic depiction of the evolution of the amnion (the innermost membrane enclosing the fetus). Its presence is denoted by "+", with mammals and snakes sharing an amniotic ancestor ("a"). The cenancestor "r" of all three terminal taxa was anamniotic ("0"), however, just like present-day frogs. We therefore have a synapomorphy of the amnion character in Ophidia and Mammalia relative to "r", but a symplesiomorphy of the amnion in Ophidia and Mammalia relative to their cenancestor "a".

The right dendrogram is a depiction of the Ophidian autapomorphy. Snakes have lost the two pairs of legs/limbs (presence denoted by "+") that are a characteristic of all Tetrapoda. There is thus a symplesiomorphy of the character in Amphibia and Mammalia relative to "r", while the Ophidia taxon presents an autapomorphy in its absence of legs/limbs. Note that, phyletically, snakes still belong to the Tetrapoda phylon since the cenancestor "r" of all three terminal taxa was a tetrapod (endowed with four limbs).

4.3.2. *Cladistic maximum parsimony (CMP) and character compatibility (CC) methods*

CMP methods search for the MP (maximum parsimony or most parsimonious) tree(s), which minimizes the sum *S* of absolute dissimilarities between all pairs of adjacent nodes in the tree.[o] Another way of explaining the CMP method is that it creates trees in which the states of all characters are defined, even those at internal nodes. The trees which are consistent with the smallest sum *S* are considered the most parsimonious and are therefore retained.

To find the most parsimonious tree, CMP methods continually compare hypotheses about the possible states of ancestral nodes. They proceed by progressively looking for synapomorphies, thus indirectly minimizing the total degree of homoplasy. Since the methods need to find synapomorphies, it becomes evident that a character with states differing from one group of taxons to another group is more informative than a character which has the same state throughout all taxons, or all taxons minus one. The differences between taxons are methodologically more important than the similarities.[p]

CC methods proceed in roughly the same manner, but they look for the global MP tree(s) that can account for the largest clique (or set) of characters without having to resort to any hypothesis of homoplasy.

Common parsimony programs include PAUP*,[2] TNT,[3] and MEGA.[4] Parsimony methods are generally much slower than NTP methods, mainly due to the nature of the heuristic methods used.

4.3.3. *CMP common artefacts*

Since dissimilarities are not additive, a cladophylogram generated by a CMP or CC method is not necessarily a good indicator of relative evolutionary

[o] Note that a node is only adjacent to its direct ancestor and to its direct descendants. Since the dissimilarity in NTP methods is calculated between pairs of terminal taxa, which can never be adjacent, the values of an NTP dissimilarity semimatrix are never used in CMP.

[p] Computer algorithms proceed differently to reconstruct CMP trees. They simply calculate the length of all possible trees and retain the best results. If heuristics are used, only the subset of trees most likely to contain MP trees is calculated.

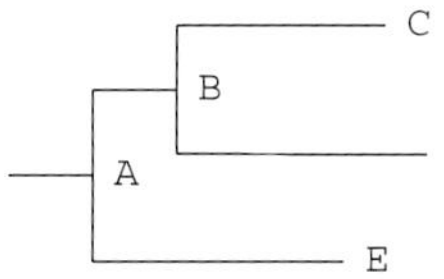

Fig. 11. An example of a CMP phylogram.

rates. Two lineages [*AD*] and [*AE*] could have evolved at precisely the same rate since they diverged from their cenancestor A, but if one lineage [*AD*] has more internal nodes along the path from node *A* to the leaf *D*, its total length will inevitably tend to be larger than the total length of lineage [*AE*]. The phylogram (Fig. 11) correctly tells us that the rate of transformation is higher in branch [*BD*] than it is in branch [*BC*], but the length of branches [*AC*] and [*AD*] is not directly comparable to that of branch [*AE*].

This size illusion effect is the same as in a small-scale (or low-resolution) map displaying roads. When switching to a larger-scale map (one with a higher resolution), more and more curves in the road become apparent. Measuring the length of a road on both maps will give different results and the one with more detail will invariably give a larger result. A road appears to become "straighter" when the resolution is decreased. The same holds true for CMP analysis: since the states of all characters at the internal nodes are reconstructed, the more nodes there are on a path, the more precise the measured evolutionary distance will become. As precision grows, the measured dissimilarity can only become larger. Thus, comparing two branches with a different number of nodes along each one is like measuring two roads on maps of different resolutions.

Another, more subtle effect is long-branch attraction.[5,6] When two or more lineages evolve rapidly, the probability of convergence arising by pure chance increases. This effect is particularly pronounced because of the small alphabet size of nucleotide sequences. The overall result is that rapidly evolving lineages — or long branches in the tree — tend to be placed phyletically close to each other. Long-branch attraction may be avoided by adding taxons that are related to those which have evolved rapidly in order to break up the longer branches.

4.4. *Probabilistic Methods*

Probabilistic methods (statistical phylogenetics) explicitly define a probabilistic model of phylogenesis. At a minimum, each character is attributed a matrix of probabilities for the transformation from one state to another, at any point on the tree. More complex models include variations of mutation rates between sites, explicitly modeled insertion/deletion events,[7,8] and even horizontal gene transfer.

It is then possible to estimate the statistical likelihood of an evolutionary scenario of transformation of the characters, from root to leaves, for each conceivable tree (topology and branch lengths).

Two main classes of algorithms are used to evaluate the likelihood of a given tree. The first one consists of maximum likelihood (ML) methods, which are generally combined with more or less straightforward hill-climbing algorithms. The second class relies on a Bayesian framework and is frequently coupled with Markov chain Monte Carlo (MCMC) algorithms; these Bayesian inferences are also probabilistic in nature, but they proceed by refining a model according to the available data. On occasion, Metropolis-coupled Markov chain Monte Carlo (MC)[3] methods are also used.[q]

Using a probabilistic model has many advantages. Heterogeneous rates of evolution between branches and between sites as well as homoplasy are dealt with explicitly because they are inherent to the models. Statistical evaluation of the results is simple, because the probability of the best calculated tree actually having arisen under the chosen model of evolution can be given directly. Furthermore, probabilistic methods can be used to verify whole new classes of hypotheses, which are beyond the scope of classic methods. Which model of evolution best fits the data? Does the data fit a model of evolution which allows for recombination? If so, which parts of the sequence underwent recombination? All of these questions become possible because likelihoods under various models can be compared and models can be optimized for the dataset being studied.

On the downside, model-based methods generally are very computationally intensive and — as for CMP methods — no guarantee can be

[q] For a detailed review of these methods, their advantages, and their drawbacks, see Holder and Lewis.[9]

given that the heuristics used have actually found the best tree.[6] Indeed, consecutive runs of the same tree reconstruction program on the same data may yield quite different results. Also, care must be taken regarding the duality between model improvement and tree reconstruction: often, exactly the same datasets are used in both steps, making model overfitting a potential problem.

Popular ML methods include PHYML,[10] Tree-Puzzle,[11] and RAxML.[12,13] The most common software implementing Bayesian methods is MrBayes.[14] More elaborate programs exist that can, for instance, perform phylogenetic reconstruction and sequence alignment simultaneously.[15]

Due to the complexity of computing the likelihood function, probabilistic methods are generally slow, even slower than CMP methods.

4.5. *Searching for an Optimal Tree in a Large, Populated Space*

Phylogenetic methods form a complex domain of research and development with numerous technical problems such as preprocessing and postprocessing of the data, inferrence of the soundness of results (statistical significance), etc. The scope of this introductory chapter only allows us to mention some of the more classical ones.

Of the methods mentioned above, only simple clustering methods and NJ give a tree structure directly. The remaining methods primarily score a given tree and use some form of search algorithm to find the tree with the best score. Several approaches are possible. The conceptually simplest approach is to perform a brute-force search in which every conceivable tree topology is tested. Unfortunately, this rapidly becomes unfeasible as the number of taxons increases.

4.5.1. *The number of possible phylogenetic trees*

The search space of possible trees grows very quickly with the number of terminal nodes. The number B of strictly bifurcating unrooted tree topologies for s leaves can be calculated according to a simple formula[16,17]:

$$B(s) = \prod_{t-3}^{s} \frac{(2s-5)!}{2^{s-4}(s-4)}. \tag{1}$$

From a topological point of view, the root of the tree can be regarded as an additional leaf. Thus, the number of rooted tree topologies B_r is given by

$$
\begin{aligned}
B_r(s) &= \prod_{t=3}^{s+1}(2t-5) \\
&= \prod_{t=2}^{s}(2(t+1)-5) \\
&= \prod_{t=2}^{s}(2(t-3) = \frac{(2s-3)!}{2^{s-2}(s-2)!}.
\end{aligned} \tag{2}
$$

This is the same as the simple multiplication of odd numbers shown in Table 1. It is quite clear that performing an exhaustive search of all possible tree topologies for a dataset containing more than 12 to 13 leaves is not a realistic approach, even with modern computers.

4.5.2. *The branch-and-bound algorithm*

A clever algorithm called branch-and-bound is able to reduce the search space and still guarantee that the globally best tree will be found. This is done simply by abandoning any tree-building avenue where the current

Table 1. Multiplication of odd numbers.

s	$B(s)$ || $B_r(t)$			t
3	1	=	1	2
4	1×3	=	3	3
5	$1 \times 3 \times 5$	=	15	4
6	$1 \times 3 \times 5 \times 7$	=	105	5
7	$1 \times 3 \times 5 \times 7 \times 9$	=	945	6
8	$1 \times 3 \times 5 \times 7 \times 9 \times 11$	=	10 395	7
9	$1 \times 3 \times 5 \times 7 \times 9 \times 11 \times 13$	=	135 135	8
10	$1 \times 3 \times 5 \times 7 \times 9 \times 11 \times 13 \times 15$	=	2 027 025	9
11	$1 \times 3 \times 5 \times 7 \times 9 \times 11 \times 13 \times 15 \times 17$	=	34 459 425	10
12	$1 \times 3 \times 5 \times 7 \times 9 \times 11 \times 13 \times 15 \times 17 \times 19$	=	654 729 075	11
13	$1 \times 3 \times 5 \times 7 \times 9 \times 11 \times 13 \times 15 \times 17 \times 19 \times 21$	=	13 749 310 575	12
…	…	=	…	…

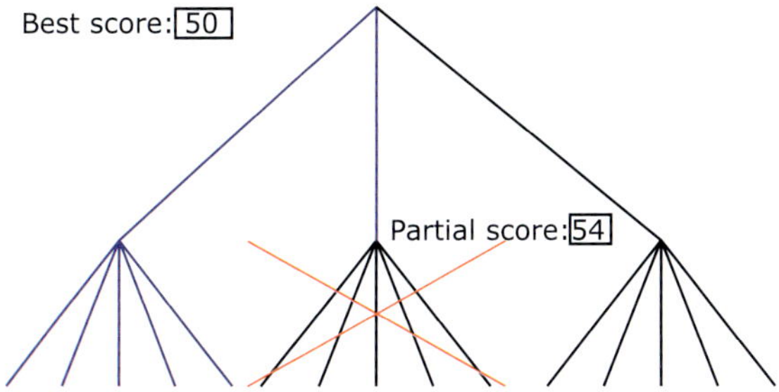

Fig. 12. The branch-and-bound algorithm. The leaves of this decision tree represent all possible complete phylogenetic tree topologies with 5 taxons. The internal nodes represent incomplete phylogenetic tree topologies. The left subtree has already been explored and an optimal tree length of 50 has been found (shorter is better). Since the partial topology represented by the node in the center of the figure already has a length of 54, the entire subtree (of the decision tree) can only contain tree topologies with lengths greater than 54. Since the best complete tree so far has a length of 50, it is not necessary to calculate the length of the phylogenetic tree topology at each of the five leaves in the central part of the decision tree.

incomplete tree is already longer than the shortest complete tree in memory.

The search space of all possible phylogenetic trees can be represented in the form of a decision tree (see Fig. 12). There is only one possible topology for a three-taxon tree. Thus, if taxons are added one by one to a phylogenetic tree, after the tree has reached a size of three taxons, a decision must be made where to place each consecutive taxon as it is added. This process is mirrored by a path from root to leaf in the decision tree. Each internal node of the tree (including the root) represents a decision on where to place the next taxon in the phylogenetic tree. Depending on the chosen position, the corresponding branch in the decision tree is followed. Thus, the leaves of the decision tree represent all possible tree topologies.

Since the inclusion of additional taxons to a partial tree can only make the tree longer (and thus decrease the score), a lower bound for tree score is given by the score of a partial tree corresponding to an internal node of the decision tree. If this score is already worse than that of the best complete tree found so far, then the subtree of the decision tree can be eliminated from the search space in a single step.

4.5.3. *Heuristic methods in tree reconstruction*

The branch-and-bound algorithm is able to increase the practical maximum number of sequences or taxons up to 20 (depending on the degree of homoplasy in the data). Once this threshold is passed, heuristic methods become necessary, as was the case with the multiple-alignment programs mentioned in Sec. 3.2. Otherwise, depending on the complexity of the data (the degree of homoplasy, for instance), CMP, CC, and MLE (maximum likelihood estimation) methods can be impossibly slow in providing a globally optimal solution.

Heuristic methods do not search the entire search space for a global optimum, but nevertheless attempt to find the globally best tree. They must be able to search through parts of the solution space efficiently. Generally, the heuristics consist in searching through the search space, starting from random locations and continuing in a direction where better results are found. Once a local optimum is reached, results become worse in all directions.

The methods used range from simple hill-climbing optimization approaches to the complex MCMC and $(MC)^3$ algorithms. Creating a decision tree as for the branch-and-bound algorithm is no longer possible; therefore, tree rearrangement techniques are used to modify a given tree in the process of optimizing it, including when a local optimum has been found.

It must be stressed that the programs return the best result found, but have no way of knowing whether they have reached the global optimum or not. Thus, all heuristic approaches are somewhat sensitive to local optima and no guarantee can be given that the best tree is indeed found. This problem is common to all nonexhaustive combinatorial search algorithms, whether they are applied to phylogenetics or not.

4.5.4. *A rapid maximum likelihood method: RAxML*

RAxML is one of the fastest probabilistic methods available to date. It uses a combination of specially designed heuristics and simple program optimization. More specifically, it entails[12]

- efficient storage of intermediate solutions (topologies and branch lengths) using rearrangement operators;

- use of lazy subtree rearrangement (LSR), a technique which avoids having to recalculate the complete likelihood function after tree rearrangement;
- dynamic adaptation of rearrangement distance. The LSR technique allows larger and smaller degrees of change during tree rearrangement. This parameter is optimized at the beginning of the search;
- low-level optimizations of the likelihood functions for several models of evolution; and
- efficient implementation of a CMP method to create a starting tree for the ML search.

4.5.5. *Creating consensus trees*

Often, when searching for the "best tree", a number of equally good trees are found. This is especially true for CMP and CC analyses executed on large molecular datasets, where several MP trees may suggest completely contradictory evolutionary scenarios. Obviously, one could consider all of these MP trees individually as alternative solutions; but when several become several thousands, there is a practicality problem.

So, when there are too many MP trees, it becomes necessary to create a consensus tree from all or part of the MP trees. This can be done by drawing a tree containing only the branches common to a certain percentage of MP trees. When the percentage is 50%, 66%, or 100%, the consensus is referred to as majority consensus, semistrict consensus, and strict consensus, respectively.

A somewhat more complicated consensus definition is the Adams consensus. Each MP tree is traced from the root to the leaves, and at each bifurcation the two subsets of terminal taxa are determined. If there is an overlap of these subsets in all MP trees, then it is retained in the Adams consensus.[r] The method is impractical if the number of trees is large.

---

[r] Consider trees I and II in Fig. 13. The strict consensus would have an unresolved tetrachotomy, forming the rake-like tree III. Let us see what the Adams consensus looks like. Starting from the root and arriving at the first intersection of tree I, we obtain subsets {*a*} and {*b*, *c*, *d*}. Likewise, for tree II, we obtain the subsets {*a*, *b*, *c*} and {*d*}. An overlap or nonempty intersection between {*b*, *c*, *d*} and {*a*, *b*, *c*} exists. This subset {*b*, *c*} is retained in the Adams consensus, as shown in tree IV.

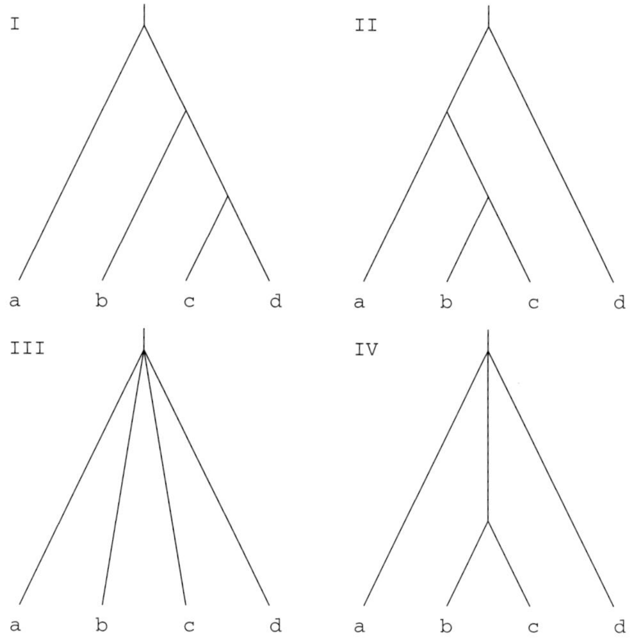

Fig. 13. Tree III is the strict consensus of trees I and II. Tree IV is the Adams consensus of trees I and II.

4.5.6. *Estimating tree robustness*

Because probabilistic methods are based on a probabilistic model, it is possible to analyze the validity of a result statistically. The ability to do so is not inherent to the other methods. However, if we make the assumption that all characters evolve independently of each other and that they all follow the same law of distribution (which need not be known), it is still possible to apply statistics to their outcome.

Some characters are under more selective pressure than others in evolution. A mutation in one region of a gene may result in a complete rearrangement of the three-dimensional structure of a protein; whereas in another region, it would have a biochemically negligible effect such as the change of a single amino acid in a loop structure. However, the correct weighting for characters cannot be known beforehand. Therefore, a large number of random weighting schemes are tested.

If phylogenetic analysis is performed with each of the weighting schemes and the same tree is produced each time, a high degree of confidence can be attributed to the result. On the other hand, if changing a single weight completely alters the outcome, care should be taken when interpreting the results.

How then is this random weighting done? The simplest technique is called jackknife. For a dataset containing n characters, n trees are calculated by eliminating each character in turn from the original dataset. In other words, the weight of a single character is set to 0 in each dataset and all other characters are given identical weights. The fraction of jackknife trees that are identical to the global tree solution is given.

A slightly more complicated technique, taking more calculating time but producing more convincing estimates of robustness, is called the bootstrap. Here, a large number of artificial datasets are randomly created, typically 100 or more. Each artificial or bootstrap dataset is created by drawing n characters randomly from the original dataset of size n, with replacement. One can picture a box containing the n characters: after a character is drawn from the box, it is put back in for the next draw. This way, some characters may be drawn several times, while others are not drawn at all. Subsequently, a bootstrap tree is calculated for each bootstrap dataset. Then, for each branch of the original tree in turn (the one obtained with the original dataset), the bootstrap proportion of that branch is calculated as the percentage of bootstrap trees which contain the branch. The bootstrap proportion may be considered as a measure of confidence in the branch. The confidence level can therefore vary throughout the tree.

There are other methods of testing the robustness of individual branches, but they are specific to a certain reconstruction technique. MCMC methods, for instance, test many intermediate trees during their search phase. It is possible to benefit from these intermediate trees in order to obtain confidence levels for individual branches of the final phylogenetic tree. However, since the sample of intermediate trees is strongly autocorrelated, much larger samples are required than for bootstrap methods.[9]

Similarly, in CMP and CC, a method called branch-decay analysis can be used. As with Bayesian methods, large numbers of trees are calculated while the solution space is explored, with only the shortest MP trees being retained. Since tree length in CMP methods is defined as the minimal number of mutations in a given topology, the result will always be an integer. Likewise, in CC methods, the size of the largest clique will always be an integer. Thus, in these methods, if intermediary trees are not immediately discarded, a strict consensus tree can be produced for a set consisting of both the MP trees and the trees which are just one unit longer than optimal (longer tree for CMP, smaller clique for CC).[s] Any branch of the final tree that disappears in this consensus is considered a weak branch and given a decay index of one. This method can be extended for trees which are *z* units longer than optimal, yielding branches with a decay index of *z*, until all of the branches have been given a decay index. A high decay index indicates a high confidence level.

4.6. *Recommended, Acceptable, and Unacceptable Groupings of Taxons*

The central objective of phylogenetics is the grouping of taxa, on the basis of evolutionary criteria, in order to reconstruct an evolutionary tree. As we have seen earlier (Sec. 2.4), an erroneous hypothesis of homology could have been made. Clearly, such an error would entail mistaken groupings of taxa. But groupings are not necessarily all wrong or all true. There are nuances in plausibility, and some categories of grouping are considered to be *a priori* of higher or lower theoretical quality according to the classification rules of taxonomy, depending on the circumstances and the classification objectives.

A grouping that contains the complete set of terminal nodes descendant from a common ancestor is called holophyletic, and forms a holophylon. This is a somewhat utopian concept, since we cannot really be sure that all descendants of a given ancestor are known. A grouping of terminal taxa in a dendrogram is called monophyletic if it contains all

[s] There is no need to keep all intermediary trees in memory until the end of the search. Since a strict consensus is used, only the preliminary consensus tree for each score needs to be conserved.

taxa displayed in the tree, which are descendants of an internal node (therefore, this node is the cenancestor of the terminal taxa in the grouping in question); this grouping forms a monophylon. Accordingly, a holophylon is a special, complete kind of monophylon.

It is possible to define a group in which some descendants of its cenancestor are not included. In other words, a systematic choice is made (consciously or not) to exclude some siblings from a group. Such groupings come in three different varieties, some of which are more acceptable than others. They are discussed in the remainder of this section.

4.6.1. *Paraphylon — Acceptable with caveat*

When a grouping of symplesiomorphic taxa — assembled on the basis of some shared ancestral characteristic(s) — excludes from its midst some siblings, i.e. descendants from the cenancestor of this group, it is called paraphyletic and constitutes a paraphylon. These siblings have been excluded either because they do not share the plesiomorphy of the group (some ancestral characteristic(s) which they have lost) or because they have developed their own apomorphy (a unique trait), which sets them apart.

Though phyletically incorrect, paraphylons are generally considered acceptable if they are helpful from a classification point of view. A classical example is the exclusion of birds and mammals from present-day reptiles.

On the basis of a number of "primitive" traits, present-day reptiles are defined as including the tuatara (*Sphenodon*, a "living-fossil" rhynchocephalian), Chelonia (turtles), crocodilians, as well as Squamata (lizards and snakes), and excluding birds and mammals. However, nowadays, crocodilians are generally considered to be phyletically closer to birds than to any other existing reptile (see right dendrogram in Fig. 14); accordingly, the *Reptilia* taxon is a paraphylon. Nevertheless, it is still a taxonomically useful grouping because it marks the evolutionary and anagenetically important boundary between the two taxa of fully homoiothermic amniotes (*Mammalia* and *Aves*) and poikilothermic amniotes (*Reptilia*).

There is another classical example of useful paraphylon. As opposed to "primitive", cartilaginous, sea-dwelling fish (*Chondrichthyes*), bony

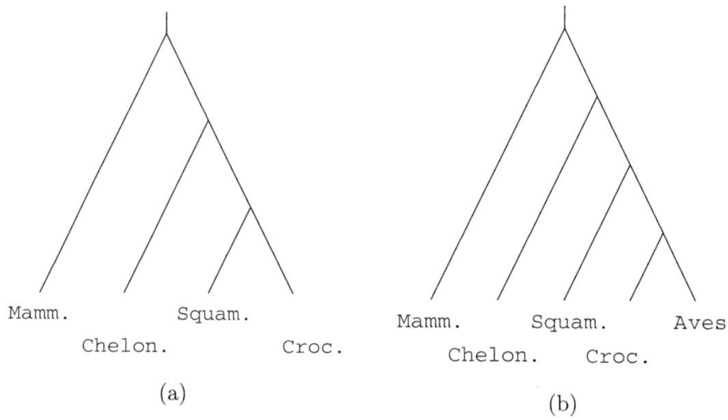

Fig. 14. (a) The grouping of Chelonia (turtles), Squamata (snakes and lizards), and Crocodilia as reptiles forms a monophylon; but this grouping cannot be considered holophyletic, since the Aves taxon (birds) is missing (see (b)).

fish (*Osteichthyes*) are defined to include *Actinopterygii*, *Crossopterygii* (the living-fossil coelacanth), as well as *Dipnoi* (lung-fish), and to exclude Tetrapoda, which are bony but not fish. However, lungfish and coelacanths are phyletically nearer to tetrapods than either is to actinopterygian fish, so the *Osteichthyes* taxon is a paraphylon. Nevertheless, it is a taxonomically useful grouping, marking the evolutionary and anagenetically important boundary between fish with a more or less bony skeleton and strictly cartilaginous *Chondrichthyes*.

4.6.2. *Convergence and reversion polyphylons — Unacceptable*

A polyphylon is a grouping of taxa based on some evolved characteristic(s) which is indeed shared, but has appeared at several distinct times in evolution. As seen in Sec. 2.4, a homoplasy is the presence of a character state in various taxa that is not due to common ancestry. Accordingly, a phylogenetic grouping based on this character is polyphyletic. While there are lots of cases of homoplasy in nature, groupings based on them should be avoided: polyphylons are phyletically unacceptable.

Referring to the definition of homoplasy, one can differentiate between convergence and reversion polyphylons. Let us consider the return-to-the-sea example from Sec. 2.4. If all marine vertebrates were grouped together in a reversion polyphylon, on the basis of their common hydrodynamic morphology, one would mask the fact that the marine mammals have returned to the sea: the land-dwelling adaptation of ancient mammals as well as the following readaptation to sea of marine mammals are the two masked events. The marine mammals are wrongly presented as not having evolved at all morphologically, while they have in fact evolved a lot! They are definitely not fish!

Alternatively, if one grouped all present-day active-flight vertebrates (birds and bats), as we have seen this would mean creating a convergence polyphylon, a grouping which would conceal the fact that their cenancestor did not fly at all! The fundamental fact that active flight has developed twice and independently so in the two lineages would be completely masked. In a way, a convergence polyphylon is even more absurd than a reversion polyphylon because the very character that has been used to combine the two lineages into a group is not present at all in their cenancestor!

5. Uses of Phylogenetics in Molecular Biology

So far, we have seen many sides of phylogenetics, ranging from the hypothesis of homology to the problems of homoplasy and varying evolutionary rates. We have briey discussed three main classes of algorithms used for the reconstruction of phylogenetic trees. In this section, we will shed some light on the role of phylogenetics in molecular biology and genomics.

5.1. *Prediction of Gene Function*

A most fruitful use of phylogenetics is in predicting gene function. With the sequencing of many genomes, including the human one, focus is shifting towards the identification of genes and the determination of gene function. While gene function was often known at the time of sequencing before the large genome projects began, today many sequences are only surmised to be genes due to the structure of their sequence, and functions are attributed by comparison with genes in other species.

Given a recent gene duplication event, even if one copy has degenerated and lost its function, it will still have a gene-like structure. Gene identification algorithms will classify it as a novel gene and automatic systems will attempt to determine its function. It is only by creating a paralogous phylogenetic tree of the gene family that the error can be detected. Furthermore, since after gene duplication constraints on the nonfunctional copy are virtually nonexistent, this copy will evolve more rapidly than a functional one. It is therefore important to use robust tree reconstruction methods. A simple NTP clustering method will not give a useful result.

If, on the other hand, an attempt is made to determine the function of a gene by analogy to genes in another organism without creating a detailed phylogenetic tree beforehand, problems will also arise. One such method of function attribution called COG (Cluster of Orthologous Genes)[18] proceeds as follows. Gene *X* from species 1 and gene *Y* from species 2 are considered orthologous if *X* has a higher similarity to *Y* than to any other gene from species 2, and vice versa. In other words, the top-scoring match in a BLAST search of *X* against known sequences in species 2 returns *Y* at the top of the list, and a BLAST search of *Y* against all known sequences from species 1 returns *X* at the top of the list. Once genes are considered orthologous, their functions are deemed to be similar and the unknown gene can be attributed its putative function.

This process is fundamentally flawed. Inferring orthology on the sole criterion of highest degree of similarity is not possible. If two paralogous genes are present in an ancestor and a different one of the pair is deleted in each of two descendants, it is likely that the COG method will show the two surviving genes to be orthologs. Moreover, this is a best-case scenario, in the sense that datasets are assumed to be complete. If one of the genomes has not been completely sequenced, matters are still worse since the real ortholog may be present in the genome but absent from the database. The only way to resolve the issue of orthology is to create a phylogenetic tree of a gene family with as high a number as possible of different species present. While two descendants of a given ancestor may each contain one copy of a pair of paralogs, it is likely that both paralogs will have survived in a third or fourth species.

5.2. Advanced Phylogenetic Analyses and New Directions

In this chapter, we have considered each nucleotide to be a separate character and each changed nucleotide to be the result of at least one independent evolutionary event. In reality, however, large portions of genes may be deleted, inserted, or moved in a single event, and this can occur across the boundaries of a gene. Advanced methods of phylogenetic inference can detect such events and situate them within a tree.

Furthermore, there are some areas of biology where the standard tree model for evolution does not hold true. This is especially true for viruses. Specific techniques such as split decomposition analyses have been developed to analyze such cases. Epidemiologists can learn a great deal about the origin of new diseases by applying these tools. As an example, the SARS epidemic is thought to be the result of a past recombination of mammalian and avian viruses.[19]

Other advanced methods of tree analysis allow more information to be gleaned from an evolutionary scenario. Protein domain shuffling, evolutionary diversification of gene families, and prediction of structure-function relationships are just some of the possibilities.

The diminishing cost of sequencing and the ever-greater availability of molecular sequence data are leading to a number of novel approaches in phylogenetics. Here are some of the new directions being explored:

- Methods have been developed to allow the construction of a plausible tree from thousands of homologous sequences, explicitly taking into account homoplasy and lineage-dependent rate heterogeneity.[12,20,21]
- With the sequencing of complete genomes, evolution of complete genomes can be studied and phylogenetics can take a totally new direction. The presence or absence of complete (orthologous) genes can be considered a phylogenetic character. Matrices of the presence/absence of genes can be used as input for phylogenetic analysis.[22]
- Studies have attempted to reconstruct large portions of ancestral genomes.[7,8]

- Large-scale sequencing of key genes from unclassified micro-organisms has led to phylogenetic studies in which sequences are not attributed to specific species. Rather, a threshold of sequence identity is used to classify sequences into species before phylogenetic reconstruction begins.[23] This approach may lead to rapid surveys of ecosystems if sequences attibuted to known species are added to the dataset, or may yield estimates of biodiversity on their own. For micro-organisms, the major advantage of such techniques lies in the fact that unculturable species can be included.[24–26]
- Allelic variations have been included in large-scale phylogenetic studies (such data are normally used in population studies).[27]

6. Phylogenetics Resources

Due to the highly volatile nature of websites and their diversity, it is difficult to give Internet references here. Joseph Felsenstein's site (http://evolution.genetics.washington.edu/phylip.html) does, however, seem to be fairly stable. It is the home of PHYLIP (PHYLogeny Inference Package), a classic suite of phylogenetic computer programs provided for free with the C sources. It is also a good portal to many other theoretical and practical reference sites on phylogenetics, and maintains a listing of available phylogenetics software.

References

1. Semple C, Steel M. (2003) *Phylogenetics.* Oxford Lecture Series in Mathematics and Its Applications. New York, NY: Oxford University Press.
2. Swofford DL. (2003) *PAUP*. Phylogenetic Analysis Using Parsimony (*and other Methods). Version 4.0b10.* Sunderland, MA: Sinauer Associates.
3. Meier R, Ali FB. (2005) Software review. The newest kid on the parsimony block: TNT (Tree analysis using New Technology). *Syst Entomol* **30**: 179–82.
4. Kumar S, Tamura K, Nei M. (2004) Mega3: integrated software for molecular evolutionary genetics analysis and sequence alignment. *Brief Bioinform* **5**: 150–63.
5. Bergsten J. (2005) A review of long-branch attraction. *Cladistics* **21**: 163–93.

6. Felsenstein J. (1978) Cases in which parsimony or compatibility methods will be positively misleading. *Syst Zool* **27**: 401–10.
7. Blanchette M, Green ED, Miller W, Haussler D. (2004) Reconstructing large regions of an ancestral mammalian genome *in silico*. *Genome Res* **14**: 2412–23.
8. Ma J, Zhang L, Suh BB *et al.* (2006) Reconstructing contiguous regions of an ancestral genome. *Genome Res* **16**: 1557–65.
9. Holder M, Lewis PO. (2003) Phylogeny estimation: traditional and Bayesian approaches. *Nat Rev Genet* **4**: 275–84.
10. Guindon S, Gascuel O. (2003) A simple, fast, and accurate algorithm to estimate large phylogenies by maximum likelihood. *Syst Biol* **52**: 696–704.
11. Schmidt HA, Strimmer K, Vingron M, von Haeseler A. (2002) TREE-PUZZLE: maximum likelihood phylogenetic analysis using quartets and parallel computing. *Bioinformatics* **18**: 502–4.
12. Stamatakis A. (2006) RAxML-VI-HPC: maximum likelihood-based phylogenetic analyses with thousands of taxa and mixed models. *Bioinformatics* **22**: 2688–90.
13. Stamatakis A, Ludwig T, Meier H. (2005) RAxML-III: a fast program for maximum likelihood-based inference of large phylogenetic trees. *Bioinformatics* **21**: 456–63.
14. Ronquist F, Huelsenbeck JP. (2003) MrBayes 3: Bayesian phylogenetic inference under mixed models. *Bioinformatics* **19**: 1572–4.
15. Suchard MA, Redelings BD. (2006) BAli-Phy: simultaneous Bayesian inference of alignment and phylogeny. *Bioinformatics* **22**: 2047–8.
16. Cavalli-Sforza LL, Edwards AWF. (1967) Phylogenetics estimation. Models and estimation procedures. *Am J Hum Genet* **19**: 233–57.
17. Felsenstein J. (1978) The number of evolutionary trees. *Syst Zool* **27**: 27–33.
18. Tatusov RL, Galperin MY, Natale DA, Koonin EV. (2000) The COG database: a tool for genome-scale analysis of protein functions and evolution. *Nucleic Acids Res* **28**: 33–6.
19. Stavrinides J, Guttman DS. (2004) Mosaic evolution of the severe acute respiratory syndrome coronavirus. *J Virol* **78**: 76–82.
20. Bittar G. (2002) The Anâtaxis phylogenetic method. *Arch Sci Geneve* **55**: 1–20.
21. Sonderegger BP. (2007) The Anâtaxis phylogenetic reconstruction algorithm. PhD thesis. University of Geneva, Geneva, Switzerland. URL http://www.unige.ch/cyberdocuments/theses2007/SondereggerBP/these.pdf/.
22. Dessimoz C, Cannarozzi G, Gil M *et al.* (2005) Oma, a comprehensive, automated project for the identification of orthologs from complete genome data: introduction and first achievements. In: Istrail S, Pevzner P, Waterman M (eds.). *Comparative Genomics*: *RECOMB 2005 International Workshop*. Lecture Notes in Computer Science, Vol. 3678. Berlin, Germany: Springer, pp. 61–72.
23. Pons J, Barraclough TG, Gomez-Zurita J *et al.* (2006) Sequence-based species delimitation for the DNA taxonomy of undescribed insects. *Syst Biol* **55**: 595–609.

24. Ley RE, Bäckhed F, Turnbaugh P *et al.* (2005) Obesity alters gut microbial ecology. *Proc Natl Acad Sci USA* **102**: 11070–5.
25. Ley RE, Harris JK, Wilcox J *et al.* (2006) Unexpected diversity and complexity of the Guerrero Negro hypersaline microbial mat. *Appl Environ Microbiol* **72**: 3685–95.
26. Robertson CE, Harris JK, Spear JR, Pace NR. (2005) Phylogenetic diversity and ecology of environmental archaea. *Curr Opin Microbiol* **8**: 638–42.
27. Joly S, Bruneau A. (2006) Incorporating allelic variation for reconstructing the evolutionary history of organisms from multiple genes: an example from *Rosa* in North America. *Syst Biol* **55**: 623–36.

Chapter 12

Phylogenetic Tree Building Methods

Manuel Gil and Gaston H. Gonnet

1. Introduction

Phylogenetic tree construction, from given data, is a well-defined problem that is taken as a basic building block in bioinformatics. A phylogenetic tree is a hierarchical representation of some source data. The leaves of a tree represent present-day objects (genes, species), and the internal nodes represent common ancestors. A tree is generally n-ary, i.e. the internal nodes can have any number of descendants, although there is no loss of generality representing them as binary trees. From now on, we will consider all our trees binary. A tree can be rooted or unrooted. The root is the place of the common ancestor of all leaves. Most tree reconstruction methods are not able to determine the root and produce unrooted trees. It can be shown[1] that the number of unrooted leaf-labeled tree topologies with n leaves is

$$3 \times 5 \times 7 \times \cdots \times (2n - 5).$$

While for five leaves there are 105 trees, already for 50 leaves the number explodes to about 2.84×10^{74}.[a] The goal of phylogenetic tree construction is to find the tree in the tree space that describes the relation between a given set of input data points as best as possible. It is normally difficult or impossible to obtain data for the internal nodes of the tree.

[a] The number of atoms in the universe is estimated to be about 10^{80}.

We only have data for the leaves so that the branching process is implicitly assumed not to be observable and its reconstruction is our main goal.

This chapter is organized into three sections, progressing from the general to the specific. In the first section, we give some overall features used to classify phylogeny reconstruction methods and present some popular methods. The second section treats a particular class, the optimality-based methods. We present a general strategy to search for the optimal tree(s) for a given criterion. One particular optimality-based method is the least squares method. In the third section, we describe the least squares tree building program developed in our research group. The section also includes a time comparison with the corresponding implementation in PAUP*[2] and a detailed description of a new heuristic. We conclude this chapter with an outlook on the assessment of the accuracy of constructed trees.

2. Types of Tree Building Methods

Tree building methods can be classified according to several criteria. A criterion can refer not only to the algorithmic or statistical method used to infer a tree from given input data, but also to the input data itself. Maximum likelihood (ML) methods, for example, get their name from the statistical approach taken; whereas the term "distance methods" refers to the type of input data. Certain criteria are related to each other. For instance, ML and parsimony methods are both character-based and optimize for a defined score, so we can expect that the two methods use, to some extent, similar algorithmic approaches. In this section, we first list prevalent criteria used to classify phylogeny methods. Subsequently, we present some widely used tree building methods in light of these criteria.

- *Type of input data.* Sequence-based phylogeny methods normally use a multiple sequence alignment (MSA) as input. The two-step approach, MSA construction followed by phylogeny construction, is a very common one. Nevertheless, it is well known that MSA and phylogeny construction are closely related tasks; this can be seen in the many MSA construction programs (e.g. CLUSTAL W,[3] MAFFT,[4] MUSCLE[5]) that use, as a means to an end, a guide tree obtained by some crude method like UPGMA[6] (Unweighted pair group method using arithmetic averages). The relatedness between the two problems has been a motivation to come up with phylogeny methods that operate on unaligned sequences and solve the extended problem of combined MSA and tree inference (see, for example, Refs. 7–9). Sequence-based methods are based on similarities in the amino acid or base sequence of genes.

Evolution also operates on whole genomes by the inversion, transposition, and duplication of (groups of) genes. Gene content methods use the presence/absence profiles of orthologous genes to construct trees. Together with the sequence-based methods, they can be classified as character-based methods. In recent years, interest in genome rearrangement tree building algorithms that use the order of genes on a chromosome as an input has flourished. Popular implementations include MGR[10] and GRAPPA.[11]

Distance methods rely on a measure of the pairwise evolutionary distances between the objects (sequences, genomes, species) being classified. The distances should reflect the leaf-to-leaf path lengths of an underlying tree. Typically, they are estimated from character or gene order data in a statistical framework under some model of evolution, but they can also be obtained from other processes such as DNA–DNA hybridization.

Sequence and distance methods are the most established ones and can be used to build gene as well as species trees, whereas gene content and gene order methods lead to species trees only.

- *Model assumptions on the data.* All tree building methods make, explicitly or implicitly, assumptions about the data. For character-based methods, the assumptions often refer to the evolutionary processes under which the data arose. For example, sequence-based methods commonly use a first-order Markovian model of sequence evolution.[b] Such models of evolution are not used to classify tree building methods.

The ones used are rather related to the underlying tree structure. Some character and distance methods, for example, assume a molecular clock. Under a molecular clock, the underlying phylogenetic tree has equal root-to-leaf path lengths for all lineages. Such a tree is called an ultrametric tree. Another example concerns distance methods. As mentioned above, evolutionary distances have to be estimated and are therefore uncertain. Nevertheless, there are phylogenetic methods that assume exact distances called additive tree construction methods. Distance methods that take the inference uncertainty into account, on the other hand, use a model to characterize the error. Least squares tree construction is an example of such a method; Sec. 4 is devoted to the corresponding method developed in our research group. A property assumed implicitly by essentially all distance methods is additivity. In the evolution from an object A to B and then from B to C, additive distances have the property $d_{AB} + d_{BC} = d_{AC}$. To be able to derive branch lengths from leaf-to-leaf distances (sums of many branches), the distance function has to be additive. Distance estimates of nonadditive distances are not suitable for tree reconstruction.[c]

- *Algorithmic methods.* Algorithmic methods can be divided into two classes. In the first class, the tree inference and the definition of a preferred tree are combined into a single statement. Such methods usually perform some sort of clustering and have the advantage of being fast. They exist for character and distance input data. They are widely used for the latter and are in this case typically designed to infer the correct tree, given that the distances are

[b] The first-order Markov property implies that each character mutates with a probability that depends only on itself and not on the history of previous mutations. Additionally to the Markov property, the sites in a sequence are in almost all models assumed to mutate independently of each other.

[c] The distance metric related to the percent identity (essentially one minus the fraction of identical residues in an alignment) is an example of a nonadditive distance.

additive or ultrametric, and exact. These restricted requirements are, however, never met in practice, so the trees returned are suboptimal. UPGMA and neighbor joining (NJ) are examples of these methods.

The second class is in principle better suited to deal with real data. The goal for these methods is to find the best tree for an explicit optimality criterion, thereby separating the problem of evaluating and searching trees. Ideally, we would want to score all tree topologies to find the optimal one; however, as shown in Sec. 1, the number of tree topologies grows rapidly with the number of leaves so that a complete enumeration becomes impractical already for, say, 15 leaves.[d] In all cases, searching the tree space makes the problem difficult. Given a topology, finding the branch lengths is generally easy. Section 3 is devoted to the description of a heuristic approach that can be taken to tackle this problem.

In our opinion, the first class is poorly defined. By lacking an optimization criterion, one can never be sure whether a tree is poorly constructed (algorithm is not good enough) or the data is not good enough. Hence, we recommend algorithms that have a precise optimization goal.

- *Statistical method.* The school of statistics used is a further way to classify tree building methods. There are two common approaches for the statistical analysis of empirical data and parameter estimation: the frequentist and the Bayesian approaches. They are divided on the fundamental definition of probability. In the frequentist's definition, probability is seen as the long-run expected frequency of the occurrence of events; whereas the Bayesian definition views probability as a measure of a state of knowledge. Implicit in the frequentist's view is that a parameter ϕ is a fixed quantity in nature that we wish to measure. In the Bayesian approach, the existence of a true value of ϕ is not necessarily assumed. To Bayesians, parameters are random variables, not constants. For example, in the frequentist's view of a coin-tossing

[d] All of the nontrivial tree building can be viewed formally as an NP-complete optimization problem (see Refs. 12–14).

experiment, the probability that a coin will come up heads is ϕ, because as the coin is flipped more and more times, the observed proportions of heads become on average closer and closer to ϕ. In the Bayesian view, a coin does not pose as a constant ϕ; instead, the probability that a coin should come up heads is viewed as a current state of knowledge, like, for example, the physics of the situation and a number of test flips. Cox[15] gives a short comparative introduction to the two schools.

One can also differentiate between parametric and nonparametric statistics. In parametric statistical modeling, one assumes that the variables being determined belong to known parametric probability distributions such as, for example, the normal distribution. In contrast, nonparametric models do not rely on assumptions that the data are a realization of a particular distribution. A histogram is an example of a nonparametric estimate of a probability distribution.

Given the number of possible approaches to tree building, it is no surprise that there is a great and increasing number of tree building methods. In the following, we review some popular ones.

- *Unweighted pair group method using arithmetic averages* (*UPGMA*). UPGMA is a classic, greedy, bottom-up clustering algorithm. It is a very fast distance method and produces ultrametric trees. At the beginning of the algorithm, each object is in its own cluster. At each step, the two closest clusters are merged and the distances are updated. The distance between clusters U and V is taken to be the average of all distances between pairs of objects in U and V. UPGMA does not use a statistical method. It is designed to construct the correct tree given that the input data are ultrametric and exact. If the data do not behave like a molecular clock, then UPGMA may produce the wrong tree, even for simple quartets and exact distances. Here is an example. Given the pairwise distances of the tree UPGMA

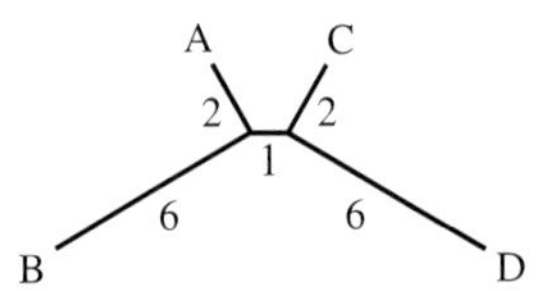

clusters, in the first step, the leaves A and C as they have the smallest distance ($d_{AC} = 2 + 1 + 2 = 5$), and therefore produces the wrong tree. The wrong grouping of long branches due to methodological problems is known as long branch attraction (LBA). LBA is not exclusively a problem of UPGMA; any tree building method using a too simplistic model of evolution or nonadditive distances can be susceptible to LBA. An elaborated review of LBA is Bergsten.[16]

- *Least squares* (*LS*). Besides ultrametricity, UPGMA assumes input distances that exactly reflect a tree. In practice, distances have an estimation error; as a consequence, they cannot be mapped exactly on a tree. The goal then is to find a tree (topology and branch lengths) that satisfies them as best as possible according to some optimality criterion. Let ε_{ij} be the normalized discrepancy for the distance between leaves i and j, defined as

$$\varepsilon_{ij} = \frac{T_{ij} - d_{ij}}{\sigma_{ij}},$$

where T_{ij} is the leaf-to-leaf path length on the tree; d_{ij} the input distance; and σ^2_{ij}, its variance. A very good optimality criterion is to minimize the norm $||\varepsilon||$. A common choice is to use the Euclidean norm $||\varepsilon||_2 = \sum_{ij}\varepsilon^2_{ij}$, which leads to a weighted least squares (WLS) tree. This has a statistical justification. Under a model where the errors in d_{ij} are independently and normally distributed with expected value zero and variance σ^2_{ij}, the tree with minimal $||\varepsilon||_2$ is the maximum likelihood distance tree.[c] Maximum likelihood is described later.

Sometimes, the variances are not known and modeled as a function of the distances. In the Fitch and Margoliash method,[17] for instance, it is assumed that the variances are proportional to the squared distances. If all of the variances are assumed to be equal, the method is called ordinary least squares (OLS).

---

[c] This is usually not the case, as distance estimates are correlated if they share branches on their leaf-to-leaf paths of the true tree. In this case, the optimal method is generalized least squares.

- *Minimum evolution* (*ME*). The ME method[18] searches among all trees for the one with the smallest sum of branch lengths. The branch lengths for a particular topology are determined by the OLS method. The neighbor joining algorithm[19] described next is a greedy approximation of ME.
- *Neighbor joining* (*NJ*). NJ, like UPGMA, performs a bottom-up clustering. Unlike UPGMA, it does not assume a molecular clock. The basic idea of NJ is to join clusters that are not only close to one another, but also far from the rest. The algorithm starts with the star topology.

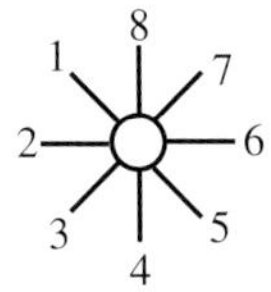

It then considers all trees obtained by grouping pairs of leaves

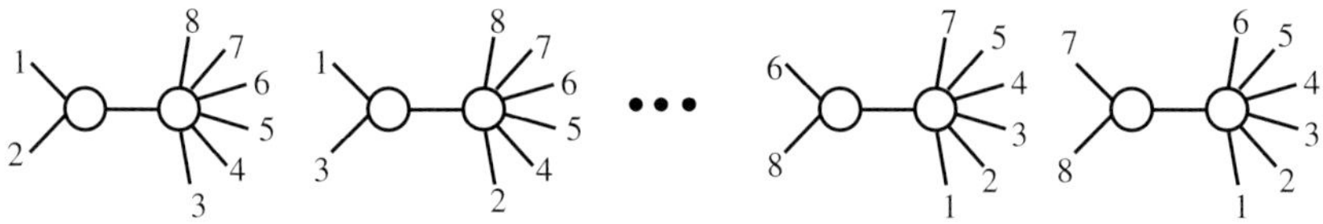

and maps the input distances by the OLS method on each of them, choosing the candidate with the smallest sum of branch lengths. This corresponds to a greedy ME step. We assume, for the sake of the example, that the pair (1,2) has been chosen. The pair (1,2) is joined and the distances between the new node (1,2) and each of the remaining leaves are computed. We omit the details of this step.

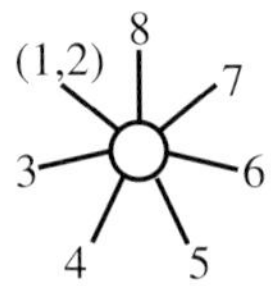

The pair (1,2) can be treated as a leaf now, and NJ repeats the procedure with the reduced star until the tree is fully resolved.

NJ reconstructs the correct tree if the input distances are additive and exact. In cases where the distances are estimated, it can be viewed as a greedy approximation to the ME method. BIONJ[20] and Weighbor[21] are weighted versions of the NJ algorithm; both methods take into account the estimation error of the input distances.

- *Parsimony*. Parsimony is a nonparametric, optimality-based character method. It selects the tree that needs the minimum number of character changes to explain the evolution of the leaves. By being nonparametric, the method does not rely on an explicit model of character evolution. Felsenstein[22] demonstrated that parsimony can lead to LBA. He used a simple model of evolution and a quartet with similar characteristics as the one given in the subsection on UPGMA (two long branches separated by a short one) to show that the trees inferred by parsimony tend to group the two long branches together.
- *Maximum likelihood* (*ML*). ML estimation belongs to the frequentist school of statistics and is a method to fit the parameters (P) of a mathematical model (M) to given data (D). The probability of the data given the model $Pr(D|M, P)$ is central to the ML method. When the model (including its parameters) is kept constant, the probability over all possible data sets sums to one. The ML method takes a different view. It considers $Pr(D|M, P)$ a function of the parameters. The data and the form of the model are kept constant. In this case, $Pr(D|M, P)$ is called the likelihood function for P, often written as $L(P) = Pr(D|M, P)$. The ML method simply chooses the parameters $\hat{P}$ which maximize $L(P)$, i.e. $\hat{P} = \text{argmax}_P\, L(P)$. In other words, the parameters of the model are chosen that make the input data most likely.

  For phylogenetic tree building, the parameters to be estimated are at least the tree topology and the corresponding branch lengths, but they can also include other model parameters like, for example, shape parameters of site rate distributions. ML can, in principle, be applied to any type of data as long as a model can be specified such that it could have been generated. We have mentioned above that, under certain assumptions, LS tree construction leads to ML distance trees. Nevertheless, when we speak of ML tree reconstruction, we normally refer to sequence-based methods. In this case, the data consist of a multiple sequence alignment and the preferred substitution model is a Markovian one. PAML,[23] PHYML,[24] and RAxML[25] are well-known implementations of the sequence-based ML method.

The same substitution model used for sequence-based ML is often used on pairs of sequences to estimate pairwise distances used by LS tree methods. During this conversion, a reduction of information takes place. As a consequence, LS methods are expected to be statistically less efficient than ML methods, which use the sequence information directly. The advantage of the distance approach is that it is substantially faster, allowing fast reconstruction of very large trees.

The tradeoff between speed and topological accuracy is also an algorithmic issue. Greedy methods are faster than optimality-based ones, but they generally produce inferior trees. In the next section, we will see how greedy methods can be used to produce starting trees, which are then improved using heuristics and an optimality criterion.

3. Tree Building with an Objective Function

In this section, we treat the problem of finding the best tree according to some optimality criterion. We have seen in Sec. 1 that an exhaustive search of the tree space is not possible, even for a small number of leaves. A general strategy to solve difficult problems is to use a constructor-heuristic-evaluator (CHE) approach. A CHE procedure constructs an initial solution with the constructor, and then applies heuristics to improve it; the result is then evaluated to decide whether it is retained or not. The following is a high-level description of the CHE approach in pseudocode:

```
sol: = constructor(problem)
repeat
  new: = heuristic(sol, problem)
  if evaluator(new) > evaluator(sol) then
     sol: = new
  endif
until finish-criteria
```

The evaluator scores a given tree with the method's optimality criterion. Examples of evaluators are $||\varepsilon||_2$ for WLS trees, $L(P)$ for ML trees, and the number of necessary changes for parsimony. We will not discuss evaluators any further. In the following two subsections, we will discuss the constructor part of the CHE and look at different kinds of heuristics to search a tree space.

3.1. *Constructors*

A constructor builds an initial tree from the input data and is therefore itself a tree building method. It should be fast and provide a reasonable start for the optimization part. The greedy clustering algorithms mentioned in the previous section have these properties. In principle, a completely random tree is also a constructor, but it is usually so poor that most heuristics are unable to recover. A constructor having some randomness is even a better idea because (a) better trees can be obtained by optimizing from different starting points, and (b) different starting trees can be optimized independently of each other and hence we can perform the time-consuming optimization in parallel. Ideally, we should be able to control the level of randomness of the constructor with a parameter, so that at one end we have total randomness (poor trees, great diversity) and at the other end we have deterministic trees (reasonably good ones, no diversity). One possibility to achieve this goal is to randomize greedy algorithms.

We illustrate this tunable randomization on the example of UPGMA. We will modify UPGMA so that at every joining step, instead of taking the nearest neighbor, it makes a random choice. Let p be the randomness parameter, where $p = 0$ means total randomness and $p = 1$ corresponds to determinism. At each step, the choices are made with the following probabilities:

- Pr {join nearest neighbors} $= p$
- Pr {join 2nd nearest neighbors} $= p(1 - p)$
- Pr {join 3rd nearest neighbors} $= p(1 - p)^2$
- ...
- if all fails, make a uniform random choice among all neighbors.

So, $p = 0$ guarantees a purely random join at every step, which will produce a random tree; while $p = 1$ corresponds to UPGMA. There is a clear trade-off between the quality of the initial tree and the diversity (and hence exploration of a large space of trees). Extensive simulations, beyond the scope of this chapter, show that choosing $p = O(1/\log(n))$ gives a very good trade-off.

3.2. *Heuristics*

The heuristics we present in this section include four that have also been described by Felsenstein[26]: nearest-neighbor interchange (NNI), subtree pruning and regrafting (SPR), tree bisection reconnection (TBR), and tree fusing. Additionally, we describe the heuristics n-optim (an extension of NNI) and reduce best-fitting subtree (RBFS). RBFS is new to the best of our knowledge.

3.2.1. *n-optim*

This is a local heuristic that identifies a pattern subtree and attempts to improve the quality by permuting it. We call it n-optim, or a pattern subtree with n leaves. Felsenstein's NNI corresponds to 4-optim. For a given quartet with the subtrees labeled by A, B, C and D

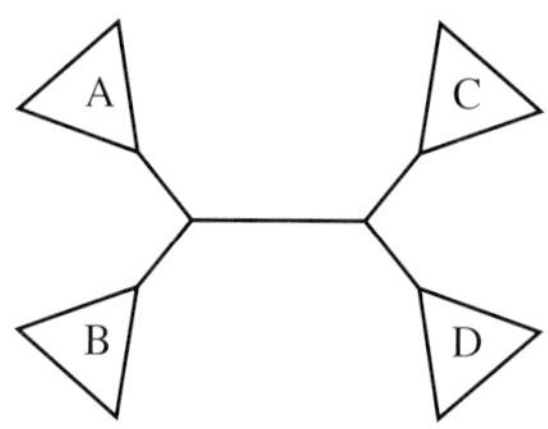

4-optim operates by considering the two other rearrangements (symmetries are not relevant):

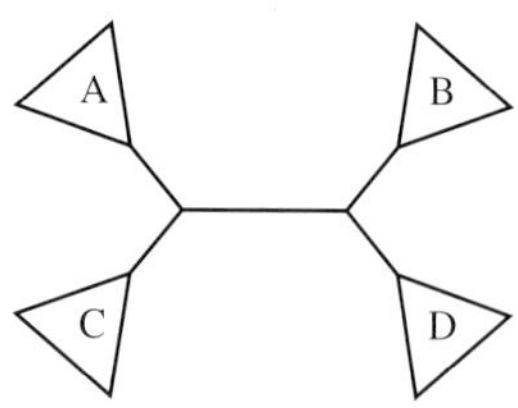

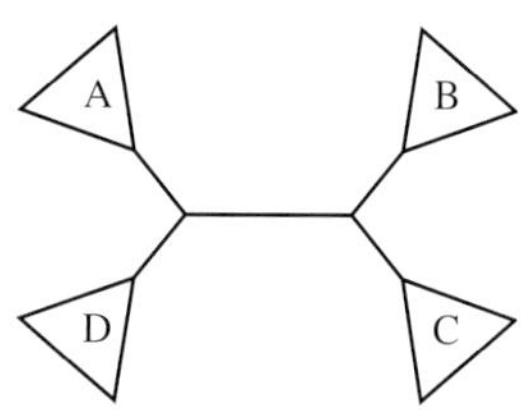

If any of the alternative orderings has a better score, we adopt it and continue with the new tree. There are $2n - 3$ internal edges in an unrooted tree with n leaves, where n of them are attached to leaves. This leaves $n - 3$ internal edges on which we can apply the test. Once the heuristic has been successful in at least one place, it is usual to retry it in all branches.

5-optim and 6-optim operate on the following patterns, respectively:

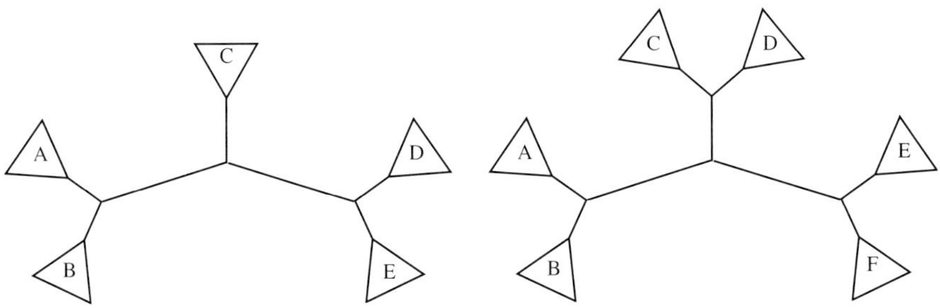

While these patterns are more powerful in searching the tree space and finding better trees, the coding required to analyze all of the alternatives becomes the limiting factor. 5-optim requires 15 cases and can be applied, in some cases, in three different ways on each internal node. Complete analysis of 6-optim requires 105 cases.

3.2.2. *Subtree pruning and regrafting (SPR)*

This heuristic considers each subtree of the tree, removes it from the tree, and attempts to insert it in a different position. The number of subtrees in an unrooted tree is three times the number of internal nodes, so 3n – 6. The total number of regrafting attempts is $O(n^2)$.

3.2.3. *Tree bisection reconnection (TBR)*

TBR takes the basic idea of SPR further. An interior branch of the tree is removed. The two resulting subtrees are then reconnected at a branch of one and a branch of the other. There are $O(n^2)$ possible reconnections. Consequently, the total number of possible TBR rearrangements is $O(n^3)$.

NNI, SPR, and TBR can be viewed as examining neighbors of a given tree in the tree space. It can be shown that NNI is a special case of SPR, which in turn is a special case of TBR. This means that NNI covers a subset of the neighbors visited by SPR.

3.2.4. *Reduce best-fitting subtree (RBFS)*

This heuristic is different from the ones presented above. The idea of RBFS is as follows. A score is computed for each subtree, which indicates whether the subtree is fitting well or not. A subtree with a high score means that its contribution to the error is small (and hence our efforts should be concentrated in other parts of the tree). Those subtrees which score the best are reduced to a single leaf. Now, the local search heuristics or other more comprehensive heuristics can be applied to the reduced tree. In Sec. 4.3, we show in detail how this idea (in particular, the scoring and reduction steps) can be applied to WLS tree construction.

3.2.5. *Tree fusing (TF)*

The final heuristic falls a bit outside the general scheme in the sense that it uses more than one tree to produce a better candidate. The idea is simple: identify the largest subtrees over the same leaves in different trees which do not have the same topology. If such a pair is found, most likely swapping them will produce an improvement in one of the new trees. A strong motivation to use this heuristic is that randomized constructors can produce many candidate trees, so we have a large population of good trees that can be used.

The heuristics described in this chapter, and in general, are likely to converge to a local minimum and "get stuck" at this minimum. The larger the neighborhood they explore (and hence the more time they consume), the more likely they will not end up in a small local minimum. There is a trade-off between making the heuristics explore large neighborhoods in the hope of jumping out of a local minimum and running several optimizations by restarting from a random solution using a randomized constructor. Our experience indicates that it is more efficient to look for local minima not too hard and in exchange start from many random initial points. As a simple rule of thumb, start n optimizations for a tree with n leaves. As mentioned above, the optimization starting at different random points can easily be done in parallel.[f]

We close this section with an important observation. When the objective function can be computed bottom-up, the NNI and k-optim heuristics offer an interesting alternative. Let us assume that what is meant by bottom-up computation is that the objective function information of every subtree can be stored in the subtree in constant space and that it can be recomputed in constant time from the information from its two descendant nodes.[g] This is the case for parsimony and ML. To take advantage of this, we will store in every internal node the three records/values[h] of each of its subtrees. Once this is done, the objective function for each NNI rearrangement can be computed in constant time. When the tree is improved, some of these records/values have to be recomputed. The strategy pays off because the number of times we evaluate the optimality for heuristics vastly exceeds the times that we are successful with an interchange.

4. Least Squares Tree Construction

In this section, we present the WLS tree building program Least Squares Tree (LST) developed by our research group. The first subsection gives a high-level description of the program. In the second subsection, we compare the running times of our implementation with the times of the corresponding program in PAUP*. The last subsection is a "specialist" section where we describe in detail the application of the RBFS heuristic.

4.1. *Least Squares Tree Function in Darwin*

Our function LST is implemented in Darwin.[27] Darwin is a programming environment with its own interpreted language and a growing library of

[f] In the jargon of parallel computing, this is called "embarrassingly parallel".
[g] Constant size and space are in relation to the size of the tree.
[h] In an unrooted tree, every internal node can be grown bottom-up from three directions.

functions for sequence management and analysis, statistics, numerics, graphics, parallel execution, etc. The closed-source kernel is coded in C language, including time-critical library functions (as LST). The library functions that are programmed in the interpreted language are open source[i] under the MIT License.

LST is a CHE procedure; that is, it constructs an initial tree and applies heuristics until the evaluator shows that the tree cannot be improved any further. The input to LST is a distance and a variance matrix, and the evaluator is the Euclidean norm of the weighted errors between the input and the actual tree. Thus, the function it minimizes is

$$\sum_{ij} \frac{\left(T_{ij} - d_{ij}\right)^2}{\sigma_{ij}^2}.$$

A missing distance between two objects A and B can be indicated with $\sigma^2_{AB} = \infty$. LST uses a randomized version of WPGMA (Weighted pair group method using arithmetic averages). WPGMA is like UPGMA, except that the computation of the distances to a joined subtree is done taking the variances into account. The new variances also have to be computed. When joining subtrees/nodes A and B

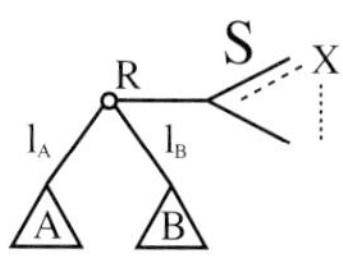

we use the following formulas: compute l_A and l_B from the equations

$$l_A + l_B = d_{AB}$$

$$l_A - l_B = \sum_{X \in S} \frac{d_{AX} - d_{BX}}{\sigma_{AX}^2 + \sigma_{BX}^2} \Bigg/ \sum_{X \in S} \frac{1}{\sigma_{AX}^2 + \sigma_{BX}^2}$$

[i] The source code of the library and executables of the kernel (for Linux, Mac OS X, Solaris, and Irix) are available for download at `http://www.cbrg.ethz.ch/`.

and then compute

$$d_{RX} = \left(\frac{d_{AX} - l_A}{\sigma^2_{AX}} + \frac{d_{BX} - l_B}{\sigma^2_{BX}} \right) \sigma^2_{RX},$$

where

$$\sigma^2_{RX} = \left(\frac{1}{\sigma^2_{AX}} + \frac{1}{\sigma^2_{BX}} \right)^{-1}.$$

The global strategy of LST is based on the following empirical observations and facts:

(a) The 4-optim heuristic is about four times less expensive to run than 5-optim or 6-optim. Consequently, it should be run first.
(b) It is a bad idea to restart the iteration of a heuristic as soon as an improvement is found. It is better to complete the iteration over all edges and, if there was an improvement, redo it. (If there are too many improvements, restarting after every improvement results in a $O(n^2)$ behavior.)
(c) We found it better to interleave 4-optim iterations with 5-optim and 6-optim iterations, i.e. run cyclically 4-, 5-, and 6-optim until three in a row do not produce an improvement.

The kernel function has two convenient additional capabilities:

(a) In a single call, a number of trees can be generated and optimized and only the best is returned. This saves all of the checking and setup times for internal structures.
(b) An initial topology can be given, from which the optimization can start.

Once a pass of 4-5-6-optim over all possible edges does not produce any improvement, the NNI cannot be continued. In this case, the kernel heuristic does not try other heuristics and returns. The standard procedure for building trees is to run the kernel optimization over 50 trees generated by the randomized WPGMA, and then apply the RBFS heuristic recursively over the best tree.

4.2. *Time Comparison with PAUP*

We have compared the running times of LST with the corresponding function *hsearch* contained in PAUP* 4.0 beta 10 on simulated data.

4.2.1. *Methods*

The input problems consisted of pairwise distances from trees with $n \in \{500, 1000\}$ leaves. The distances were obtained in the following way: We generated a random tree with uniformly distributed $U(0.1..l_{max})$ point accepted mutation (PAM) branch lengths. The maximum branch length l_{max} is a parameter in the simulation that controls the difficulty of the problems. A random sequence of 500 amino acids was generated and mutated along the random tree. We assumed a Markovian model of evolution using PAM matrices and introduced gaps of Zipfian distributed length.[28] From the sequences at the leaves of the tree, we computed pairwise distance estimates by ML.

The distances were then fed to LST and *hsearch*. We chose to use the default parameters for *hsearch* and to impose a maximal running time of 50 hours (by setting the parameter time limit = 180 000). LST was run in two modes. In the first mode, the parameter Trials was set to one, which constitutes the lowest possible optimization level. In the second mode, we set Trials = 50 and used the RBFS heuristic.

The simulations were run on an AMD Athlon, 1800 MHz processor running a Red Hat Linux 2.4.21. The times reported in the results are user times measured with the time command included in Linux.

4.2.2. *Results*

Table 1 gives the results for (a) 500 leaves and (b) 1000 leaves. A row corresponds to a particular l_{max}. It shows the average running time over 20 input problems, how often LST and *hsearch* returned the same tree,

Table 1. Time and performance comparison between LST and *hsearch* for (a) 500 leaves and (b) 1000 leaves.

(a) 500 leaves

	Trials = 1			Trials = 50 + RBFS		
l_{max}	Time [s]	Equal	Better	Time [s]	Equal	Better
20	9.30 ± 0.31	20	0	610 ± 24	20	0
100	9.63 ± 0.32	3	0	661 ± 25	14	2
150	9.34 ± 0.34	0	5	729 ± 35	3	17

(b) 1000 leaves

	Trials = 1			Trials = 50 + RBFS		
l_{max}	Time [s]	Equal	Better	Time [s]	Equal	Better
20	79.9 ± 3.3	2	16	5601 ± 177	1	17
50	80.5 ± 3.1	0	20	5699 ± 185	0	20
75	79.3 ± 3.1	0	14	5958 ± 242	0	20
100	80.4 ± 3.1	0	11	6564 ± 229	0	19

For each maximum branch length l_{max}, LST and hsearch were run on 20 problems. Trials = 1 is LST's lowest optimization level, and Trials = 50 + RBFS constitutes a high one. *hsearch* was always run for 50 hours. Time is the average running time of LST over the 20 problems. The column Equal shows how often the two programs found a tree with the same score, and the column Better shows how often LST found a tree with a better score.

and how often LST found a tree with a higher score. The timings for *hsearch* are not shown. It ran for 50 hours on each problem.

We first look at the problems with 500 leaves. In all cases, it took LST around 10 seconds to come up with a solution with Trials = 1 and around 10 minutes with Trials = 50 + RBFS. For the easiest problems (l_{max} = 20), the two programs constructed the same tree in all 20 problems and for both of LST's optimization levels. For l_{max} = 100 and l_{max} = 150, Trials = 1 was not enough for LST to be able to compete with *hsearch*. Using Trials = 50 + RBFS for l_{max} = 100 led to a comparable performance of the two programs; and for l_{max} = 150, *hsearch* never returned a better scoring tree than LST. In 17 out of 20 cases, LST found a better tree in around 10 minutes, than *hsearch* did in 50 hours.

We now look at the 1000-leaf trees. Already with Trials = 1, which took less than 2 minutes to run, LST found a better solution than *hsearch* for more than half of the 20 problems. This was the case for all l_{max} tested. Running LST with Trials = 50 + RBFS took less than 2 hours. In this case, LST was superior to *hsearch* on at least 17 of the input problems. Please notice that default parameters were used for *hsearch*, and that we did not study to which extent tweaking the parameters would change the results.

4.3. *RBFS Heuristic*

In Sec. 3.2, we have presented the general idea of the RBFS heuristic. In this section, we show how this idea can be applied to WLS tree building. To formulate RBFS, we need an index to score a given subtree and a way to substitute a well-scoring subtree by a leaf. The two components are described next.

4.3.1. *Subtree index*

Consider a subtree S in the following tree T:

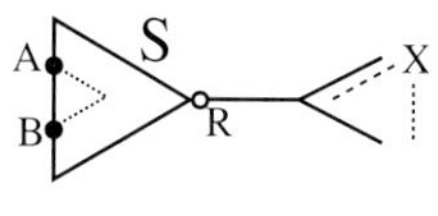

The index we present here measures the contribution of a subtree S to the total error made in T. It is composed of the sum of two parts. The first part measures the fitting of distances inside the subtree S:

$$E_1 = \sum_{A,B \in S} \frac{(T_{AB} - d_{AB})^2}{\sigma_{AB}^2},$$

where A and B are different leaves in the subtree S, T_{AB} is the distance between A and B in S, and d_{AB} and σ_{AB}^2 are the source data.

The second part measures how well leaves X outside S fit with respect to all of the leaves in S. This is

$$E_2(X) = \sum_{A \in S} \frac{(T_{AR} + T_{RX} - d_{AX})^2}{\sigma_{AX}^2}.$$

For each X, we will assume that we are free to choose its optimal distance to R. (If this is not the case, it is the "fault" of some other part of the tree, not the "fault" of S.) T_{RX} is then set so that $E_2(X)$ is minimal (which is equivalent of setting T_{RX} by LS for each S and each X). Combining the two parts (thereby considering all leaves X outside S), we get as an intermediate result the total sum of the weighted squared errors:

$$E'_S = E_1 + \sum_{X \in \bar{S}} E_2(X).$$

The number of degrees of freedom in the computation of E'_S is

$$p = \underbrace{\binom{k}{2} + k(n-k)}_{\text{\# distance relations}} - \underbrace{\big((2k-3) + 1 + (n-k)\big)}_{\text{\# branches to be set}},$$

where $k = |S|$ is the number of leaves in the subtree S. Clearly, $k > 1$ for this to make sense. Normalizing E'_S by the degrees of freedom, we obtain the final index

$$E_s = \frac{E'_s}{p} = \frac{2\left(E_1 + \sum_{X \in \bar{S}} E_2(X)\right)}{(2n-k-4)(k-1)}.$$

For each subtree, we evaluate E_S and choose the one with the smallest E_S. This is our best-fitting subtree, and we will now assume that it can be replaced by a leaf and hence decrease the complexity of the tree building. This subtree replacement can be repeated many times either until the tree is small enough or until some other criteria based on the magnitude of E_S is met.

4.3.2. *Subtree substitution*

The only remaining part is to replace the subtree S with a new leaf in such a way that all of the distance and variance information of S is preserved. Let R^* be the leaf that replaces S (which we will view as positioned at the root of the subtree S).

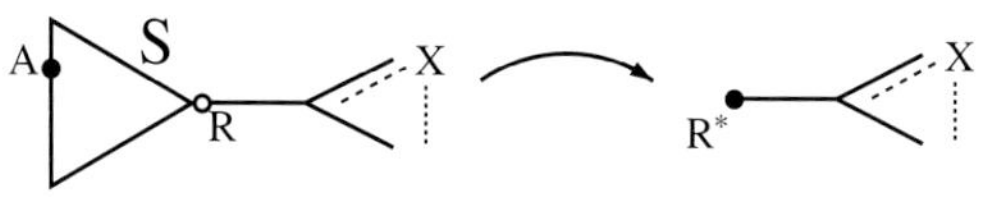

Then,

$$d_{XR^*} = \sigma^2_{XR^*} \sum_{A \in S} \frac{d_{AX} - T_{AR}}{\sigma^2_{AX}},$$

Where

$$\sigma^2_{XR^*} = \left(\sum_{A \in S} \frac{1}{\sigma^2_{AX}} \right).$$

Once the tree with S replaced by R^* is resolved (maybe recursively), R^* is replaced by S.

5. Outlook

In this chapter, we have given an overview of phylogenetic tree building methods and have delved into the ones that optimize a score. In doing so, we have focused solely on the construction of trees. A second goal after the construction is to provide confidence statements about the constructed trees. From a statistical point of view, the trees are estimates and as such have to be viewed as random variables. One difficulty is that a tree is a much more complex variable than, for example, a real valued random variable. It is composed of the discrete structure of the topology and the continuous values of the branch lengths. Once a tree (or a set of trees) is constructed, we want to make confidence statements about the branching patterns and the branch lengths, and moreover be able to compare different trees or build confidence sets of trees. An elaborate treatment of these topics is beyond the scope of this chapter. Instead, we conclude by giving some relevant references for further reading.

Probably the best known tool to measure the robustness of a constructed tree with respect to small changes in the data is bootstrapping. It is often used to obtain confidence values for the edges on a tree and can be applied in conjunction with almost all tree building methods. See Holmes[29] for a survey. However, statistical interpretation of the bootstrapping supports is unclear. Alternatively, in the context of likelihood-based methods, likelihood ratio tests[30,31] and Bayesian branch supports[32] are used.

In Sec. 3, several heuristics were described to search a tree space. During such a search, many topologies are visited and scored. Two scores that are different in absolute numbers need not be different in statistical terms. Thus, another important and nontrivial issue besides branch testing is the statistical comparison of different competing trees (e.g. the best scoring tree vs. the second-best scoring tree, or trees built on different genes). Goldman *et al.*[33] have reviewed statistical tests for the comparison of topologies and present their correct use.

References

1. Semple C, Steel M. (2003) *Phylogenetics.* New York, NY: Oxford University Press.
2. Swofford DL. (2003) *Phylogenetic Analysis Using Parsimony (*and Other Methods), Version 4.* Sunderland, MA: Sinauer Associate.
3. Thompson JD, Higgins DG, Gibson TJ. (1994) CLUSTAL W: improving the sensitivity of progressive multiple sequence alignment through sequence weighting, position-specific gap penalties and weight matrix choice. *Nucleic Acids Res* **22**: 4673–80.
4. Katoh K, Kuma K, Toh H, Miyata T. (2005) MAFFT version 5: improvement in accuracy of multiple sequence alignment. *Nucleic Acids Res* **33**(2): 511–8.

5. Edgar RC. (2004) MUSCLE: a multiple sequence alignment method with reduced time and space complexity. *BMC Bioinformatics* **5**: 113.
6. Sokal RR, Michener CD. (1958) Statistical method for evaluating systematic relationships. *Univ Kans Sci Bull* **38**: 1409–38.
7. Sankoff D. (1975) Minimal mutation trees of sequences. *SIAM J Appl Math* **28** 35–42.
8. Redelings BD, Suchard MA. (2005) Joint Bayesian estimation of alignment and phylogeny. *Syst Biol* **28**(3): 401–18.
9. Lunter G, Miklos I, Drummond A *et al.* (2005) Bayesian coestimation of phylogeny and sequence alignment. *BMC Bioinformatics* **6**: 83.
10. Bourque G, Pevzner P. (2002) Genome-scale evolution: reconstructing gene orders in the ancestral species. *Genome Res* **12**(1): 26–36.
11. Moret BM, Wyman S, Bader D *et al.* (2001) A new implementation and detailed study of breakpoint analysis. *Pac Symp Biocomput* (2001): 583–94.
12. Day W, Johnson D, Sankoff D. (1986) The computational complexity of inferring rooted phylogenies by parsimony. *Math Biosci* **81**: 33–42.
13. Day WHE. (1987) Computational complexity of inferring phylogenies from dissimilarity matrices. *Bull Math Biol* **49**(4) 461–7.
14. Chor B, Tuller T. (2006) Finding a maximum likelihood tree is hard. *J ACM* **53**(5): 722–44.
15. Cox DR. (2006) Frequentist and Bayesian statistics: a critique. In: *Proceedings of PHYSTAT 05*. London, UK: Imperial College Press, pp. 3–6.
16. Bergsten, J. (2005) A review of long-branch attraction. *Cladistics* **21**(2): 163–93.
17. Fitch WM, Margoliash E. (1967) Construction of phylogenetic trees. *Science* **155**: 279–84.
18. Rzhetsky A, Nei M. (1993) Theoretical foundation of the minimum evolution method of phylogenetic inference. *Mol Biol Evol* **10**: 1073–95.
19. Saitou N, Nei M. (1987) The neighbor-joining method: a new method for reconstructing phylogenetic trees. *J Mol Evol* **4**: 406–25.
20. Gascuel O. (1997) BIONJ: an improved version of the NJ algorithm based on a simple model of sequence data. *Mol Biol Evol* **14**(7): 685–95.
21. Bruno WJ, Socci ND, Halpern AL. (2000) Weighted neighbor joining: a likelihood-based approach to distance-based phylogeny reconstruction. *Mol Biol Evol* **17**(1): 189–97.
22. Felsenstein J. (1987) Cases in which parsimony and compatibility methods will be positively misleading. *Syst Zool* **27**: 401–10.
23. Yang Z. (2007) PAML 4: Phylogenetic analysis by maximum likelihood. *Mol Biol Evol* **24**(8): 1586–91.
24. Guindon S, Gascuel O. (2003) A simple, fast, and accurate algorithm to estimate large phylogenies by maximum likelihood. *Syst Biol* **52**(5): 696–704.
25. Stamatakis A. (2006) RAxML-VI-HPC: maximum likelihood-based phylogenetic analyses with thousands of taxa and mixed models. *Bioinformatics* **22**(21): 2688–90.

26. Felsenstein J. (2004) Inferring Phylogenies. Sunderland, MA: Sinauer Associate.
27. Gonnet GH, Hallet MT, Korostensky C, Bernardin L. (2000) Darwin v. 2.0: an interpreted computer language for the biosciences. *Bioinformatics* **16**(2): 101–3.
28. Benner SA, Cohen MA, Gonnet GH. (1993) Empirical and structural models for insertions and deletions in the divergent evolution of proteins. *J Mol Biol* **229**(4): 1065–82.
29. Holmes S. (2003) Bootstrapping phylogenetic trees: theory and methods. *Stat Sci* **18**(2): 241–55.
30. Gaut BS, Lewis PO. (1995) Success of maximum likelihood phylogeny inference in the four-taxon case. *Mol Biol Evol* **12**: 152–62.
31. Anisimova M, Gascuel O. (2006) Approximate likelihood-ratio test for branches: a fast, accurate, and powerful alternative. *Syst Biol* **55**(4): 539–52.
32. Huelsenbeck JP, Larget B, Miller RE, Ronquist F. (2002) Potential applications and pitfalls of Bayesian inference of phylogeny. *Syst Biol* **51**(5): 673–88.
33. Goldman N, Anderson JP, Rodrigo AG. (2000) Likelihood-based tests of topologies in phylogenetics. *Syst Biol* **49**(4): 652–70.

Chapter 13

Bioinformatics for Evolutionary Developmental Biology

Marc Robinson-Rechavi

1. Introduction

Evolutionary developmental biology ("evo-devo") has emerged as one of the most exciting areas of biology. It is providing, for the first time, some answers to long-standing questions, such as the origin of novelty, sources of phenotypic variation, or the relationships between diverse animal body plans. It is also forcing us to reconsider apparently known answers, such as the relation between microevolution and macroevolution or the definition of homology.[1–6] In addition to developmental and evolutionary biology, diverse fields of research have contributed to this success as well, including classical zoology, paleontology, and molecular genetics.[7,8] Most recently, genome projects from diverse organisms have also provided new insight into the evolution of developmentally important genes.[9–12]

Thus, evo-devo is by its very nature an interdisciplinary science. In this chapter, we explore the interface with another interdisciplinary field, bioinformatics. While some bioinformatic studies have implications for evo-devo, and some evo-devo studies (especially the analysis of genome sequences) make use of bioinformatics, this interface has been rather neglected up to now (see also Mabee[13]).

The study of whole-genome duplication is one field of evo-devo in which the use of bioinformatics can be (extremely) helpful. Although suggestions of the importance of duplication in evolution have been recurrent,[14] the main evidence in support of this theory came in the 1990s from the study of evo-devo, most notably from the discovery of four Hox gene complexes in human and mouse compared to only one Hox complex in many invertebrates such as the fruit fly.[15] This was interpreted as being consistent with a hypothesis of two rounds of genome duplication at the origin of vertebrates, as suggested by the work of Susumo Ohno,[16] who provided the classical framework for the study of genome duplication. Ohno suggested that mutations of existing genes cannot generate new functions without risking loss of the original function; whereas duplication creates redundancy of this original function, allowing one copy to diverge and adopt a new function. Thus, the divergence of orthologs (homologs diverging after speciation) would be conservative, whereas the divergence of paralogs (homologs diverging after duplication) would allow for the evolution of novelty. This idea has since become mainstream in comparative genomics (e.g. Koonin[17]). Ohno also suggested that these duplications could be linked to the "complexity" of lineages such as vertebrates, an appealing idea in light of the duplications of Hox complexes (key regulators of animal development), but one which has proven difficult to test, since the position of mammals as the apex of such "complexity" was short-lived with the discovery of seven Hox complexes in the zebrafish.[18]

Genome duplication, although rare, has emerged as an important factor in genome evolution. Genome-scale evidence was first obtained, surprisingly, from the simple yeast *Saccharomyces cerevisiae*, with the discovery that the yeast genome was tiled by nonoverlapping duplicated blocks.[19] Further studies in yeast established comparative mapping on duplicated and nonduplicated species as the best way to prove and date whole-genome duplication.[20] In addition to yeast, evidence for ancient whole-genome duplications has notably been found in *Arabidopsis thaliana*,[21,22] cereals,[23] teleost fish,[24,25] and paramecium.[26] Finally, a combination of comparative mapping and phylogeny has provided support for Ohno's[16] hypothesis of two whole-genome duplications at the origin of vertebrates.[27–29] More recent tetraploids are also found in various vertebrate lineages.[30]

We will first present results on the use of sequence analysis, notably phylogenetics, which shed some light on questions from evo-devo. In the second part, we will present ongoing research to model more complex anatomical and developmental data, which could provide the foundation for a bioinformatic platform for evo-devo studies.

2. The Easy Part: The Evolution of Gene Sequences

2.1. *Rapid Overview of Bioinformatics Involved*

The basic task in relating sequence evolution to developmental biology is finding genes of interest, listing all of their homologs, and determining their phylogenetic relationships. To study their evolution, we may also be interested in functional classification as well as in evidence of selective pressure. For example, a change in selective pressure on some sites may be evidence of a change in function of the protein.

The first task, identifying genes of interest, might be approached in two ways: we may start with candidate genes and search for their homologs, or we may start by determining homologs genome-wide and use the result of this analysis to select genes of interest. In both cases, a key step is identifying homologs by sequence similarity. This is a topic abundantly treated elsewhere, but we need to note here that many genes of interest in evo-devo are characterized by short conserved domains which may be difficult to identify in standard scans using, for example, BLASTP.[31] Transcription factors such as the Hox, bZIP, and bHLH transcription factors are thus best identified by the careful use of hidden Markov models (e.g. Amoutzias *et al.*[32]), and are often missed in large-scale scans for homologs. An interesting exception to this is the nuclear hormone receptor superfamily, whose members can be readily identified thanks to their ligand-binding domain of approximately 200 amino acids.[33,34]

The topic of genome duplication raises that of the distinction between orthologs and paralogs.[17,35] The theoretically correct way to do this is through phylogenetic inference, although numerous alternative methods have been proposed. We will not treat these methods in detail here, but note that we systematically use likelihood methods (e.g. PhyML[36]). While the most common use of phylogenies in such studies is to start with the target genes and analyze their phylogeny, in some cases we perform the reverse. For this, we rely on existing databases of gene trees, such as TreeFam[37] or the databases of the Pôle Bioinformatique Lyonnais (PBIL),[38–40] combined with tree reconciliation tools.[41] The latter allow us to specify a topology and search for all gene trees (or subtrees) that match it. In this way, we can identify all genes that were retained in duplicate after whole-genome duplication in fish, but not duplicated in tetrapodes, specifying also species or lineages in which gene loss is allowed or forbidden.

2.2. *The Importance of Duplication and Loss*

2.2.1. *Why don't flies have retinoic acid receptors?*

Nuclear hormone receptors (or nuclear receptors, NRs) are transcription factors that are specific to Metazoa (animals). They include receptors of major hormones, such as steroids or thyroid hormones. NRs play important roles in many central biological processes, notably development (reviewed in Laudet and Gronemeyer[42]). A typical example is the group of retinoic acid receptors (RARs), which includes RARα, RARβ, and RARγ in humans. RARs mediate the regulation of anteroposterior expression of Hox genes in vertebrate development by retinoic acid. Whereas the Hox genes are largely conserved between mammals and flies, no ortholog of RAR is found in the *Drosophila* genome, and neither are present orthologs of classical steroid receptors (ER, AR, PR, MR, and GR) nor of thyroid hormone receptors (TRs). Moreover, orthologs of these genes are not found in nematode genomes either. The steroid receptors are even absent from the genome of the sea squirt, *Ciona intestinalis*. In fact, while most of the 48 human nuclear receptors have known ligands (hormones or fatty acids), most of the 21 fly nuclear receptors are so-called "orphans", without a known ligand. These observations led to the suggestion that most liganded nuclear receptors were vertebrate innovations.[9,43]

When a sufficient sampling of animal genomes became available, we took advantage of the conserved structure of nuclear receptors to search for all homologs, and performed a global phylogenetic analysis of the superfamily.[44] By combining this gene tree with the known species phylogeny, we can date not only duplications, but losses as well. On a rooted tree, events that are closer to the tips are more recent, and events closer to the root are more ancient. This does not provide an absolute dating (in years), but it does provide a relative dating. If we start with human RARs (Fig. 1), we see that all speciations among vertebrates

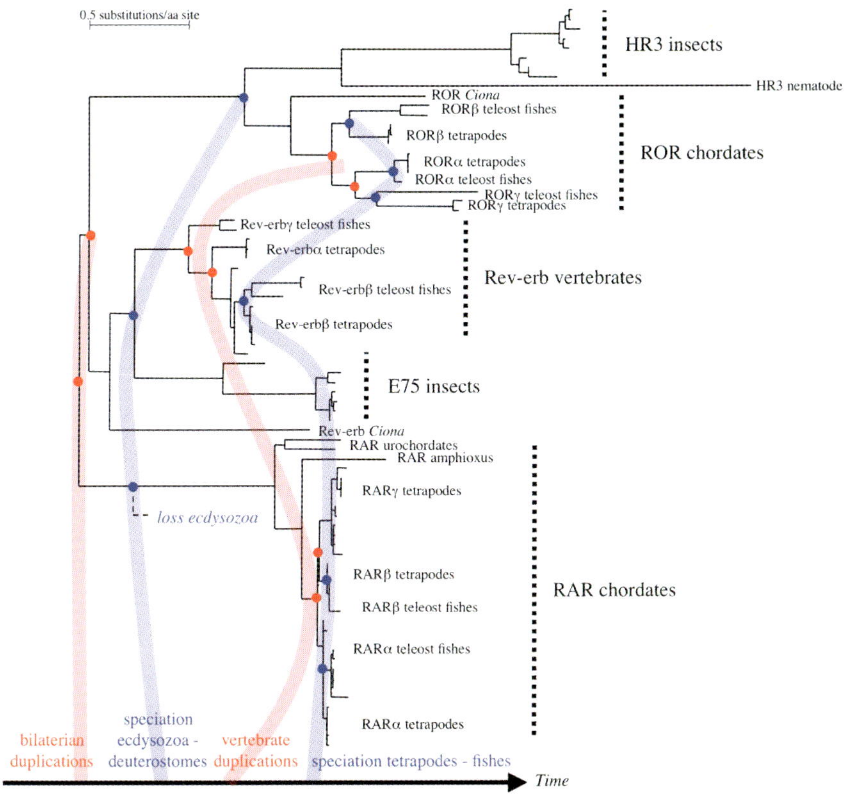

Fig. 1. Simplified phylogenetic tree of three groups of nuclear receptors: maximum likelihood phylogeny (PhyML,[36] JTT model, four rate categories, gamma shape parameter, and proportion of invariant estimated) of 78 nuclear receptors from the groups NR1B (RAR), NR1D (Rev-erb, E75), and NR1F (ROR, HR3). Branch length is proportional to substitutions per amino acid site. The relative timing of key speciation events is shown in blue, and the relative timing of duplication events is shown in red.

(e.g. tetrapode/fish) happened more recently than the duplications which gave rise to RARα, RARβ, and RARγ, but that these duplications occurred after the speciation between *Ciona* and vertebrates. Thus, these duplications date to the origin of vertebrates. If we go back in time before the split between the *Ciona* and vertebrate RAR orthologs, we do not find a speciation node, but an older gene duplication that gave rise to RARs and other nuclear receptors. When did this older duplication occur? The ROR/HR3 subtree includes not only speciations among chordates and vertebrate-specific duplications, like RARs, but also a speciation node between chordates on the one hand and insects and nematodes on the other — the speciation between ecdysozoans and deuterostomes that occurred at the origin of bilaterian animals.[45,46] In the tree, this speciation is clearly more recent than the duplication leading to RARs, RORs, and other nuclear receptors. We can therefore assume the following order of events (Fig. 1): duplications leading to proto-RAR, proto-ROR, proto-Rev-erb, and other NRs, followed by speciation between ecdysozoans and deuterostomes. Then, to explain the lack of any RAR orthologs in the sequenced ecdysozoan genomes, we must infer that this gene was lost in ecdysozoans, and that RAR does not appear as a vertebrate innovation but rather as an ecdysozoan loss. Of note, an RAR ortholog has also been identified in the sea urchin genome,[47] a deuterostome but not a chordate.

In the analysis of the whole superfamily, we find a recurrent pattern: vertebrate innovations are invertebrate losses.[44] Symmetrically, insect-specific genes (e.g. E78) are ancestral bilaterian genes lost in the chordate lineage. This pattern has been spectacularly confirmed for steroid receptors, with the cloning and characterization of an estrogen receptor ortholog from a mollusk.[48,49] Indeed, ancestral sequence reconstruction shows that the ancestor of bilaterian animals probably had a receptor activated by estrogen.

2.2.2. *Why do humans have three retinoic acid receptors?*

Flies may not have retinoic acid receptors, but humans have three. RARα, RARβ, and RARγ are paralogs, kept from the genome duplications at the origin of vertebrates. All three bind all-trans retinoic acid, and activate transcription of target genes. Yet they are not redundant. Not only have the three copies been kept over 400 million years of vertebrate evolution, but knock-out (KO) experiments show paralog-specific phenotypes, mostly affecting development.[50] Like other nuclear receptors, RARs have a DNA-binding domain and a ligand-binding domain. The latter contains about 270 amino acids in RARs, and is composed of 12 α-helices. Of these, 25 amino acids in the hydrophobic ligand-binding pocket make direct contact with the ligand. The binding pockets of the human RAR paralogs differ in three of these 25 positions. They also differ in their *in vitro* binding to different synthetic retinoids, and in the resulting transactivation of target genes.

To gain further insight into the evolution of these differences, we compared all chordate RARs, including vertebrates, amphioxus, and tunicates.[51] The phylogeny confirms the dating of the duplications at the origin of vertebrates, with single-copy orthologs in amphioxus and tunicates. The amino acid sequence of the ligand-binding domain of the RAR immediately predating the duplications was predicted using maximum likelihood, and synthesized. As expected, all homologs and the predicted ancestral protein transactivate with all-trans retinoic acid, with similar EC50 values. On the other hand, the ancestral, amphioxus, and tunicate RARs do not transactivate in the presence of retinoids that are specific to human RARα or RARγ; they do bind the RARβ-specific retinoid, with strong transactivation in amphioxus RAR and with weaker transactivation in tunicate RAR and predicted ancestral RAR. Targeted mutations of the amphioxus RAR show that the three positions identified in human are indeed key to the differences in specificity. These results suggest that RARβ is closest to the ancestral function, with changes evolving by point mutations in the ligand-binding pocket of RARα and RARγ. This is confirmed by limited proteolytic experiments, which show that all RARs do bind the β-specific retinoid, even when it is not sufficient for transactivation, but do not bind the other specific retinoids. Interestingly, *in situ* hybridization shows that the expression pattern of amphioxus RAR is most similar to that of RARβ. Thus, it appears that, after whole-genome duplications, one copy remained similar to the ancestral function in terms of both sequence and expression pattern, while the other two acquired derived characteristics.[51]

This study focused on differences between paralogs, due to duplication. However, it is worth noting that we also found differences between orthologs. For example, both zebrafish and Xenopus RARγ transactivate with both the α- and γ-specific retinoids (as established in mammals), and indeed have one amino acid in helix H3 that is identical to mammalian RARα, but not to RARγ.[51] A more general consideration of nuclear receptors shows that functional differences between orthologs are not rare when distant organisms are compared. For example, vertebrate Rev-erbs are orphan receptors involved in circadian cycle regulation;[52] but E75, their insect ortholog (Fig. 1), is a heme receptor involved in ecdysone regulation.[53] Thus, function may change not only after duplication, but also between species (see also Markov *et al.*[54]).

2.2.3. *Biased gene loss after whole-genome duplication*

After whole-genome duplication, duplicate copies of genes may evolve in different manners, gaining or losing functions (reviewed in Semon and Wolfe[55]). But the most common fate is certainly loss of one of the copies. Although this may not seem very exciting in itself, the contrast between which genes are lost and which ones are kept as duplicates has emerged as one of the most important features of whole-genome duplication. The rate of gene loss has been estimated at 88% in about 80 million years since genome duplication in yeasts,[20] 70% in ≤86 million years in *Arabidopsis*,[56] and 79% in about 61–67 million years in cereals.[23] By comparing only genes that were mapped to chromosomes in the fish *Tetraodon* and in humans, and whose evolutionary fate could be determined by phylogenetic analysis, we obtained a figure of 85% of gene loss after whole-genome duplication in teleost fishes,[57] despite the greater age of the event. These similar figures are best explained if most losses occur rapidly after duplication,[58] so that subsequent evolution does not change the figure significantly.

In an important study, Davis and Petrov[59] showed that slowly evolving genes are more likely to be found duplicated. The bias is similar in yeast and in nematodes, and is maintained over evolutionary time, indicating that gene retention was also biased after the whole-genome duplication in yeast. An important aspect of the work of Davis and Petrov[59] was to use estimates of selective pressure that are phylogenetically independent of the duplication. We conducted a similar study in fish,[57] and found that these conclusions also applied to a more ancient genome duplication in a vertebrate lineage. The selective pressure was measured by the number of the number of nonsynonymous substitutions per site (d*N*) to the number of synonymous substitutions per site (d*S*) between the human and mouse orthologs of genes that either lost one copy (singletons) or were kept as duplicates after the fish whole-genome duplication (Fig. 2). Using only human-mouse d*N* to measure the rate of evolution of the encoded proteins, we found that nonduplicated orthologs of gene pairs retained after duplication evolve 30% slower. This is comparable to observations for nematode (25%) and yeast (50%) genes.

What is the relevance of these observations to evo-devo? First, if whole-genome duplication has really played a key role in the establishment of the developmental diversity of fish[60,61] (but see Donoghue and

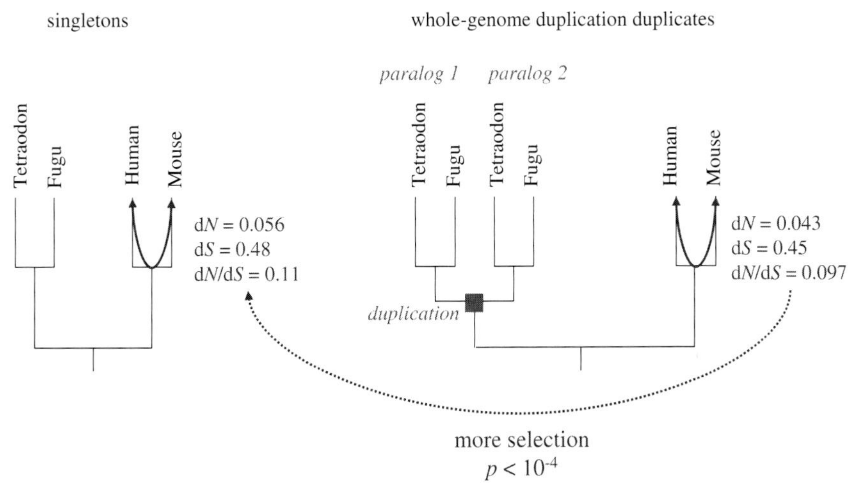

Fig. 2. Comparison of selection pressure on duplicated and singleton genes. Schematic phylogenetic classification of genes according to duplication and loss. d*N* is the number of nonsynonymous substitutions per site, and d*S* is the number of synonymous substitutions per site. The arrow represents the unpaired *t*-test between d*N*/d*S* values.

Purnell[62]), we need to understand its mechanisms as well as possible. Gene retention may also be biased relative to function, and enrichments in communication and developmental genes have been reported in insects, yeasts,[63] and *Arabidopsis*.[64] In fish, we found an excess of genes annotated with terms related to development and signaling functions,[57] which supports the putative link between genome duplication and developmental innovations.

Second, once the importance of biased gene retention after duplication is established, this provides a convenient measure of selective pressures on the genome. We have used this to test developmental constraints on genome evolution. Two models have been proposed for developmental constraints on morphological evolution. The first, dating back to pre-Darwinian observations by von Baer,[1,65] is that there is a progressive divergence of morphological similarities between vertebrate embryos. More general characters would form in early development, which would be highly constrained; while species-specific characters would form in late development, which would be more open to innovation. An alternative model was proposed more recently[66,67]: the "hourglass model", which is based on the observation of large morphological diversity in very early development (e.g. blastula). This model assumes a constrained stage in middle development, around the vertebrate pharyngula stage. To evaluate the impact of the constraints postulated by both models on the genome, we investigated the pattern of expression during development according to gene retention after whole-genome duplication. We expect genes that are kept in double to be highly expressed at developmental stages that are open to evolutionary innovation, but not at stages that are highly constrained. These should be characterized by conservatism: no duplication and no loss of highly expressed genes (Fig. 3). We also expect a high cost of gene loss (e.g. lethal phenotype for gene KO) in more constrained stages.

By combining a zebrafish time series microarray experiment (E-TABM-33 accession from ArrayExpress) with phylogenetic definition of retention or loss after duplication in fish (Fig. 2), we find that genes kept in duplicate are lowly expressed in early development, and then increase regularly their expression to reach a maximum in late development.[68] In principle, this could be the result of biased evolution after

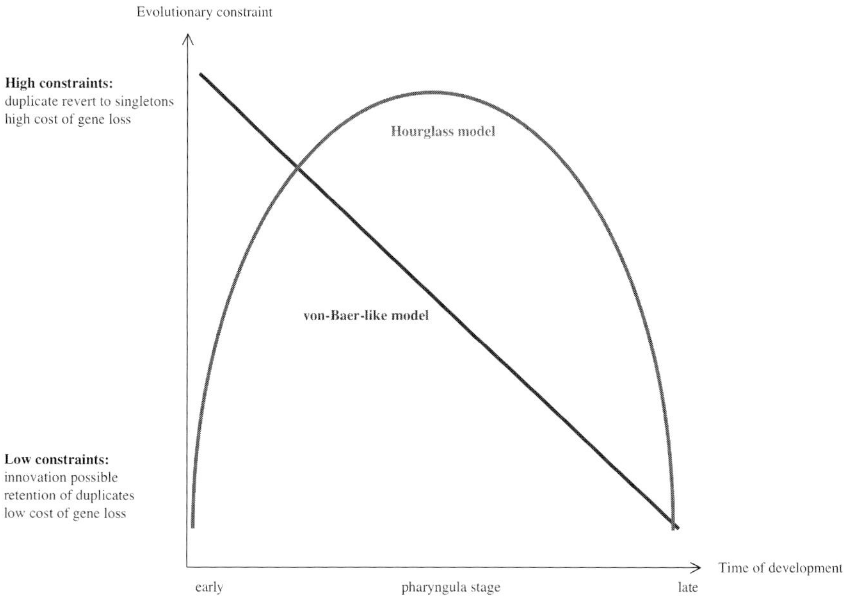

Fig. 3. Schematic predictions of two models of developmental constraints on evolution.

duplication (e.g. duplicate genes could acquire lower expression in early development) as well as of biased retention and loss. To check this, we used expressed sequence tag (EST) data to determine expression in mouse development: the orthologs of genes kept in duplicate in fish are lowly expressed in early mouse development. Moreover, the data fit a simple linear correlation, while excluding the parabola curve expected from the hourglass model (Fig. 3). The same results are obtained when we investigate biased gene retention after whole-genome duplications at the origin of vertebrates. Finally, gene KO phenotypes are also consistent with decreasing constraints over development in both zebrafish and mouse.

Taken together, these results show that (a) timing of expression during development is a strong and conserved constraint on genome evolution; (b) gene duplication is restricted when phenotype is constrained; and (c) at the genomic level, the traditional von Baer-like model provides a better fit than the hourglass model.

In conclusion of this section, both gene duplication and gene loss must be understood in order to clarify the relationship between genome evolution and developmental (and hence, morphological) evolution.

3. Developing Bioinformatic Tools for Evo-Devo

3.1. *Defining Homology for Bioinformatics*

Homology is one of the most fundamental concepts in biology. It has also generated abundant terminological and conceptual discussions (e.g. Hall[69]). It was originally defined based on the similarity of anatomical structures, with emphasis on their relations, thus clarifying that a bird wing and a mammalian forelimb are homologous. The term later acquired a historical dimension: homologous structures are assumed to derive from the same ancestral structure, to which they owe their similarity, whereas analogous structures have converged to similarity from different ancestral starting points. This definition can create complex situations, since the bat wing and the bird wing are analogous as wings, having converged from an ancestral walking limb; but they are homologous as forelimbs, since they derive from the forelimb of an ancestral tetrapode. Moreover, different fields of research have developed different operational definitions[70]: historical homology, defined by phylogenetic continuity; morphological homology, defined by structural similarity; or biological homology, defined by similar developmental constraints. Morphologists also define serial homology between organs repeated along the axis of the same organism, such as vertebrae (discussed in McKitrick[71]). Since the original evolutionary formulations of homology, two fields of research have hugely influenced our view of biological diversity and evolution: molecular evolution and evo-devo.

At the molecular level, homologs between species have been discovered whose origin predates the divergence of bacteria and eukaryotes; while inside each species, genomes include many families of homologous genes. To clarify this situation, three main types of molecular homology are distinguished: orthology, or divergence by speciation; paralogy, or divergence by sequence duplication; and xenology, or divergence after gene transfer between species.[35] Such molecular homology is probably the only type that has been well formalized in bioinformatics up to now.

Evo-devo is dependent on definitions of homology both for anatomical structures and for genes, and this field has also brought important new information relative to homology. Some of this new information has raised new questions, which are best exemplified by the now-classical case of animal eyes — if insect eyes and vertebrate eyes are organized in fundamentally different ways, they are probably analogs; but if they are determined during development by orthologous genes, are they not homologs? Such cases have led to the controversial proposal that the presence of key orthologous genes in development suffices to define homology (discussed in Wray and Abouheif[72]). An attempt to solve this issue is the new term "homocracy", which is defined as sharing the expression of the same patterning genes.[73] Homocratic structures may or may not be homologous; homologous structures are often homocratic, but this is not a logical necessity.

When defining homology in development, we must also take into account heterochrony, which is defined as changes in the relative timing of development during evolution. For each organ, homology should be defined at specific developmental stages, which differ between organs. For example, the heart develops later in primates relative to rodents, while the ear develops earlier, thus changing the timing of development of these organs relative to each other.[74] This makes it impossible to define homology between developmental stages as a whole in many cases, and greatly complicates automatic comparison of embryos between species. Moreover, developmental "stages" as defined in the literature are somewhat arbitrary divisions of a continuous process. The limits of these divisions may not be consistent between species, even without heterochrony.

Despite the importance of homology, little attention has been paid to its careful implementation in bioinformatics. Notably, few ontologies contain any notion of homology. Ontologies are formal representations of a field of knowledge, including terms, definitions and relations between the terms (e.g. hydrolase "is_an" enzyme). They have become an important tool in bioinformatics, responding to the need to formalize many complex descriptions in biology. Some ontologies do provide homology relationships. PATIKA,[75] a pathway ontology, includes a "homology" relation; in practice, it is used to manage paralogy inside gene families. Protein homology is defined as a "synonym" in the MoleculeRole Ontology of the INOH Pathway Database. To our knowledge, the most detailed implementation of the concept of homology is found in the Sequence Ontology[76]: it includes a "homologous_to" relation, which is the only child of "similar_to"; and has three children, "orthologous_to", "paralogous_to", and "non_functional_homolog_to". The latter is an interesting formalization of the relation between a gene and a pseudogene. The child relation to "similar_to" shows that a morphological definition of homology was chosen, whereas the Cell Ontology uses a definition of historical homology.[77] However, in the Cell Ontology, the relation is not implemented explicitly. Instead, homology is the default for the same term in different species (e.g. "muscle_cell" in human and fly); otherwise, several lineage-specific terms are created, as in "pigment_cell_(sensu_Vertebrata)" and "pigment_cell (sensu_Nematoda_and_Protostoma)". A similar approach is used in the Plant Ontology.[78] In several ontologies, homology is not defined as a type of relation, but is discussed in the definitions. For example, good discussions of anatomical homology, including serial homology and analogy, appear in definitions of the ontologies of mosquito[79] or corn.[80]

3.2. *Modeling Homology Relationships*

To conduct evo-devo studies computationally, we need to define homology relationships between ontologies describing the anatomy and development of different species. Designing such relationships consists in finding correspondences (homology relationships) between the concepts (organs) of these ontologies. This problem is a special case of "schema matching" or "ontology alignment". Ontology alignment[81] is the process of determining correspondences between ontology concepts. Usually, this technique is used to find the common concepts present in two ontologies. In the case of anatomical ontologies, the concepts to align are not strictly common, but rather related: a homology relationship is not an equivalence relationship. For this reason, classical ontology alignment approaches cannot be applied here, as these methods would be misled by the existence of elements with the same names and related to the same concept but not homologous (eye of insects and of vertebrates, for instance), or homologous elements with different names (caudal fin and upper limb, for instance). This is why, in our modified ontology alignment technique (Parmentier and Robinson-Rechavi, unpublished), an expert has to manually validate the putative homologs.

Our process is a supervised one: at each step, some homology relationships are proposed to the expert, who may validate them or not. Computations are made based on these decisions, and new propositions are made to the expert.

The algorithm starts with a list of pairs that have identical names. This is based on the assumption that two structures that have the same name are likely homologous. For example, "optic cup" of ZFIN (zebrafish)[82] and "optic cup" of EHDA (human) will be paired, but "optic cup" of ZFIN will not be initially paired with "optic nerve" of EHDA. The score of similarity between terms is upweighted by the proportion of common words, and downweighted by the frequency of these words (frequent words are less informative, e.g. "endoderm"). Moreover, scores are propagated between pairs that are neighbors in both ontologies. For example, the score of the "optic cup" pair is added to the score of the "eye" pair, as the optic cup is part of the eye.

Each pair is proposed to the expert, in descending order of scores. The expert may validate or invalidate the hypothesis of homology, or delay decision. The expert may choose to evaluate any number of pairs before triggering an iteration, in which computations are performed. Computation creates or extends homology groups. The new homology information is propagated through the ontologies. The underlying idea is that if two concepts A and B are homologous, then one of the subconcepts of A is probably homologous to one of the subconcepts of B. Of note, validated homology contributes a significantly higher score than name similarity. Propagation is downweighted by the number of subconcepts to avoid generating many false positives (i.e. all of the children of "whole body"). Evaluation of pairs, ordered by total score (base score + propagated score), is iterated until the expert decides to terminate or when no more pairs are proposed.

Our method is implemented in Homolonto (Parmentier and Robinson-Rechavi, unpublished), a software that we have developed in Java. Compared to manual alignment of the ontologies, Homolonto reduces time considerably, with high sensitivity. Thus, aligning the zebrafish (ZFIN; 2087 terms) and Xenopus (Xenbase; 480 terms) ontologies took 1 month by hand, but 2 days using Homolonto. The first 213 pairs proposed to the expert (i.e. 1 day of work) were valid at 80%, and contained 91% of all true positives.

3.3. *Bgee, a Database for Gene Expression Evolution*

To be useful for evo-devo and other comparative studies, the homology relationships determined using Homolonto are implemented in a database, Bgee. This "dataBase for Gene Expression Evolution" is being developed to facilitate comparisons of gene expression between animal species (http://bgee.unil.ch).[83]

To enable the comparison of large-scale gene expression patterns, Bgee must answer three conditions:

(a) precise description of the anatomy and developmental stages of each species, stored in a computer-understandable way. This is done using existing ontologies, such as ZFIN[82];
(b) comparison criteria between anatomies, developmental stages, and genes. For anatomy, this is done using Homolonto. For development, we have developed a small ontology of "metastages" that are common to all bilaterian animals, such as "blastula part_of embryo". For genes, we use homology predictions from other sources (i.e. Ensembl[84]); and
(c) integration of expression data in order to know in which anatomical features (spatial mapping) and at which developmental stages (temporal mapping) genes are expressed.

The relationships between these types of information are represented in a very simplified manner in Fig. 4.

Concerning developmental stages, we came to the conclusion that it is not possible to detail homologous stages in a similar way to organs.

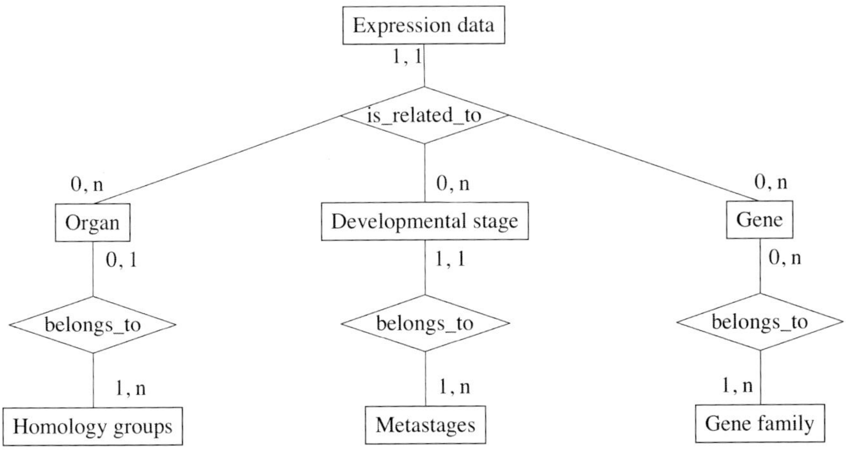

Comparisons between species:
Homologous genes expressed in homologous organs at equivalent stages

Fig. 4. Schematic representation of the relationships between types of information in the Bgee database. Expression data are central to relating genes to anatomical and developmental terms in each species.

We therefore developed a simplified ontology of metastages. Despite the resulting loss of accuracy, it allows comparison of gene expression patterns, taking into account developmental time.

Concerning expression data, we face two challenges: integrating heterogeneous data types,[85,86] and transforming quantitative data (level of expression) into the qualitative information that is standard in typical developmental studies ("expressed" or not). The reason for integrating heterogeneous expression data is that they complement each other in terms of coverage. For example, EST libraries typically present an incomplete picture of the transcriptome, but they are available for many species and allow good identification of closely related paralogs. Oligonucleotide microarrays are much more complete, but different experiments are difficult to compare and nonmodel species are not covered. Both ESTs and microarrays are usually annotated to coarse anatomical and developmental descriptions (e.g. "adult brain"), whereas *in situ* hybridizations can provide very detailed accounts of gene expression. However, *in situ* hybridization provides limited genome coverage (although see Thisse *et al.*[87]), is not applicable to humans, and can be more challenging to treat automatically than other transcriptome data types.[86,88]

Briefly, the basic approach chosen in Bgee is to recode expression data for a gene in an organ and developmental stage as "not detected", "expressed with high confidence", or "expressed with low confidence". For experiments based on tag counting, such as ESTs, SAGE, or MPSS, we have considered a gene as expressed with a high confidence if the 95% confidence interval does not include zero.[89] For microarray data, a gene is considered expressed if the normalized signal is significantly above the background signal.[90] In the future, we plan to add information on whether a gene is significantly more expressed in one condition than another (e.g. more expressed in muscle than brain), and integrate other types of data.

The database is developed with MySQL, and currently includes four vertebrate species. The website allows users to retrieve information on gene expression by querying the database for keywords or gene identifiers, or browsing anatomical or developmental ontologies. In addition to species-specific or gene-specific views, users may view all gene families expressed in homologous organs between chosen species, and the complete expression information of a gene family across species. All queries may be constrained by data type, data quality, and keywords or identifiers.

Bgee is a promising tool to enable evo-devo studies on a larger scale. We hope that it will also be useful to put functional genomics studies in a comparative context, and provide a platform for integration of anatomical homology information into bioinformatics.

4. Conclusion

The intersection of bioinformatics and evo-devo is still relatively small, but holds a large potential for bringing the tools of high-throughput biology to illuminate our understanding of some of the most fundamental questions in biology: the origin of novelty, the role of constraints, the importance of loss vs. gain, and the extent of conservation between distant taxa. In this chapter, we have discussed briefly two aspects of the integration of bioinformatics and evo-devo: sequence-based analysis, and modeling homology relationships. A third approach should be mentioned in conclusion: gene regulatory networks. The characterization of gene regulatory networks controlling development is still a recent field,[91] and data have often proven costly to produce even in one species. It is to be hoped that, in the future, sufficient functional data will be available from a wider array of species. This should allow mechanistic yet large-scale studies of the control of morphological diversity by the genome.

Acknowledgments

I thank Frédéric Bastian, Gilles Parmentier, and Julien Roux for their help in preparing this chapter. Research in the laboratory of MRR was supported by the EU program Crescendo, the Swiss National Science Foundation, Etat de Vaud, the Swiss Institute of Bioinformatics, and the Decrypthon program.

References

1. Gould SJ. (1977) *Ontogeny and Phylogeny*. Cambridge, MA: The Belknap Press of Harvard University Press.
2. Gould SJ. (2002) *The Structure of Evolutionary Theory*. Boston, MA: Belknap Press.
3. Carroll S. (2005) *Endless Forms Most Beautiful: The New Science of Evo Devo and The Making of the Animal Kingdom*. New York, NY: W. W. Norton & Co.
4. Sanetra M, Begemann G, Becker MB, Meyer A. (2005) Conservation and co-option in developmental programmes: the importance of homology relationships. *Front Zool* **2**: 15.
5. Theissen G. (2005) Birth, life and death of developmental control genes: new challenges for the homology concept. *Theory Biosci* **124**: 199–212.
6. Brakefield PM. (2006) Evo-devo and constraints on selection. *Trends Ecol Evol* **21**: 362–8.
7. Raff RA. (2007) Written in stone: fossils, genes and evo-devo. *Nat Rev Genet* **8**: 911–20.
8. Breuker CJ, Debat V, Klingenberg CP. (2006) Functional evo-devo. *Trends Ecol Evol* **21**: 488–92.
9. Dehal P, Satou Y, Campbell RK *et al.* (2002) The draft genome of *Ciona intestinalis*: insights into chordate and vertebrate origins. *Science* **298**: 2157–67.

10. Sea Urchin Genome Sequencing Consortium; Sodergren E, Weinstock GM, Davidson EH *et al.* (2006) The genome of the sea urchin *Strongylocentrotus purpuratus*. *Science* **314**: 941–52.
11. King N, Westbrook MJ, Young SL *et al.* (2008) The genome of the choanoflagellate *Monosiga brevicollis* and the origin of metazoans. *Nature* **451**: 783–8.
12. Canestro C, Yokoi H, Postlethwait JH. (2007) Evolutionary developmental biology and genomics. *Nat Rev Genet* **8**: 932–42.
13. Mabee PM. (2006) Integrating evolution and development: the need for bioinformatics in evo-devo. *Bioscience* **56**: 301–9.
14. Taylor JS, Raes J. (2004) Duplication and divergence: the evolution of new genes and old ideas. *Annu Rev Genet* **38**: 615–43.
15. Holland PW, Garcia-Fernandez J, Williams NA, Sidow A. (1994) Gene duplications and the origins of vertebrate development. *Dev* (Suppl): 125–33.
16. Ohno S. (1970) *Evolution by Gene Duplication*. Heidelberg, Germany: Springer-Verlag.
17. Koonin EV. (2005) Orthology, paralogs and evolutionary genomics. *Annu Rev Genet* **39**: 309–38.
18. Amores A, Force A, Yan YL *et al.* (1998) Zebrafish hox clusters and vertebrate genome evolution. *Science* **282**: 1711–4.
19. Wolfe KH, Shields DC. (1997) Molecular evidence for an ancient duplication of the entire yeast genome. *Nature* **387**: 708–13.
20. Kellis M, Birren BW, Lander ES. (2004) Proof and evolutionary analysis of ancient genome duplication in the yeast *Saccharomyces cerevisiae*. *Nature* **428**: 617–24.
21. Grant D, Cregan P, Shoemaker RC. (2000) Genome organization in dicots: genome duplication in *Arabidopsis* and syntheny between soybean and *Arabidopsis*. *Proc Natl Acad Sci USA* **97**: 4168–73.
22. Simillion C, Vandepoele K, Van Montagu MCE *et al.* (2002) The hidden duplication past of *Arabidopsis thaliana*. *Proc Natl Acad Sci USA* **99**: 13627–32.
23. Paterson AH, Bowers JE, Chapman BA. (2004) Ancient polyploidization predating divergence of the cereals, and its consequences for comparative genomics. *Proc Natl Acad Sci USA* **101**: 9903–8.
24. Jaillon O, Aury J-M, Brunet F *et al.* (2004) Genome duplication in the teleost fish *Tetraodon nigroviridis* reveals the early vertebrate proto-karyotype. *Nature* **431**: 946–57.
25. Woods IG, Wilson C, Friedlander B *et al.* (2005) The zebrafish gene map defines ancestral vertebrate chromosomes. *Genome Res* **15**: 1307–14.
26. Aury JM, Jaillon O, Duret L *et al.* (2006) Global trends of whole-genome duplications revealed by the ciliate *Paramecium tetraurelia*. *Nature* **444**: 171–8.
27. Dehal P, Boore JL. (2005) Two rounds of whole genome duplication in the ancestral vertebrate. *PLoS Biol* **3**: e314.

28. Nakatani Y, Takeda H, Kohara Y, Morishita S. (2007) Reconstruction of the vertebrate ancestral genome reveals dynamic genome reorganization in early vertebrates. *Genome Res* **17**: 1254–65.
29. Putnam NH, Hellsten U, Yu JS *et al.* (2008) The amphioxus genome and the evolution of the chordate karyotype. *Nature* **453**: 1064–71.
30. Otto SP, Whitton J. (2000) Polyploid incidence and evolution. *Annu Rev Genet* **34**: 401–37.
31. Altschul SF, Madden TL, Schaffer AA *et al.* (1997) Gapped BLAST and PSI-BLAST: a new generation of protein database search programs. *Nucleic Acids Res* **25**: 3389–402.
32. Amoutzias GD, Veron A, Weiner AJ *et al.* (2007) One billion years of bZIP transcription factor evolution: conservation and change in dimerization, and DNA-binding site specificity. *Mol Biol Evol* **24**: 827–835.
33. Escriva Garcia H, Laudet V, Robinson-Rechavi M. (2003) Nuclear receptors are markers of animal genome evolution. *J Struct Funct Genomics* **3**: 177–84.
34. Robinson-Rechavi M, Laudet V. (2003) Bioinformatics of nuclear receptors. In: Russell DW, Mangelsdorf DJ (eds.). *Methods in Enzymology*. New York, NY: Academic Press, pp. 93–118.
35. Fitch WM. (2000) Homology: a personal view on some of the problems. *Trends Genet* **16**: 227–31.
36. Guindon S, Gascuel O. (2003) A simple, fast, and accurate algorithm to estimate large phylogenies by maximum likelihood. *Syst Biol* **52**: 696–704.
37. Li H, Coghlan A, Ruan J *et al.* (2006) TreeFam: a curated database of phylogenetic trees of animal gene families. *Nucleic Acids Res* **34**: D572–80.
38. Duret L, Mouchiroud D, Gouy M. (1994) HOVERGEN: a database of homologous vertebrate genes. *Nucleic Acids Res* **22**: 2360–5.
39. Perrière G, Combet C, Penel S *et al.* (2003) Integrated databanks access and sequence/structure analysis services at the PBIL. *Nucleic Acids Res* **31**: 3393–9.
40. Perriere G, Duret L, Gouy M. (2000) HOBACGEN: database system for comparative genomics in bacteria. *Genome Res* **10**: 379–85.
41. Dufayard JF, Duret L, Penel S *et al.* (2005) Tree pattern matching in phylogenetic trees: automatic search for orthologs or paralogs in homologous gene sequence databases. *Bioinformatics* **21**: 2596–603.
42. Laudet V, Gronemeyer H. (2002) *The nuclear receptors factsbook*. London, UK: Academic Press.
43. Escriva H, Delaunay F, Laudet V. (2000) Ligand binding and nuclear receptor evolution. *Bioessays* **22**: 717–27.
44. Bertrand S, Brunet FG, Escriva H *et al.* (2004) Evolutionary genomics of nuclear receptors: from twenty-five ancestral genes to derived endocrine systems. *Mol Biol Evol* **21**: 1923–37.

45. Adoutte A, Balavoine G, Lartillot N *et al.* (2000) The new animal phylogeny: reliability and implications. *Proc Natl Acad Sci USA* **97**: 4453–6.
46. Dunn CW, Hejnol A, Matus DQ *et al.* (2008) Broad phylogenomic sampling improves resolution of the animal tree of life. *Nature* **452**: 745–9.
47. Marletaz F, Holland LZ, Laudet V, Schubert M. (2006) Retinoic acid signaling and the evolution of chordates. *Int J Biol Sci* **2**: 38–47.
48. Thornton JW, Need E, Crews D. (2003) Resurrecting the ancestral steroid receptor: ancient origin of estrogen signaling. *Science* **301**: 1714–7.
49. Keay J, Bridgham JT, Thornton JW. (2006) The *Octopus vulgaris* estrogen receptor is a constitutive transcriptional activator: evolutionary and functional implications. *Endocrinology* **147**: 3861–9.
50. Mark M, Ghyselinck NB, Chambon P. (2006) Function of Retinoid Nuclear Receptors: lessons from genetic and pharmacological dissections of the retinoic acid signaling pathway during mouse embryogenesis. *Annu Rev Pharmacol Toxicol* **46**: 451–80.
51. Escriva H, Bertrand S, Germain P *et al.* (2006) Neofunctionalization in vertebrates: the example of retinoic acid receptors. *PLoS Genet* **2**: e102.
52. Duez H, Staels B. (2008) Rev-erb[alpha] gives a time cue to metabolism. *FEBS Lett* **582**: 19–25.
53. Reinking J, Lam MMS, Pardee K *et al.* (2005) The *Drosophila* nuclear receptor E75 contains heme and is gas responsive. *Cell* **122**: 195–207.
54. Markov G, Lecointre G, Demeneix B, Laudet V. (2008) The "street light syndrome", or how protein taxonomy can bias experimental manipulations. *Bioessays* **30**: 349–57.
55. Semon M, Wolfe KH. (2007) Consequences of genome duplication. *Curr Opin Genet Dev* **17**: 505–12.
56. Bowers JE, Chapman BA, Rong J, Paterson AH. (2003) Unravelling angiosperm genome evolution by phylogenetic analysis of chromosomal duplication events. *Nature* **422**: 433–8.
57. Brunet FG, Crollius HR, Paris M *et al.* (2006) Gene loss and evolutionary rates following whole-genome duplication in teleost fishes. *Mol Biol Evol* **23**: 1808–16.
58. Scannell DR, Byrne KP, Gordon JL *et al.* (2006) Multiple rounds of speciation associated with reciprocal gene loss in polyploid yeasts. *Nature* **440**: 341–5.
59. Davis JC, Petrov DA. (2004) Preferential duplication of conserved proteins in eukaryotic genomes. *PLoS Biol* **2**: e55.
60. Meyer A, Van de Peer Y. (2005) From 2R to 3R: evidence for a fish-specific genome duplication (FSGD). *Bioessays* **27**: 937–45.
61. Volff JN. (2005) Genome evolution and biodiversity in teleost fish. *Heredity* **94**: 280–94.
62. Donoghue PCJ, Purnell MA. (2005) Genome duplication, extinction and vertebrate evolution. *Trends Ecol Evol* **20**: 312–9.

63. Jordan IK, Wolf YI, Koonin EV. (2004) Duplicated genes evolve slower than singletons despite the initial rate increase. *BMC Evol Biol* **4**: 22.
64. Maere S, De Bodt S, Raes J *et al.* (2005) Modeling gene and genome duplications in eukaryotes. *Proc Natl Acad Sci USA* **102**: 5454–9.
65. von Baer KE. (1828) *Ueber Entwicklungsgeschichte der Thiere: Beobachtung und Reflexion.* Königsberg, Germany: Bornträger.
66. Duboule D. (1994) Temporal colinearity and the phylotypic progression: a basis for the stability of a vertebrate Bauplan and the evolution of morphologies through heterochrony. *Dev Suppl*: 135–42.
67. Raff RA. (1996) *The Shape of Life: Genes, Development, and the Evolution of Animal Form.* Chicago, IL: University of Chicago Press.
68. Roux J, Robinson-Rechavi M. (2009) Developmental constraints on vertebrate genome evolution. *PLoS Genet* (in press).
69. Hall B. (1999) *Homology: The Hierarchical Basis of Comparative Biology.* New York, NY: John Wiley & Sons.
70. Abouheif E. (1997) Developmental genetics and homology: a hierarchical approach. *Trends Ecol Evol* **12**: 405–8.
71. McKitrick MC. (1994) On homology and the ontological relationship of parts. *Syst Biol* **43**: 1–10.
72. Wray GA, Abouheif E. (1998) When is homology not homology? *Curr Opin Genet Dev* **8**: 675–80.
73. Nielsen C, Martinez P. (2003) Patterns of gene expression: homology or homocracy? *Dev Genes Evol* **213**: 149–54.
74. Jeffery J, Bininda-Emonds O, Coates M, Richardson M. (2005) A new technique for identifying sequence heterochrony. *Syst Biol* **54**: 230–40.
75. Demir E, Babur O, Dogrusoz U *et al.* (2004) An ontology for collaborative construction and analysis of cellular pathways. *Bioinformatics* **20**: 349–56.
76. Eilbeck K, Lewis SE. (2004) Sequence Ontology annotation guide. *Comp Funct Genomics* **5**: 642–7.
77. Bard J, Rhee S, Ashburner M. (2005) An ontology for cell types. *Genome Biol* **6**: R21.
78. Jaiswal P, Avraham S, Ilic K *et al.* (2005) Plant Ontology (PO): a controlled vocabulary of plant structures and growth stages. *Comp Funct Genomics* **6**: 388–97.
79. Topalis P, Koutsos A, Dialynas E *et al.* (2005) AnoBase: a genetic and biological database of anophelines. *Insect Mol Biol* **14**: 591–7.
80. Vincent PLD, Coe JEH, Polacco ML. (2003) *Zea mays* Ontology — a database of international terms. *Trends Plant Sci* **8**: 517–20.
81. Shvaiko P, Euzenat J. (2007) *Ontology Matching.* Berlin, Germany: Springer Verlag.
82. Sprague J, Bayraktaroglu L, Clements D *et al.* (2006) The Zebrafish Information Network: the zebrafish model organism database. *Nucleic Acids Res* **34**: D581–5.

83. Bastian F, Parmentier G, Roux J *et al.* (2008) Bgee: integrating and comparing heterogeneous transcriptome data among species. In: *International Workshop on Data Integration in the Life Sciences (DILS) 2008*. LNBI series, Vol. 5109. Evry, France: Springer, pp. 124–131.
84. Hubbard TJ, Aken BL, Beal K *et al.* (2007) Ensembl 2007. *Nucleic Acids Res* **35**: D610–7.
85. Kuo WP, Liu F, Trimarchi J *et al.* (2006) A sequence-oriented comparison of gene expression measurements across different hybridization-based technologies. *Nat Biotechnol* **24**: 832–40.
86. Lee CK, Sunkin SM, Kuan C *et al.* (2008) Quantitative methods for genome-scale analysis of *in situ* hybridization and correlation with microarray data. *Genome Biol* **9**: R23.
87. Thisse B, Heyer V, Lux A *et al.* (2004) Spatial and temporal expression of the zebrafish genome by large-scale *in situ* hybridization screening. *Methods Cell Biol* 77: 505–19.
88. Ye J, Chen J, Li Q, Kumar S. (2006) Classification of *Drosophila* embryonic developmental stage range based on gene expression pattern images. *Comput Syst Bioinformatics Conf*: 293–8.
89. Audic S, Claverie J-M. (1997) The significance of digital gene expression profiles. *Genome Res* 7: 986–95.
90. Wu ZJ, Irizarry RA, Gentleman R *et al.* (2004) A model-based background adjustment for oligonucleotide expression arrays. *J Am Stat Assoc* **99**: 909–17.
91. Davidson EH, Erwin DH. (2006) Gene regulatory networks and the evolution of animal body plans. *Science* **311**: 796–800.

Section IV

MODELING OF BIOLOGICAL SYSTEMS

Chapter 14

Spatiotemporal Modeling and Simulation in Biology

Ivo F. Sbalzarini

1. Introduction

Describing the dynamics of processes in both space and time simultaneously is referred to as spatiotemporal modeling. This is in contrast to describing the dynamics of a system in time only as is, for example, usually done in chemical kinetics or pathway models. Solving spatiotemporal models in a computer requires spatiotemporal computer simulations. While computational data analysis allows unbiased and reproducible processing of large amounts of data from, e.g. high-throughput assays, computer simulations enable virtual experiments *in silico* that would not be possible in reality. This greatly expands the range of possible perturbations and observations. Computational experiments allow studying systems whose complexity prohibits manual analysis, and they make accessible time and length scales that cannot be reached by lab experiments. Examples of the latter include molecular dynamics (MD) studies in structural biology and studies in ecology or evolutionary biology. In virtual experiments, all variables are controllable and observable. We can thus measure everything and precisely control all influences and cross-couplings. This allows disentangling coupled effects that could not be separated in real experiments, greatly reduces or eliminates the need for indirect control experiments, and facilitates interpretation of the results. Finally, computational models do not involve living beings, thus enabling experiments that would not be possible in reality due to ethical reasons. Although we focus on applications of spatiotemporal computer simulations in biology, the employed concepts and methods are more generally valid.

Resolving a dynamic process in space greatly increases the number of degrees of freedom (variables) that need to be tracked. Consider, for example, a biochemical heterodimerization reaction. This reaction can be modeled by its chemical kinetics using three variables: the concentrations of the two monomers and the concentration of dimers. Assume now that monomers are produced at certain locations in space and freely diffuse from there. Their concentration thus varies in space in such a way that it is higher close to the source and lower farther away, which greatly increases the number of variables we have to track in the simulation. If we are, say, interested in the local concentrations at 1000 positions, we already have to keep track of 3000 variables. Moreover, the reactions taking place at different points in space are not independent. Each local reaction can influence the others through diffusive transport of monomers and dimers. The complexity of spatiotemporal models thus rapidly increases. In fact, there is no theoretical limit to the number of points in space that we may use to resolve the spatial patterns in the concentration fields. Using infinitely many points corresponds to modeling the system as a continuum.

A number of powerful mathematical tools are available to efficiently deal with spatiotemporal models and to simulate them. While it is not possible within the scope of this chapter to describe each of them in detail, we will give an overview with references to specialized literature. We then review in detail one particular method that is well suited for applications in biology. But before we start, we revisit some of the motivations and particularities of spatiotemporal modeling in the life sciences.

In spatiotemporal modeling, nature is mostly described in four dimensions: time plus three spatial dimensions. While time and the presence of reservoirs (integrators) are essential for the existence of dynamics, three-dimensional (3D) spatial aspects also play important roles in many biological processes. Think, for example, of predators hunting their prey in a forest, of blood flowing through our arteries, of the electromagnetic fields in the brain, or of such an unpleasant phenomenon as the epidemic spread of a disease. In all of these examples, and many others, the spatial distributions of some quantities play an essential role. Models and simulations of such systems should thus account for and resolve these distributions. When determining the location of an epileptic site in the brain, it is, for instance, of little value to know the total electric current density in the whole brain — we need to know precisely where the source is. These examples extend across all scales of biological systems, from the above-mentioned predator–prey interactions in ecosystems over morphogenesis[1–5] and intracellular processes to single molecules. Think, for example, of conformational changes in proteins. Examples at the intracellular level include virus entry[6,7] and transport,[8–10] intracellular signaling,[11,12] the diffusion of proteins in the various cellular compartments,[13–15] or the fact that such compartments exist in the first place.

Spatial organization is important, as the same protein can have different effects depending on the intracellular compartment in which it is located. The most prominent example is probably cytochrome C, which is an essential part of the cell respiration chain in the mitochondria, but triggers programmed cell death (apoptosis) when released into the cytoplasm.[16] Another example is found in the role of transmembrane signaling during morphogenesis. Differences in protein diffusion constants are not large enough to produce Turing patterns,[1] and the slow transport across intercompartment membranes is essential.[17] Examples of spatiotemporal processes at the multi-cellular level include tumor growth[18–20] and cell-cell signaling,[21] including phenomena such as bacterial quorum sensing, the microscopic mechanism underlying the macroscopic phenomenon of bioluminescence in certain squid.[22]

Given the widespread importance of spatiotemporal processes, it is not surprising that a number of large software projects for spatiotemporal simulations in biology have been initiated. Examples in computational cell biology[23,24] include E-Cell, MCell, and the Virtual Cell.

2. Properties of Biological Systems

Simulating spatially resolved processes in biological systems, such as geographically structured populations, multicellular organs, or cell organelles, provides a unique set of challenges to any mathematical method. One often hears that this is because biological systems are "complex".

Biochemical networks, ecosystems, biological waves, heart cell synchronization, and life in general are located in the high-dimensional, nonlinear regime of the map of dynamical systems,[25] together with quantum field theory, nonlinear optics, and turbulent flows. None of these topics are completely explored. They are at the limit of our current understanding and will remain challenging for many years to come. Why is this so, and what do we mean by "complex"?

Biological systems exhibit a number of characteristics that render them difficult. These properties frequently include one or several of the following:

- high-dimensional (or infinite-dimensional in the continuum limit);
- regulated;
- delineated by complex shapes;
- nonlinear;
- coupled across scales and subsystems;
- plastic over time (time-varying dynamics); and/or
- nonequilibrium.

Due to these properties, biological systems challenge existing methods in modeling and simulation. They are thus particularly well suited to drive the development of new methods and theories. The challenges presented by spatiotemporal biological systems have to be addressed on several fronts simultaneously: numerical simulation methods, computational algorithms, and software engineering.[26] Numerical methods are needed that can deal with multi-scale systems[27–31] and topological changes in complex geometries. Computer algorithms have to be efficient enough to deal with the vast number of degrees of freedom, and software platforms must be available to effectively and robustly implement these algorithms on multiprocessor computers.[32]

2.1. *Dimensionality and Degrees of Freedom*

The large number of dimensions (degrees of freedom) is due to the fact that biological systems typically contain more compartments, components, and interaction modes than traditional engineering applications such as electronic circuits or fluid mechanics.[26] In a direct numerical simulation, all degrees of freedom need to be explicitly tracked. In continuous systems, each point in space adds additional degrees of freedom, leading to an infinite number of dimensions. Such systems have to be discretized, i.e. the number of degrees of freedom needs to be reduced to a computationally feasible amount, which is done by selecting certain representative dimensions. Only these are then tracked in the simulation, approximating the behavior of the full, infinite-dimensional system. Discretizations must be consistent, i.e. the discretized system has to converge to the full system if the number of explicitly tracked degrees of freedom goes to infinity.

Discrete biological systems already have a finite number of degrees of freedom and can sometimes be simulated directly. If the number of degrees of freedom is too large, as is e.g. the case when tracking the motion of all atoms in a protein, we do, however, again have to reduce them in order for simulations to be feasible. This can be done by collecting several degrees of freedom into one and only tracking their collective behavior. These so-called "coarse graining" methods greatly reduce the computational cost and allow simulations of very large, high-dimensional systems such as patches of lipid bilayers with embedded proteins,[33,34] or actin filaments.[35] Coarse graining thus allows extending the capabilities of molecular simulations to time and length scales of biological interest.

2.2. *Regulation*

In biological systems, little is left to chance, which might seem surprising given the inherently stochastic nature of molecular processes, environmental influences, and phenotypic variability. These underlying fluctuations are, however, in many cases a prerequisite for adaptive deterministic behavior, as has been shown, for example, in gene regulation networks.[36] In addition to such indirect regulation mediated by bistability and stochastic fluctuations, feedback and feed-forward loops are ubiquitous in biological systems. From signal transduction pathways in single cells to Darwinian evolution, regulatory mechanisms play important roles. Results from control theory tell us that such loops can alter the dynamic behavior of a system, change its stability or robustness, or give rise to multi-stable behavior that enables adaptation to external changes and disturbances.[36] Taking all of these effects into account presents a grand challenge to simulation models not only because many of the hypothetical regulatory mechanisms are still unknown or poorly characterized.

2.3. *Geometric Complexity*

Biological systems are mostly characterized by irregular and often moving or deforming geometries. Processes on curved surfaces may be coupled to processes in enclosed spaces; and surfaces frequently change their topology, such as in fusion or fission of intracellular compartments. Examples of such complex geometries are found on all length scales and include the prefractal structures of taxonomic and phylogenetic trees,[37] regions of stable population growth in ecosystems,[38] pneumonal and arterial trees,[39] the shapes of neurons,[40] the cytoplasmic space,[41] clusters of intracellular vesicles,[42] electric currents through ion channels in cell membranes,[43] protein chain conformations,[44] and protein structures.[45] Complex geometries are not only difficult to resolve and represent in the computer, but the boundary conditions imposed by them on dynamic spatiotemporal processes may also qualitatively alter the macroscopically observed dynamics. Diffusion in complex-shaped compartments such as the endoplasmic reticulum (ER; Fig. 1) may appear anomalous, even if the underlying molecular diffusion is normal.[46–49]

2.4. *Nonlinearity*

Common biological phenomena such as interference, cooperation, and competition lead to nonlinear dynamic behavior. Many processes, from repressor interactions in gene networks over predator–prey interactions in ecosystems to calcium waves in cells, are not appropriately described by linear systems theory as predominantly used and taught in physics and engineering. Depending on the number of degrees of freedom, nonlinear systems exhibit phenomena not observed in linear systems. These phenomena include bifurcations, nonlinear oscillations, and chaos and fractals. Nonlinear models are intrinsically hard to solve. Most of them are impossible to solve analytically; and computer simulations are hampered

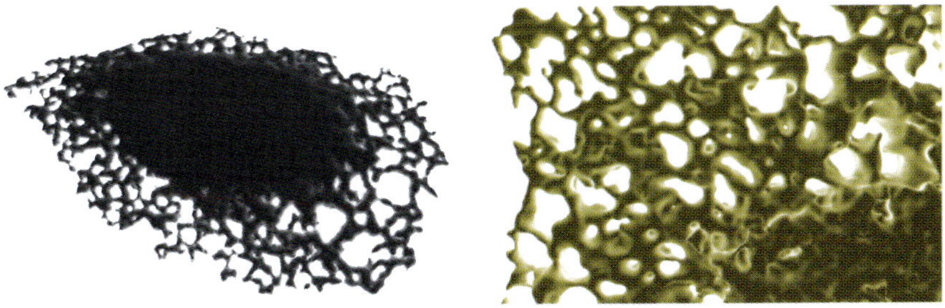

(a) (b)

Fig. 1. (a) Shaded view of a 3D computer reconstruction of the geometry of an endoplasmic reticulum (ER) of a live cell.[49] (b) Close-up of a reconstructed ER, illustrating the geometric complexity of this intracellular structure.

by the fact that common computational methods, such as normal mode analysis, Fourier transforms, or the superposition principle, break down in nonlinear systems because a nonlinear system is not equal to the sum of its parts.[25]

2.5. *Coupling Across Scales*

Coupling across scales means that events on the microscopic scale such as changes in molecular conformation can have significant effects on the global, macroscopic behavior of the system. This is certainly the case for many biological systems — bioluminescence due to bacterial quorum sensing[22] for example, or the effect on the behavior of a whole organism when hormones bind to their receptors. Such multi-scale systems share the property that the individual scales cannot be separated and treated independently. There is a continuous spectrum of scales with coupled interactions that impose stringent limits on the use of computer simulations. Direct numerical simulation of the complete system would require resolving it in all detail everywhere. Applied to the simulation of a living cell, this would mean resolving the dynamics of all atoms in the cell. A cell consists of about 10^{15} atoms, and biologically relevant processes such as protein folding and enzymatic reactions occur on the time scale of milliseconds. The largest molecular dynamics (MD)[50] simulations currently done consider about 10^{10} atoms over one nanosecond. In order to model a complete cell, we would need a simulation about 100 000 times larger, running over a millionfold longer time interval. This would result in a simulation at least 10^{11} times bigger than what can currently be done. This is certainly not feasible and will remain so for many years to come. Even if one could simulate the whole system at full resolution, the results would be of questionable value. The amount of data generated by such a simulation would be vast, and the interesting macroscopic phenomena that we are looking for would mostly be masked by noise from the small scales. In order to treat coupled systems, we thus have to use multi-scale models[28–31] and formulations at the appropriate level of detail.

2.6. *Temporal Plasticity*

While the analysis of high-dimensional, nonlinear systems is already complicated as such, the systems themselves also frequently change over time in biological applications. In a mathematical model, this is reflected by jumps in the dynamic equations or by coefficients and functions that change over time. During its dynamics, the system can change its behavior or switch to a different mode. For example, the dynamics of many processes in cells depend on the cell cycle, physiological processes in organisms alter their dynamics depending on age or disease, and environmental changes affect the dynamic behavior of ecosystems. Such systems are called "plastic" or "time-varying". Dealing with time-varying systems, or equations that change their structure over time, is an open issue in numerical simulations. Consistency of the solution at the switching points must be ensured in order to prevent the simulation method from becoming unstable.

2.7. *Nonequilibrium*

According to the second law of thermodynamics, entropy can only increase. Life evades this decay by feeding on negative entropy.[51] The discrepancy between life and the fundamental laws of thermodynamics has puzzled scientists for a long time. It can only be explained by assuming that living systems are not in equilibrium. Most of statistical physics and thermodynamics has been developed for equilibrium situations and, hence, does not readily apply to living systems. Phenomena such as the establishment of cell polarity or the organization of the cell membrane can only be explained when accounting for nonequilibrium processes such as vesicular recycling.[52] Due to our incomplete knowledge of the theoretical foundations of nonequilibrium processes, they are much harder to understand. Transient computer simulations are often the sole method available for their study.

3. Spatiotemporal Modeling Techniques

Dynamic spatiotemporal systems can be described in various ways, depending on the required level of detail and fidelity. We distinguish three dimensions of description: phenomenological vs. physical, discrete vs. continuous, and deterministic vs. stochastic. The three axes are independent and all combinations are possible. Depending on the chosen system description, different modeling techniques are available. Figure 2 gives an overview of the most frequently used ones as well as examples of dynamic systems that could be described with them.

3.1. *Phenomenological vs. Physical Models*

Phenomenological models reproduce or approximate the overall behavior of a system without resolving the underlying mechanisms. Such models are useful if one is interested in analyzing the reaction of the system to a known perturbation, without requiring information about how this reaction is brought about. This is in contrast to physical models, which faithfully reproduce the mechanistic functioning of the system. Physical models thus allow predicting the system behavior in new, unseen situations, and they give information about how things work. Physical models are based on first principles or laws from physics.

3.2. *Discrete vs. Continuous Models*

The discrete vs. continuous duality relates to the spatial resolution of the model. In a discrete model, each constituent of the system is explicitly

	continuous	discrete
deterministic	PDEs	interacting particles
	diffusion	molecular dynamics
stochastic	SDEs	random events
	reaction-diffusion with low molecule numbers	population dynamics

Fig. 2. Most common modeling techniques for all combinations of continuous/discrete and deterministic/stochastic models. The techniques for physical and phenomenological models are identical, but in the former case the models are based on physical principles. Common examples of application of each technique are given in the shaded areas.

accounted for as an individual entity. Examples include MD simulations,[50] where the position and velocity of each atom are explicitly tracked and atoms are treated as individual, discrete entities. In a continuous model, a mean field average is followed in space and time. Examples of such field quantities are concentration, temperature, and charge density.

In continuous models, we distinguish two types of quantities. On the one hand, quantities whose value in a homogeneous system does not depend on the averaging volume are called "intensive". Examples include concentration or temperature. If 1 L of water at 20°C is divided into two half-liter glasses, the water in each of the two glasses will still have a temperature of 20°C, even though the volume is halved. The temperature of the water is independent of the volume of water, hence making temperature an intensive property. On the other hand, quantities whose value in a homogeneous system depends on the volume are called "extensive". These are quantities such as mass, heat, or charge. Neither of the two half-liters of water has the same mass as the original liter. Intensive and extensive quantities come in pairs: concentration–mass, temperature–heat, charge density–charge, etc. Field quantities as considered in continuous models are always intensive, and quantities in discrete models are usually extensive. Corresponding extensive and intensive quantities are interrelated through an averaging operation. The concentration of molecules can e.g. be determined by measuring the total mass of all molecules within a given volume and dividing this mass by the volume. We imagine such an averaging volume around each point in space in order to recover a spatially resolved concentration field. If the averaging volume chosen is too small, entry and exit of individual molecules will lead to significant jumps in the average. With a growing averaging volume, the concentration may converge to a stable value. If the volume is further enlarged, the concentration may again start to vary due to macroscopic spatial gradients. This behavior is illustrated in Fig. 3. Above the continuum limit λ, the average is converged and microscopic single-particle effects are no longer significant. The value of the continuum limit is governed by the abundance of particles compared to the size of the averaging volume. If the microscopic particles are molecules such as proteins, λ is related to their mean free path. On length scales larger than the scale of field variations L, macroscopic gradients of the averaged field become apparent if the field is not homogeneous, i.e. if its value varies in space. The dimensionless ratio $\text{Kn} = \lambda/L$ is called the Knudsen number.

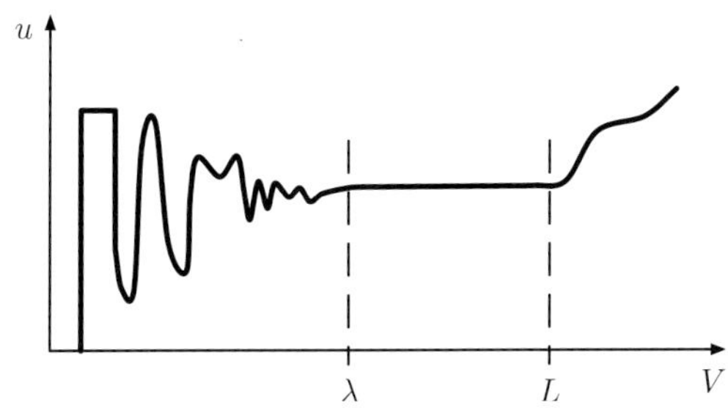

Fig. 3. The value u of a volume-averaged intensive field quantity depends on the size of the averaging volume V. For volumes smaller than the continuum limit λ, individual particles cause the average to fluctuate. In the continuum region above λ, the volume average can be stationary or vary smoothly due to macroscopic field gradients above the length scale of field variations L.

Continuous models are valid only if the microscopic and macroscopic scales are well separated, i.e. if $\mathrm{Kn} \ll 1$; for any spatial distribution with $\mathrm{Kn} \gg 1$, discrete models are the only choice since each particle is important and no continuum region exists. Between these two cases lies the realm of mesoscopic models.[53]

Continuous deterministic models are characterized by smoothly varying (on length scales $>L$) field quantities whose temporal and spatial evolution depends on some derivatives of the same or other field quantities. The fields can, for example, model concentrations, temperatures, or velocities. Such models are naturally formulated as unsteady partial differential equations (PDEs),[54,55] since derivatives relate to the existence of integrators, and hence reservoirs, in the system. The most prominent examples of continuous deterministic models in biological systems include diffusion models, advection, flow, and waves. Discrete deterministic models are characterized by discrete entities interacting over space and time according to deterministic rules. The interacting entities can, e.g. model cells in a tissue,[5] individuals in an ecosystem, or atoms in a molecule.[50] Such models can mostly be interpreted as interacting particle systems or automata. In biology, discrete deterministic models can be found in ecology or in structural biology.

3.3. *Stochastic vs. Deterministic Models*

Biological systems frequently include a certain level of randomness, as is the case for unpredictable environmental influences, fluctuations in molecule numbers upon cell division, and noise in gene expression levels. Such phenomena can be accounted for in stochastic models. In such models, the model output is not entirely predetermined by the present state of the model and its inputs, but it also depends on random fluctuations. These fluctuations are usually modeled as random numbers of a given statistical distribution. Continuous stochastic models are characterized by smoothly varying fields whose evolution in space and time depends on probability densities that are functions of some derivatives of the fields. In the simplest case, this amounts to a single noise term modeling, e.g. Gaussian or uniform fluctuations in the dynamics. Models of this kind are mostly formalized as stochastic differential equations (SDEs).[56]

These are PDEs with stochastic terms that can be used to model probabilistic processes such as the spread of epidemics, neuronal signal transduction,[57,58] or evolution theory. In discrete stochastic models, probabilistic effects mostly pertain to discrete random events. These events are characterized by their probability density functions. Examples include population dynamics (individuals have certain probabilities to be born, die, eat, or be eaten), random walks of diffusing molecules, or stochastically occurring chemical reactions. Several methods are also available for combining stochastic and deterministic models into hybrid stochastic-deterministic models.[59,60]

4. Spatiotemporal Simulation Methods

Depending on the modeling technique chosen for a given system (Fig. 2), different numerical methods exist for simulating the resulting model in a computer. While it is impossible to give an exhaustive list of all available methods, we will highlight the most important ones for each category of models. The same numerical methods can be used for both physical and phenomenological models.

4.1. *Methods for Discrete Stochastic Models*

Discrete stochastic models as formulated by events occurring according to certain probability distributions can be simulated using stochastic simulation algorithms (SSAs) such as Gillespie's algorithm[61,62] or the Gibson–Bruck algorithm.[63] While most of these algorithms were originally developed for temporal dynamics only, they have since been generalized to spatiotemporal models such as reaction-diffusion models.[64,65] Monte Carlo methods[66,67] provide the basis for most of these algorithms. Simulating probabilistic trajectories of the model thus amounts to sampling from the model probability distributions. In order to estimate the average trajectory or the standard deviation, ensemble averages over many simulations must be computed. This fundamentally limits the convergence properties of these methods to $\mathcal{O}(1/\sqrt{N})$, where N is the number of simulations performed.[68] Agent-based methods with probabilistic agents are also frequently used to simulate discrete stochastic models. A prominent example is Brownian agents.[69]

4.2. Methods for Discrete Deterministic Models

Simulations of discrete deterministic models are frequently implemented using methods from the class of finite automata. The most prominent examples are cellular[a] automata[70–72] and agent-based simulations.[73] In finite automata, spatially distributed computational cells (or agents) with certain attributed properties interact according to sets of deterministic rules. These interaction rules map the state (the values of the attributed properties) of the interacting cells to certain actions, which in turn change the states of the cells. Finite automata are powerful and fascinating tools, as already small sets of simple rules can give rise to very complex nonlinear model behavior. They can be used for diverse purposes such as studying behavioral aspects of interacting individuals in ecosystems, studying artificial life, simulating interacting neurons,[71] simulating social interactions,[74] or simulating pattern-forming mechanisms in morphogenesis.[5] Another important class of discrete deterministic simulations is found in MD.[50] Here, the atomistic behavior of molecules is simulated by explicitly tracking the dynamics and positions of all atoms. Atoms in classical MD are modeled as discrete particles that interact according to deterministic mechanisms such as interatomic bonds, van der Waals forces, or electrostatics.

4.3. Methods for Continuous Stochastic Models

Continuous stochastic models formulated as SDEs can be numerically simulated using a variety of stochastic integration methods,[75,76] most notably Euler–Maruyama or Milstein's higher-order method.[76] It is, however, important to keep in mind that each simulation represents just one possible realization of the stochastic process. In order to estimate means and variances, many independent simulations need to be performed

---

[a] The word "cellular" refers to computational cells in the algorithm and does not imply any connection with biological cells.

and an ensemble average computed.[76] While the topic of SDEs may seem exotic to many computational biologists, it is more widespread than one would think. Simulating, for example, a reaction-diffusion model with stochastic reactions amounts to numerically solving an SDE.[64,65]

4.4. *Methods for Continuous Deterministic Models*

Continuous deterministic models as represented by PDEs can be solved using any of the discretization schemes from numerical analysis.[77,78] The most common ones include finite difference (FD) methods,[79] finite element (FE) methods,[80,81] and finite volume (FV) methods for conservation laws.[82,83] FD methods are based on Taylor series expansions[84] of the spatial field functions and approximation of the differential operators by difference operators such that the first few terms in the Taylor expansion are preserved. FE methods express the unknown field function in a given function space. The basis functions of this space are supported on polygonal elements that tile the computational domain. Determining the unknown field function then amounts to solving a linear system of equations for the weights of the basis functions on all elements. FV methods make use of physical conservation laws such as conservation of mass or momentum. The computational domain is subdivided into disjoint volumes, for each of which the balance equations are formulated (change of volume content equals inflow minus outflow) and numerically solved.

All of these methods have the common property that they require a computational mesh — regular or irregular — that discretizes the computational domain into simple geometric structures such as lines (FD), areas (FE), or volumes (FV) with the appropriate connectivity. For complex-shaped domains as they frequently occur in biological systems (cf. Fig. 1), it can be a daunting task to find a good connected mesh that respects the boundary conditions and has sufficient regularity to preserve the accuracy and efficiency of the numerical method. Mesh-free particle methods[85–87] relax this constraint by basing the discretization on point objects that do not require any connectivity information. While particle formulations are the natural choice for discrete models, their advantages can be transferred to the continuous domain using continuum particle methods as described in Sec. 5; they are based on approximating the smooth field functions of a continuous model by integrals that are being discretized onto computational elements called particles. While the particles in discrete simulations represent real-world objects such as molecules, atoms, animals, or cells, particles in continuous methods are computational elements that collectively approximate a field quantity as outlined in Sec. 5.1.

4.5. *Representing Complex Geometries in the Computer*

Complex geometries and surfaces can be represented in the computer using a variety of methods,[88] which can be classified according to the connectivity information they require. Triangulated surfaces[89] are an example of connectivity-based representations, as they require each triangle to know which other triangles it is connected to. Establishing this connectivity information on complex-shaped surfaces is computationally expensive, so these representations are preferably used in conjunction with numerical methods that operate on the same connectivity. This is the case when using FE methods with triangular elements in simulations involving triangulated surfaces,[3,90] or FD methods in conjunction with pixelated surface representations.[91] An example of a complex triangulated surface is shown in Fig. 1(a).

Connectivity-less surface representations include scattered point clouds[92] and implicit surface representations such as level sets.[93] In level set methods, the geometry is implicitly represented as an isosurface of a higher-dimensional level function. Level sets are well suited to be used in combination with particle methods because the level function can directly be represented on the same set of computational particles.[94–96] This allows treating arbitrarily complex geometries at constant computational cost, and simulating moving and deforming geometries with no linear stability limit. The ER shown in Fig. 1(b) was, e.g. represented in the computer as a level set in order to simulate diffusion processes on its surface using particle methods.[96]

5. Introduction to Continuum Particle Methods

Particle methods are point-based spatiotemporal simulation methods that exhibit a number of favorable properties, which help address the complications of spatiotemporal biological systems (cf. Sec. 2):

- They are the most universal simulation method. Particle methods can be used for all types of models in Fig. 2, whereas most other numerical methods are limited to one or two types of models.
- They are intimately linked to the physical or biological process they represent, since particles correspond to real-world entities (in discrete models) or approximate field quantities (in continuous models). The interactions between the particles can mostly be intuitively understood as forces or exchange of mass. This prevents spurious, unphysical modes from showing up in the simulation results, a capability that has recently also been developed for FE methods.[97]
- They are well suited for simulations in complex geometries, such as the ER shown in Fig. 1,[49] and for simulations on complex curved surfaces such as intracellular membranes.[96] No computational mesh needs to be generated and no connectivity constraints satisfied. This effectively avoids the increased algorithmic complexity of mesh-based methods in complex geometries due to loss of the "nice" structure of the matrix.
- Due to their inherent regularity (particles have a finite size that defines the resolution of the method; cf. Sec. 5.1), particle methods can easily handle topological changes such as fusion and fission in the simulated geometry. Mesh-based methods need special regularization so as not to become unstable when two fusing or separating objects touch in exactly one point.
- They are inherently adaptive, as particles are only required where the represented quantity is present, and the motion of the particles automatically tracks these regions. This constitutes an important computational advantage compared to mesh-based methods, where a mesh is required throughout the computational domain.

- Particle methods are not subject to any linear convective stability condition (CFL condition[98,99]).[27,100] When simulating flows or waves with mesh-based methods, the CFL condition imposes a time step limit such that the flow or wave can never travel more than a certain number of mesh cells per time step. In particle methods, convection simply amounts to moving the particles with the local velocity field, and no limit on how far they can move is imposed as long as their trajectories do not intersect.
- Thanks to advancements in fast algorithms for N-body interactions (cf. Sec. 6), particle methods are as computationally efficient as mesh-based methods. In addition, the data structures and operators in particle simulations can be distributed across many processors, enabling highly efficient parallel simulations.[32]

Since continuum applications of particle methods are far less known than discrete ones, we focus on deterministic continuous models. For reasons of simplicity, however, we will not cover the most recent extensions of particle methods to multi-resolution and multi-scale[27–31] problems using concepts from adaptive mesh refinement,[101] adaptive global mappings,[101] or wavelets.[100]

In continuum particle methods, a particle p occupies a certain position x_p and carries a physical quantity $\boldsymbol{\omega}_p$, referred to as its strength. These particle attributes — strength and location — evolve so as to satisfy the underlying governing equations in a Lagrangian frame of reference.[27] Simulating a model thus amounts to tracking the dynamics of all N computational particles carrying the physical properties of the system being simulated. The dynamics of the N particles are governed by sets of ordinary differential equations (ODEs) that determine the trajectories of the particles p and the evolution of their properties $\boldsymbol{\omega}$ over time. Thus,

$$\begin{aligned} \frac{\mathrm{d}\boldsymbol{x}_p}{\mathrm{d}t} = \boldsymbol{v}_p(t) = \sum_{q=1}^{N} \boldsymbol{K}(\boldsymbol{x}_p, \boldsymbol{x}_q; \boldsymbol{\omega}_p, \boldsymbol{\omega}_q) \qquad p = 1, 2, \ldots, N \\ \frac{\mathrm{d}\boldsymbol{\omega}_p}{\mathrm{d}t} = \sum_{q=1}^{N} \boldsymbol{F}(\boldsymbol{x}_p, \boldsymbol{x}_q; \boldsymbol{\omega}_p, \boldsymbol{\omega}_q) \qquad p = 1, 2, \ldots, N, \end{aligned} \tag{1}$$

where $\boldsymbol{v}_p(t)$ is the velocity of particle p at time t. The dynamics of the particles are completely defined by the functions $\boldsymbol{K}$ and $\boldsymbol{F}$, which represent the model being simulated. In pure particle methods, $\boldsymbol{K}$ and $\boldsymbol{F}$ emerge from integral approximations of differential operators (cf. Sec. 5.2.1); in hybrid particle-mesh (PM) methods, they entail solutions of field equations that are discretized on a superimposed mesh (cf. Sec. 5.2.2). The sums on the right-hand side of Eq. (1) correspond to quadrature[84] (numerical integration) of some functions.

In order to situate continuum particle methods on the map of numerical analysis, we consider the different strategies to numerically solve a differential equation as outlined in Fig. 4 for the example of a simple PDE, the Poisson equation, which we wish to solve for the intensive field quantity u. One way consists of discretizing the equation onto a computational mesh with resolution h using FD, FE, or FV, and then numerically solving the resulting system $\mathbf{A}\boldsymbol{u} = \boldsymbol{f}$ of linear algebraic equations. The discretization needs to be done consistently in order to ensure that the discretized equations model the same system as the original PDE, and the numerical solution of the resulting linear system is subject to stability criteria. An alternative route is to solve the PDE analytically

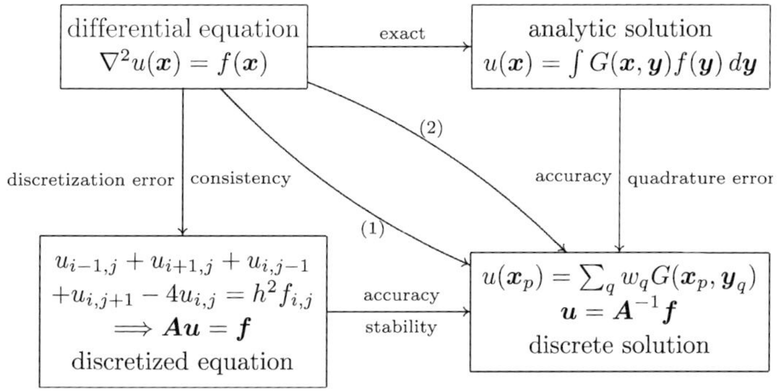

Fig. 4. Strategies to numerically solve a differential equation, illustrated on the example of the two-dimensional (2D) Poisson equation: (1) discretization of the PDE on a mesh with resolution h, followed by numerical solution of the discretized equations for the intensive property u; or (2) integral solution for the extensive property that is numerically approximated by quadrature.

using Green's function[55] $G(\boldsymbol{x}, \boldsymbol{y})$.[b] The resulting integral defines an extensive quantity that is then discretized and computed by a quadrature[84] with some weights w. The values of the weights depend on the particular quadrature rule used. For midpoint quadrature[102] and the example of Fig. 4, they would be $w_q = f(\boldsymbol{y}_q)\mathrm{d}\boldsymbol{y}$. This defines the right-hand side of Eq. (1) for this example. The advantages of the latter procedure are that the integral solution is always consistent (even analytically exact), and that numerical quadrature is always stable. The only property that remains to be concerned about is the solution's accuracy. The first way of solution is sometimes referred to as the "intensive method", and the second as the "extensive method".

5.1. *Function Approximation by Particles*

The approximation of a continuous field function $u(\boldsymbol{x})$ by particles in d-dimensional space can be developed in three steps:

- *Step 1: integral representation.* Using the Dirac δ-function identity, the function u can be expressed in integral form as

$$u(\boldsymbol{x}) = \int u(\boldsymbol{y})\,\delta(\boldsymbol{x} - \boldsymbol{y})\mathrm{d}\boldsymbol{y}. \tag{2}$$

  In point particle methods such as random walk (cf. Sec. 7.1), this integral is directly discretized on the set of particles using a quadrature rule with the particle locations as quadrature points. Such a discretization, however, does not allow recovering the function values at locations other than those occupied by the particles.
- *Step 2: integral mollification.* Smooth particle methods relax this limitation by regularizing the δ-function by a mollification kernel $\zeta_\epsilon = \epsilon^{-d}\zeta(\boldsymbol{x}/\epsilon)$, with $\lim_{\epsilon\to 0}\zeta_\epsilon = \delta$, that conserves the first $r - 1$ moments of the δ-function identity.[85] The kernel ζ_ϵ can be thought of as a cloud or blob of strength, centered at the particle location,

[b] Note that Green's function always exists, even though it may not be known in closed form in most cases.

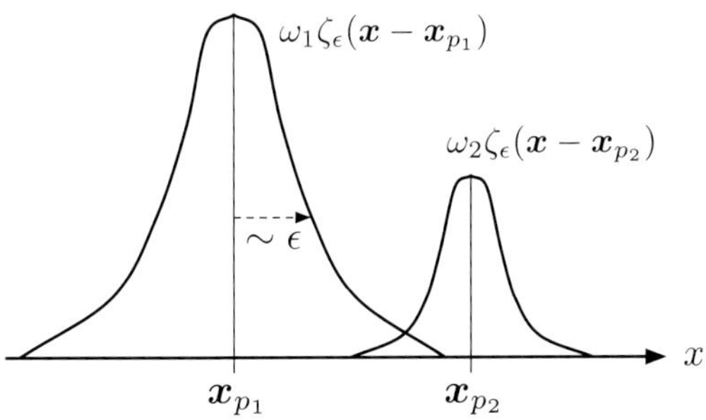

Fig. 5. Two particles of strengths ω_1 and ω_2, carrying mollification kernels ζ_ϵ.

as illustrated in Fig. 5. The core size ϵ defines the characteristic width of the kernel and thus the spatial resolution of the method. The regularized function approximation is defined as

$$u_\epsilon(\boldsymbol{x}) = \int u(\boldsymbol{y})\zeta_\epsilon(\boldsymbol{x} - \boldsymbol{y})\mathrm{d}\boldsymbol{y} \tag{3}$$

and can be used to recover the function values at arbitrary locations $\boldsymbol{x}$. The approximation error is of order ϵ^r, hence

$$u_\epsilon(\boldsymbol{x}) = u(\boldsymbol{x}) + \mathcal{O}(\epsilon^r). \tag{4}$$

As introduced above, r is the first nonvanishing moment of the mollification kernel.[27,85] For positive symmetric kernels, such as a Gaussian, $r = 2$.

- *Step 3: mollified integral discretization.* The regularized integral in Eq. (3) is discretized over N particles using the quadrature rule

$$u_\epsilon^h(\boldsymbol{x}) = \sum_{p=1}^{N} \omega_p^h \zeta_\epsilon\left(\boldsymbol{x} - \boldsymbol{x}_p^h\right), \tag{5}$$

where $\boldsymbol{x}_p^h$ and ω_p^h are the numerical solutions of the particle positions and strengths, determined by discretizing the ODEs in Eq. (1) in time. The quadrature weights ω_p are the particle strengths and depend on the particular quadrature rule used. The most frequent choice is to use midpoint quadrature,[102] thus setting $\omega_p = u(\boldsymbol{x}_p)V_p$, where V_p is the volume of particle p. Using this discretization, we obtain the function approximation

$$u_\epsilon^h(\boldsymbol{x}) = u_\epsilon(\boldsymbol{x}) + \mathcal{O}\left(\frac{h}{\epsilon}\right)^s = u(\boldsymbol{x}) + \mathcal{O}(\epsilon^r) + \mathcal{O}\left(\frac{h}{\epsilon}\right)^s, \tag{6}$$

where s depends on the number of continuous derivatives of the mollification kernel ζ_ϵ,[27,85] and h is the interparticle distance. For a Gaussian, $s \to \infty$.

From the approximation error in Eq. (6), we see that it is imperative that the distance h between any two particles is always less than the kernel core size ϵ, thus maintaining

$$\frac{h}{\epsilon} < 1 \tag{7}$$

at all times. If this "particle overlap" condition is violated, the approximation error becomes arbitrarily large, and the method ceases to be well posed.

5.2. *Operator Approximation*

Two strategies are distinguished to evaluate differential operators on particles: pure particle methods and hybrid PM methods.

5.2.1. *Pure particle methods*

In pure particle methods, differential operators on functions that are represented on particles are approximated by integral operators. The sums on the right-hand side of Eq. (1) thus represent the discretized versions of these integral operators. If we are interested in diffusion processes, for example, the relevant differential operators are ∇^2 and $\nabla \cdot (\boldsymbol{D}\nabla)$ (cf. Sec. 7). Both of them can be approximated by integral operators that allow consistent evaluation on scattered particle locations and conserve mass exactly[103,104] (cf. Sec. 7.2). The concept of this approximation method has been extended to a general, systematic framework for approximating any differential operator by a corresponding integral.[105] Following this framework, a differential operator L^β of order β applied to a continuous function $u(\boldsymbol{x})$ is equivalent to the integral operator

$$L^\beta u(\boldsymbol{x}) = \frac{1}{\epsilon^{|\beta|}} \int (u(\boldsymbol{y}) \pm u(\boldsymbol{x}))\eta_\epsilon^\beta(\boldsymbol{x} - \boldsymbol{y})\mathrm{d}\boldsymbol{y} \tag{8}$$

with a suitable scaled kernel $\eta_\epsilon^\beta(\boldsymbol{x}) = \epsilon^{-d}\eta^\beta(\boldsymbol{x}/\epsilon)$ of core size ϵ. This integral operator is then discretized over the particles using, e.g., midpoint quadrature[102] of resolution h, yielding

$$L_h^\beta u(\boldsymbol{x}_p) = \frac{1}{\epsilon^{|\beta|}} \sum_q V_q(u(\boldsymbol{x}_q) \pm u(\boldsymbol{x}_p))\eta_\epsilon^\beta(\boldsymbol{x}_p - \boldsymbol{x}_q). \tag{9}$$

Pure particle methods thus amount to evaluating direct particle–particle (PP) interactions, which means that for each particle $p = 1, 2, \ldots, N$ we have to compute a sum over all particles $q = 1, 2, \ldots, N$ [cf. Eq. (1)]. The computational complexity of this N-body problem thus nominally scales as $\mathcal{O}(N^2)$. Efficient algorithms do, however, exist to reduce it to $\mathcal{O}(N)$ in all practical cases. These algorithms will be outlined in Sec. 6. Alternatively, hybrid PM methods, as described next, can be used.

5.2.2. *Hybrid particle-mesh (PM) methods*

In hybrid PM methods, as pioneered by Harlow,[106] some (but not all) of the differential operators are evaluated on a superimposed regular Cartesian mesh.[107] This amounts to splitting the operators into separate short-range and long-range contributions. The short-range interactions are directly evaluated on the particles, whereas the long-range contributions are evaluated on the mesh. Using direct PP interactions for the short-range part allows better resolving local phenomena and retaining the favorable stability properties of particle methods in the case of convection (moving the particles is a local operation). Prominent examples of hybrid PM methods can be found in fluid dynamics and electrostatics. Both applications involve long-range interactions in order to compute the velocities (fluid dynamics) or forces (electrostatics). In a hybrid PM approach, such long-range interactions are modeled by a corresponding field equation that is then solved on the mesh. In many applications, the fields to be discretized are gradient fields, such that the corresponding long-range operator is the Laplace operator and the field equation hence is the Poisson equation (cf. Fig. 4). This equation can efficiently be solved using, e.g. FD[79] implemented in a multigrid algorithm,[108] or Poisson solvers based on fast Fourier transforms. In hybrid PM methods, the functions $\boldsymbol{K}$ and $\boldsymbol{F}$ in Eq. (1) may thus contain contributions corresponding to the solution of the field equation on the mesh. Therefore, hybrid methods require

- interpolation of the $\boldsymbol{\omega}_p$ carried by the particles from the irregular particle locations $\boldsymbol{x}_p$ onto the M regular mesh points ($\boldsymbol{\omega}_m$):

$$\boldsymbol{\omega}_m^h = \sum_{p=1}^{N} Q(\boldsymbol{x}_m - \boldsymbol{x}_p^h)\boldsymbol{\omega}_p^h \qquad m = 1, 2, \ldots, M; \tag{10}$$

- and interpolation of the field solution $\boldsymbol{F}_m$ (and, similarly, $\boldsymbol{K}_m$ if present) from the mesh to the (not necessarily same) particle locations ($\boldsymbol{F}_p$):

$$\boldsymbol{F}_p^h = \sum_{m=1}^{M} R(\boldsymbol{x}_p^h - \boldsymbol{x}_m)\boldsymbol{F}_m^h \qquad p = 1, 2, \ldots, N. \tag{11}$$

The accuracy of the method depends on the smoothness of $\boldsymbol{K}$ and $\boldsymbol{F}$, on the interpolation functions Q and R, and on the mesh-based discretization scheme employed for the solution of the field equations. In order to achieve high accuracy, the interpolation functions Q and R must be smooth to minimize local errors, and conserve the moments of the interpolated quantity to minimize far-field errors.[27] In addition, it is necessary that Q is at least of the same order of accuracy as R in order to avoid spurious contributions to $\boldsymbol{F}_p^h$.[107] This can easily be achieved by selecting the same interpolation function, W, for both operations: $Q = R = W$. Accurate interpolation functions that conserve the moments of the interpolated quantity up to a certain order can be constructed systematically.[109] Conservation of moments is an important property in the simulation of biological systems, where the laws of physics require quantities such as mass (zero-order moment), impulse (first-order moment), and angular impulse (second-order moment) to be conserved. Building this conservation right into the method constitutes an obvious advantage. One of the most commonly used moment-conserving interpolation kernels (but not the only one) is the M_4' function,[109] given by

$$M_4'(s) = \begin{cases} 1 - \frac{1}{2}(5s^2 + 3s^3) & \text{if } 0 \le s < 1 \\ \frac{1}{2}(2-s)^2(1-s) & \text{if } 1 \le s \le 2 \\ 0 & \text{if } s > 2, \end{cases} \tag{12}$$

where $s = |\boldsymbol{x}|/h = |\boldsymbol{x}_p - \boldsymbol{x}_m|/h$. Hereby, h denotes the mesh spacing and x is the distance from the particle to the respective mesh node, as illustrated in Fig. 6. The M_4' kernel is third-order accurate, exactly conserving moments up to and including the second moment. For each particle–node pair, we compute one weight $0 \le W_{pm} \le 1$ and the portion $\boldsymbol{\omega}_m = W_{pm}\boldsymbol{\omega}_p$ of the strength of particle p is attributed to mesh node

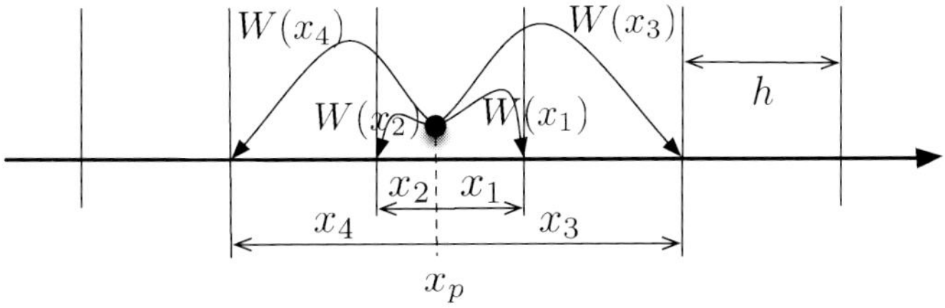

Fig. 6. Particle-to-mesh interpolation in one dimension. The interpolation weight is computed from the mesh spacing h and the distance x between the particle and the mesh node. For each particle–node pair, a different weight is computed. The particle strength is then multiplied by these weights and assigned to the mesh nodes. Mesh-to-particle interpolation works analogously and uses the same interpolation kernels.

m (Fig. 6). This is done independently for each particle and can efficiently be parallelized on vector and multi-processor computers.[32] In higher dimensions, the kernels are tensorial products of the one-dimensional (1D) kernels. Their values can thus be computed independently in each spatial direction and then multiplied to form the final interpolation weight for a given particle and mesh node: $W(x,y,z) = W_x(x)\, W_y(y)\, W_z(z)$.

Meshes are used not only to accelerate the computation of long-range interactions in hybrid PM schemes, but also to periodically reinitialize the particle locations to regular positions in order to maintain the overlap condition of Eq. (7). Reinitialization using a mesh is needed if particles tend to accumulate in certain areas of the computational domain and to disperse in others. In such cases, the function approximation would cease to be well posed as soon as the condition in Eq. (7) is violated. This can be prevented by periodically resetting the particle positions to regular locations by interpolating the particle properties to the nodes of a regular Cartesian mesh as outlined above, discarding the present set of particles, and generating new particles at the locations of the mesh nodes. This procedure is called remeshing.[85]

6. Efficient Algorithms for Particle Methods

The evaluation of PP interactions is a key component of particle methods and PM algorithms. Equation (1), however, defines an N-body problem, which is of potentially $\mathcal{O}(N^2)$ complexity to solve. It is this high computational cost that has long prevented the use of particle methods in computational science. Fortunately, this can be circumvented and the complexity can be reduced to $\mathcal{O}(N)$ in all practical cases. Together with efficient implementations on parallel computers,[32] this makes particle methods a competitive alternative to mesh-based methods.

If the functions $\boldsymbol{K}$ and $\boldsymbol{F}$ in Eq. (1) are local (but not necessarily compact), the algorithmic complexity of the sums in Eq. (1) naturally reduces to $\mathcal{O}(N)$ by considering only interactions within a certain cut-off radius r_c around each particle. This corresponds to short-range interactions where only nearby neighbors of a given particle significantly contribute. The specific value of r_c depends on the interaction law, i.e. the kernel functions $\boldsymbol{K}$ and $\boldsymbol{F}$ in Eq. (1), and has to be chosen to meet the desired simulation accuracy. The most conservative choice of r_c is given by the radius where the interaction contributions fall below the machine epsilon of the computer[84] and hence become insignificant.

For long-range interactions whose value decays as $\mathcal{O}(1/r^2)$ or slower with increasing interparticle distance r, cut-offs are not appropriate and we have to consider the full N-body problem of Eq. (1). Examples of such interactions include Coulomb forces, gravitation, or the Biot–Savart law in electromagnetism and fluid dynamics. Fast algorithms such as multipole expansions[110] (cf. Sec. 6.2) are, however, available to reduce the complexity of the corresponding pure particle method to $\mathcal{O}(N)$ also in these cases, albeit with a large prefactor. This large prefactor typically causes pure particle implementations of long-range interactions to be several orders of magnitude slower than the corresponding hybrid PM algorithm. Nevertheless, fast N-body methods are appealing from a conceptual point of view.

6.1. *Fast Algorithms for Short-Range Interactions*

Considering only the interactions within an r_c-neighborhood naturally reduces the algorithmic complexity of the PP evaluation from $\mathcal{O}(N^2)$ to $\mathcal{O}(N)$ with a prefactor that depends on the value of r_c and the local particle density. This requires, however, that the set of neighbors to interact with is known or can be determined with at most $\mathcal{O}(N)$ cost. Since particle methods do not use any connectivity information (cf. Sec. 4.4), neighborhood information is not explicitly available and it changes over time if particles move. Finding the neighbors of each particle by searching over all other particles would again render the complexity of the algorithm $\mathcal{O}(N^2)$, annihilating all benefits of a finite cut-off radius r_c. Two standard methods are available to find the interaction partners in $\mathcal{O}(N)$ time: cell lists and Verlet lists.

In cell lists, particles are sorted into equisized cubic cells whose size corresponds to the interaction cut-off r_c. Each cell contains a (linked) list of the particles residing in it. Interactions are then computed by sweeping through these lists. If particle p is to interact with all of its neighbors closer than r_c, this involves considering all other particles in the same cell

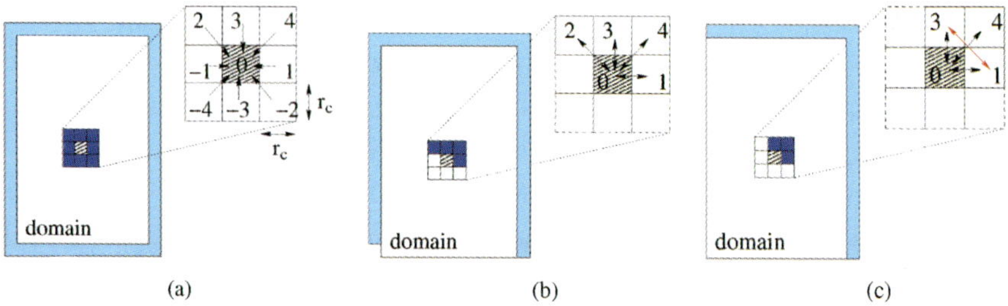

Fig. 7. Cell–cell interactions in cell list algorithms. (a) For asymmetric PP interactions, all adjacent cells have to be considered and the interactions are one-sided. (b) In traditional symmetric cell list algorithms, interactions are required on all but one boundary. (c) Introducing diagonal interactions (1–3), the cell layers for the boundary conditions (light blue; cf. Sec. 7.2.2) also become symmetric. This reduces the memory overhead and improves the efficiency of parallel implementations by reducing the communication volume. The 2D case is depicted. See text for interactions in the 3D case.

as particle p (center cell) as well as all particles in all immediately adjacent cells [Fig. 7(a)]. The shaded areas around the computational domain in Fig. 7 are needed to satisfy the boundary conditions using the method of images as outlined in Sec. 7.2.2.

For spherically symmetric interactions in 3D, cell lists contain up to $27/(4\pi/3) \approx 6$ times more particles than actually needed. Verlet lists[111] are available to reduce this overhead. For each particle p, they consist of an explicit list of all other particles with which it has to interact. This list contains the indices of all particles within a sphere around $\boldsymbol{x}_p$. The radius of this Verlet sphere has to be at least r_c, but is usually enlarged by a certain margin (skin) in order for the Verlet lists to be valid over several simulation time steps. The Verlet lists need to be rebuilt as soon as any particle has moved farther than the skin margin. Choosing the skin size is a trade-off between minimizing the lengths of the lists (and hence the number of interactions to be computed) and maximizing the time between list updates.[112] In the 3D case, Verlet list algorithms are at most $81/(4\pi(1+\text{skin})^3)$ times faster than cell list algorithms. In order to ensure overall $\mathcal{O}(N)$ scaling, Verlet lists are constructed using intermediate cell lists.

Another point of possible optimization concerns the symmetry of PP interactions. By construction of the kernel-based interactions, the effect of a particle p on another particle q is the same (with a possible sign change) as the effect of particle q on p. Looping over all particles and computing the interactions with all neighbors within the cut-off radius thus considers every interaction twice. The computational cost can be reduced by a factor of (at most) two if interactions are evaluated symmetrically. We then only loop over half of the neighbors and attribute the interaction contributions to both participating particles at once. How, then, is it possible to make sure that all interactions are considered exactly once? In cell lists, it is sufficient to loop over only those particles q in the center cell for which $q > p$, as well as over all particles in half of the neighboring cells [Fig. 7(b)]. In Fig. 7(c), diagonal interactions are introduced in order to further reduce the memory overhead for the boundary layers by 33% in the 2D case and 40% in the 3D case.[32] In parallel implementations, the diagonal interaction scheme moreover has the advantage of lower communication overhead. If the cells are numbered in ascending $x,y,(z)$, starting from the center cell with number 0, the symmetric cell–cell interactions are[32] 0–0, 0–1, 0–3, 0–4, and 1–3 in the 2D case; and 0–0, 0–1, 0–3, 0–4, 0–9, 0–10, 0–12, 0–13, 1–3, 1–9, 1–12, 3–9, 3–10, and 4–9 in the 3D case. Verlet list algorithms remain unchanged in the symmetric case, as the Verlet lists are constructed using intermediate symmetric cell lists and hence only contain unique interactions in the first place.

6.2. *Fast Algorithms for Long-Range Interactions*

In 1986, Joshua Barnes and Piet Hut[113] introduced a fast hierarchical $\mathcal{O}(N \log N)$ algorithm for N-body problems. The Barnes–Hut algorithm divides the domain into a tree of regular cuboidal cells. Each cell in the tree has half the edge length of its parent cell, and stores information about the center of mass and the total strength of all particles inside. The tree is then traversed for each particle p for which the interactions are to be evaluated. Direct PP interactions are only computed for nearby interaction partners q. If the partners are sufficiently far away, they are collectively approximated by the center of mass and the total strength of the largest possible cell that satisfies the closeness criterion

$$\frac{d}{\Delta} < \theta, \tag{13}$$

where d is the diagonal of the cell currently being considered, Δ is the distance of particle p from the center of mass of that cell, and θ is a fixed accuracy parameter ~1. This amounts to coarse-graining clusters of remote particles to single particles.

Based on the Barnes–Hut algorithm, Leslie Greengard and Vladimir Rokhlin presented the fast multipole method (FMM).[110,114,115] Their formulation uses a finite series expansion of the interaction kernel and direct cell–cell interactions in the tree. Compared to the Barnes–Hut algorithm, this further reduces the algorithmic complexity to $\mathcal{O}(N)$.

7. Particle Methods for the Simulation of Diffusion Processes

We consider the simulation of continuous spatial diffusion processes as a simple example of biological relevance.[116] Physically, the macroscopic phenomenon of diffusion is created by the collective behavior of a large (in theory, infinite) number of microscopic particles, such as molecules, undergoing Brownian motion.[116–118] From continuum theory,[119] we can define a concentration field as the mean mass of particles per unit volume at every point in space (cf. Sec. 3.2). For abundant diffusing particles, this allows formulating a continuous deterministic model for the spatiotemporal evolution of the concentration field $u(\boldsymbol{x}, t)$ in a closed, bounded domain Ω. This model is formulated as the PDE

$$\frac{\partial u(\boldsymbol{x},t)}{\partial t} = \nabla \cdot (\boldsymbol{D}(\boldsymbol{x},t)\nabla u(\boldsymbol{x},t)) \quad \text{for } \boldsymbol{x} \in \{\Omega/\partial\Omega\}, \quad 0 < t \leq T. \tag{14}$$

In this diffusion equation, $\boldsymbol{D}(\boldsymbol{x},t)$ denotes the diffusion tensor, ∇ the Nabla operator, and $\partial\Omega$ the boundary of the domain Ω.

Terminology classifies diffusion processes based on the structure of the diffusion tensor:

- If $\boldsymbol{D}$ is constant everywhere in Ω, diffusion is called "homogeneous". A $\boldsymbol{D}$ that varies in space defines "inhomogeneous diffusion".
- If $\boldsymbol{D}$ is proportional to the identity matrix, $\boldsymbol{D} = D\mathbb{1}$, diffusion is called "isotropic"; otherwise, "anisotropic". Isotropic diffusion is characterized by a flux whose magnitude does not depend on its direction, and it can be described using a scalar diffusion constant D. For isotropic, homogeneous diffusion, the diffusion equation simplifies to

$$\frac{\partial u(\boldsymbol{x},t)}{\partial t} = D\nabla^2 u(\boldsymbol{x},t) \quad \text{for } \boldsymbol{x} \in \{\Omega/\partial\Omega\}, \quad 0 < t \leq T, \tag{15}$$

where ∇^2 is the Laplace operator.

At $t = 0$, the concentration field is specified by an initial condition

$$u(\boldsymbol{x}, t = 0) = u_0(\boldsymbol{x}) \qquad \boldsymbol{x} \in \Omega.$$

The model is completed by problem-specific boundary conditions prescribing the behavior of u along $\partial\Omega$. The most frequently used types of boundary conditions are Neumann and Dirichlet conditions. A Neumann boundary condition fixes the diffusive flux through the boundary to a prescribed value f_N ($\boldsymbol{n}$ is the outer unit normal on the boundary):

$$\frac{\partial u}{\partial \boldsymbol{n}} = \nabla u(\boldsymbol{x},t) \cdot \boldsymbol{n} = f_N(\boldsymbol{x},t) \quad \text{for } \boldsymbol{x} \in \partial\Omega, \quad 0 < t \leq T;$$

whereas a Dirichlet condition prescribes the concentration f_D at the boundary:

$$u(\boldsymbol{x},t) = f_D(\boldsymbol{x},t) \quad \text{for } \boldsymbol{x} \in \partial\Omega, \quad 0 < t \leq T.$$

If the boundary function f is 0 everywhere on $\partial\Omega$, the boundary condition is called "homogeneous".

In the framework of pure particle methods, continuous diffusion models can be simulated using particles carrying mass as their extensive strength ω and collectively representing the intensive concentration field u. In the following, we review the stochastic method of random walk (RW) and the deterministic particle strength exchange (PSE) method. Using a 1D test problem, we then compare the accuracy and the convergence behavior of the two methods.

7.1. *The Method of Random Walk (RW)*

The Random Walk (RW)[116,120] method is based on the stochastic interpretation of Green's function solution[55] (cf. Fig. 4) of the diffusion equation:

$$u(\boldsymbol{x},t) = \int_{\Omega} G(\boldsymbol{x},\boldsymbol{y},t)u_0(\boldsymbol{y})\mathrm{d}\boldsymbol{y}. \tag{16}$$

In the case of d-dimensional isotropic homogeneous free-space diffusion, i.e. $\boldsymbol{D} = D\mathbb{1}$ and $\Omega = \mathbb{R}^d$, Green's function is explicitly known to be[121]

$$G(\boldsymbol{x},\boldsymbol{y},t) = \frac{1}{(4\pi Dt)^{d/2}}\exp\left[-\frac{\|\boldsymbol{x}-\boldsymbol{y}\|_2^2}{4Dt}\right]. \tag{17}$$

The RW method interprets this function as the transition density of a stochastic process.[122] In d dimensions, the method starts by either uniformly or randomly placing N particles p at initial locations $\boldsymbol{x}_p^0$, $p = 1, 2, \ldots, N$. Each particle is assigned a strength of $\omega_p = V_p u_0(\boldsymbol{x}_p^0)$, where V_p is the particle volume. This defines a point particle function approximation (cf. Sec. 5.1) to the initial concentration field $u_0(\boldsymbol{x})$. The particles then undergo a random walk by changing their positions at each positive-integer time step n according to the transition density in Eq. (17):

$$\boldsymbol{x}_p^{n+1} = \boldsymbol{x}_p^n + \mathcal{N}_p^n(0, 2D\delta t), \tag{18}$$

where $\mathcal{N}_p^n(0, 2D\delta t)$ is a vector of i.i.d. Gaussian random numbers with each component having a mean of zero and a variance of $2D\delta t$; δt is the simulation time step size. Moving the particles according to Eq. (18) creates a concentration field that, for $N \to \infty$, converges to the exact solution of the diffusion equation as given in Eq. (16). Homogeneous Neumann boundary conditions can be satisfied by reflecting the particles at the boundary. Drawing the step displacements in Eq. (18) from a multivariate Gaussian distribution readily extends the RW method to anisotropic diffusion processes.

RW is a stochastic simulation method. This Monte Carlo[66,67] character limits its convergence capabilities (cf. Sec. 4.1), since the variance of the mean of N i.i.d. random variables is given by $1/\sqrt{N}$ times the individual variance of a single random variable[68] (cf. Sec. 7.3). Moreover, the solution deteriorates with increasing diffusion constant D as the variance of the random variables becomes larger. In the case of small D ($\ll \delta t$), the motion of the particles can be masked by the sampling noise. RW thus works best for an intermediate range of diffusion constants.

7.2. *Particle Strength Exchange (PSE)*

The Particle Strength Exchange (PSE)[103,104] method is a deterministic pure particle method to simulate continuous diffusion processes in space. It is based on approximating the diffusion operator by a mass-conserving integral operator that can be consistently evaluated on the particle locations (cf. Sec. 5.2.1). The PSE scheme has been devised by Degond and Mas-Gallic for both isotropic[103] and anisotropic[104] diffusion. We illustrate the concept in the isotropic case. Anisotropic PSE is analogous and follows a similar derivation.[104]

7.2.1. *PSE for isotropic diffusion*

In free space, i.e. $\Omega = \mathbb{R}^d$, the isotropic PSE method[103] obtains an integral approximation to the Laplace operator in Eq. (15) by considering the concentration at a location $\boldsymbol{y}$ and expanding it into a Taylor series[84] around $\boldsymbol{x}$:

$$u(\boldsymbol{y}) = u(\boldsymbol{x}) + \sum_{i=1}^{r+1}\left[\frac{1}{i!}\Big((\boldsymbol{y}-\boldsymbol{x})\cdot\nabla_{\boldsymbol{x}'}\Big)^i u(\boldsymbol{x}')\right]_{\boldsymbol{x}'=\boldsymbol{x}} + \mathcal{O}\Big(\|\boldsymbol{y}-\boldsymbol{x}\|_2^{r+2}\,\|u\|_\infty\Big). \tag{19}$$

Subtracting $u(\boldsymbol{x})$ on both sides, multiplying the whole equation by a scaled kernel function $\eta_\epsilon(\boldsymbol{x}) = \epsilon^{-d}\eta(\boldsymbol{x}/\epsilon)$ of core size $\epsilon > 0$, and integrating over $\boldsymbol{y}$ yields

$$\int_{\mathbb{R}^d}(u(\boldsymbol{y})-u(\boldsymbol{x}))\eta_\epsilon(\boldsymbol{x}-\boldsymbol{y})\mathrm{d}\boldsymbol{y} = \sum_{i=1}^{r+1}\frac{1}{i!}\int_{\mathbb{R}^d}\left[\big((\boldsymbol{y}-\boldsymbol{x})\cdot\nabla_{\boldsymbol{x}'}\big)^i u(\boldsymbol{x}')\right]_{\boldsymbol{x}'=\boldsymbol{x}}\eta_\epsilon(\boldsymbol{x}-\boldsymbol{y})\mathrm{d}\boldsymbol{y} + \|u\|_\infty\,\mathcal{O}\Big(\int_{\mathbb{R}^d}\|\boldsymbol{y}-\boldsymbol{x}\|_2^{r+2}\,\eta_\epsilon(\boldsymbol{x}-\boldsymbol{y})\mathrm{d}\boldsymbol{y}\Big). \tag{20}$$

For the approximation to be consistent, we have to ask the following requirement for the kernel function η[103]:

$$\int_{\mathbb{R}^d}\prod_{i=1}^{d} x_i^{\alpha_i}\eta(\boldsymbol{x})\mathrm{d}\boldsymbol{x} = \begin{cases} 0, & \forall\boldsymbol{\alpha}\in\mathbb{N}^{\mathrm{d}},\ \ \boldsymbol{\alpha}\neq 2\boldsymbol{e}_i,\ \ 1\le\sum_{i=1}^{d}\alpha_i\le r+1 \\ 2, & \text{if } \boldsymbol{\alpha}=2\boldsymbol{e}_i,\ \ i\in\{1,2,\ldots,d\}. \end{cases} \tag{21}$$

r is the order of the approximation and $\boldsymbol{x} = (x_1, x_2,\ldots, x_d) \in \mathbb{R}^d$. $\boldsymbol{\alpha} = (\alpha_1, \alpha_2,\ldots,\alpha_d) \in \mathbb{N}^{\mathrm{d}}$ is a d-dimensional index and $(\boldsymbol{e}_1, \boldsymbol{e}_2,\ldots,\boldsymbol{e}_d)$ is the canonical basis of $\mathbb{R}^d$. In the 3D case, the above requirement can be expressed as

$$\int_{\mathbb{R}^3} x_i x_j\eta(\boldsymbol{x})\mathrm{d}\boldsymbol{x} = 2\delta_{ij} \qquad \text{for } i,j=1,2,3 \tag{22}$$

$$\int_{\mathbb{R}^3} x_1^{i_1} x_2^{i_2} x_3^{i_3} \eta(\boldsymbol{x}) \mathrm{d}\boldsymbol{x} = 0 \quad \text{if } i_1 + i_2 + i_3 = 1 \text{ or } 3 \leq i_1 + i_2 + i_3 \leq r + 1 \tag{23}$$

$$\int_{\mathbb{R}^3} \|\boldsymbol{x}\|_2^{r+2} |\eta(\boldsymbol{x})| \, \mathrm{d}\boldsymbol{x} < \infty \tag{24}$$

for $i_1, i_2, i_3 \in \mathbb{N}_0^+$, and $\delta_{ij} = 1$ if $i = j$ and 0 otherwise. The first condition normalizes the kernel function. The second one requires all moments up to order $r + 1$ to vanish, and the third one is required in order to bound the truncation error. Using the requirement in Eq. (21), the only remaining terms in Eq. (20) are

$$\epsilon^{-2} \int_{\mathbb{R}^d} (u(\boldsymbol{y}) - u(\boldsymbol{x})) \eta_\epsilon(\boldsymbol{x} - \boldsymbol{y}) \mathrm{d}\boldsymbol{y} = \nabla^2 u(\boldsymbol{x}) + \mathcal{O}(\epsilon^r), \tag{25}$$

and the integral operator that approximates the Laplacian is found as

$$\nabla_\epsilon^2 u(\boldsymbol{x}) = \epsilon^{-2} \int_{\mathbb{R}^d} (u(\boldsymbol{y}) - u(\boldsymbol{x})) \eta_\epsilon(\boldsymbol{x} - \boldsymbol{y}) \mathrm{d}\boldsymbol{y}. \tag{26}$$

While this operator is not the only possibility of discretizing the Laplacian onto particles, it has the big advantage of conserving mass exactly.[85] The approximation error is $\mathcal{O}(\epsilon^r)$, with r being the largest integer for which the condition in Eq. (21) is fulfilled.[85] Equation (26) is discretized using the particle locations as quadrature points. Thus,

$$\nabla_{\epsilon,h}^2 u^h(\boldsymbol{x}_p^h) = \epsilon^{-2} \sum_{q=1}^{N} (V_q u_q^h - V_q u_p^h) \eta_\epsilon(\boldsymbol{x}_p^h - \boldsymbol{x}_q^h), \tag{27}$$

where V_q is the volume of particle q. Inserting this discretized operator into Eq. (15), the final PSE scheme for isotropic diffusion reads

$$\begin{aligned}
\frac{\mathrm{d}\boldsymbol{x}_p^h}{\mathrm{d}t} &= 0 \\
\frac{\mathrm{d}\boldsymbol{\omega}_p^h}{\mathrm{d}t} &= V_p D\epsilon^{-2} \sum_{q=1}^{N} (V_q u_q^h - V_q u_p^h)\eta_\epsilon(\boldsymbol{x}_p^h - \boldsymbol{x}_q^h) \\
& p = 1, 2, \ldots, N.
\end{aligned} \tag{28}$$

The PSE kernel η_ϵ is local and the fast algorithms described in Sec. 6.1 can be used to reduce the computational complexity to $\mathcal{O}(N)$. In order to simulate diffusion using PSE, the strengths of all the particles change (i.e. they exchange mass) while their locations remain constant (i.e. they do not move). This is dual to the method of RW, where the particles conserve their mass but move in space. In PSE, all geometry and boundary condition processing thus only needs to be done once when initializing the particles. Combined convection-diffusion problems can be simulated by moving the particles with the convective velocity field instead of keeping them fixed.

Besides the obvious choice of using a Gaussian [cf. Eq. (17)] as the PSE kernel η, various algebraic kernels have also been derived. Algebraic kernels are computationally more efficient, since evaluating the exponential function on the floating-point unit of a computer processor takes several tens of clock cycles. In the 3D case, the following second-order accurate kernel as proposed by G.-H. Cottet (private communication, 1999) can, for example, be used:

$$\eta(\boldsymbol{x}) = \frac{15}{\pi^2} \frac{1}{|\boldsymbol{x}|^{10} + 1}. \tag{29}$$

7.2.2. *Boundary conditions in PSE*

The PSE algorithm as described so far only applies to infinite domains or to particles farther away from the boundary than r_c. For particles within an r_c-neighborhood from the boundary, we need to modify the PSE scheme in order to account for the prescribed boundary conditions.

For homogeneous boundary conditions in the case of flat (compared to the core size ϵ of the mollification kernel) boundaries, a straightforward method consists of placing mirror particles in an r_c-neighborhood outside of the simulation domain (Fig. 7). In the resulting method of images, the integral operator becomes

$$\epsilon^{-2}\int_{\mathbb{R}^d}(u(\boldsymbol{y})-u(\boldsymbol{x}))(\eta_\epsilon(\boldsymbol{x}-\boldsymbol{y})\pm\eta_\epsilon(\boldsymbol{x}+\boldsymbol{y}))\mathrm{d}\boldsymbol{y}+\mathcal{O}(\epsilon^r). \tag{30}$$

The final scheme is thus represented as

$$\frac{\mathrm{d}\omega_p^h}{\mathrm{d}t}=V_pD\epsilon^{-2}\sum_{q=1}^{N}(V_qu_q^h-V_qu_p^h)(\eta_\epsilon(\boldsymbol{x}_p^h-\boldsymbol{x}_q^h)\pm\eta_\epsilon(\boldsymbol{x}_p^h+\boldsymbol{x}_q^h)). \tag{31}$$

The positive sign between the two kernel functions applies for homogeneous Neumann boundary conditions, whereas the negative sign is to be used in the case of homogeneous Dirichlet boundary conditions. The method of images is restricted to the case of homogeneous boundary conditions. For inhomogeneous boundary conditions, the particle strengths need to be adjusted in the vicinity of the boundary.[123]

7.3. *Comparison of PSE and RW*

The accuracy of the RW and PSE methods is illustrated by using a benchmark case of isotropic homogeneous diffusion on the 1D (d = 1) ray $\Omega = [0, \infty)$, subject to the following initial and boundary conditions:

$$\begin{cases} u(x,t=0)=u_0(x)=xe^{-x^2} & x\in[0,\infty), \quad t=0 \\ u(x=0,t)=0 & x=0, \quad 0<t\le T. \end{cases} \tag{32}$$

The exact analytic solution of this problem is

$$u_{\mathrm{ex}}(x,t)=\frac{x}{(1+4Dt)^{3/2}}e^{-x^2/(1+4Dt)}. \tag{33}$$

Both RW and PSE simulations of this benchmark case are performed with varying numbers of particles in order to study spatial convergence.[49] The boundary condition at $x = 0$ is satisfied using the method of images as introduced in Sec. 7.2.2.

Figure 8 shows the RW and PSE solutions in comparison to the exact solution at a final time of $T = 10$ for $N = 50$ particles and a diffusion constant of $D = 10^{-4}$. The accuracy of the simulations for different numbers of particles is assessed by computing the final L_2 error

$$L_2 = \left[\frac{1}{N} \sum_{p=1}^{N} (u_{\mathrm{ex}}(x_p, T) - u(x_p, T))^2 \right]^{1/2} \tag{34}$$

for each N. The resulting convergence curves are shown in Fig. 9. For RW, we observe the characteristic slow convergence of $\mathcal{O}(1/\sqrt{N})$.[68] For PSE, a convergence of $\mathcal{O}(1/N^2)$ is observed, in agreement with the employed second-order kernel function. Below an error of 10^{-6}, machine precision is reached. It can be seen that the error of a PSE simulation

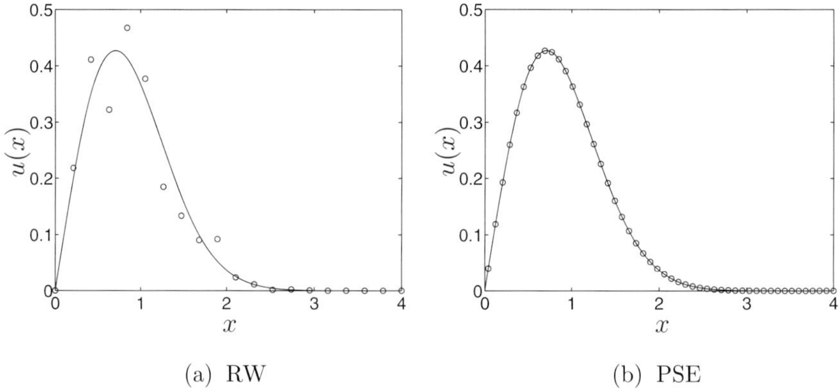

Fig. 8. Comparison of (a) RW and (b) PSE solutions of the benchmark case. The solutions at time $T = 10$ are shown (circles) along with the exact analytic solution [solid line; Eq. (33)]. For both methods, $N = 50$ particles, a time step of $\delta t = 0.1$, and $D = 10^{-4}$ are used. The RW solution is sampled in 20 intervals of width $\delta x = 0.2$. For the PSE, a core size of $\epsilon = h$ is used.

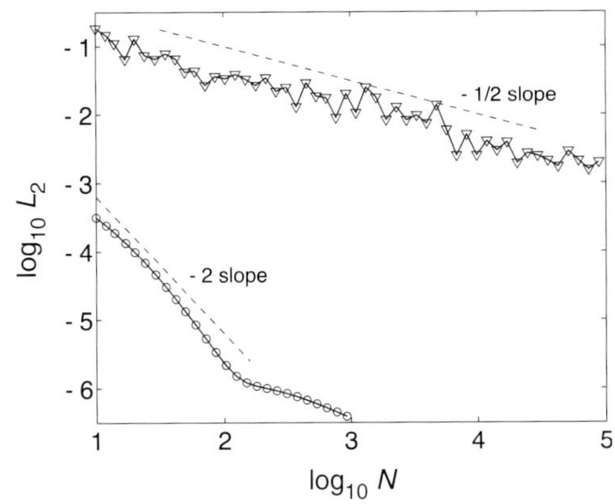

Fig. 9. Convergence curves for RW and PSE. The L_2 error versus the number of particles for the RW (triangles) and the PSE (circles) solutions of the benchmark case at time $T = 10$ are shown. For both methods, a time step of $\delta t = 0.1$ and $D = 10^{-4}$ are used. The RW solution is sampled in 20 intervals of width $\delta x = 0.2$; and for the PSE, a core size of $\epsilon = h$ is used. The machine epsilon of the computer is 10^{-6}.

is several orders of magnitude lower than the one of the corresponding RW simulation with the same number of particles. Using only 100 particles, PSE is already close to machine precision. It is evident from these results that large numbers of particles are necessary to achieve reasonable accuracy using RW in complex-shaped domains.

8. Reaction-Diffusion Processes

The reviewed particle methods for the simulation of diffusion processes can be extended to account for spatially resolved (bio)chemical reactions using either deterministic reaction kinetics or stochastic models. In the following, we consider reaction-diffusion systems governed by equations of the Fisher–KPP[124,125] type:

$$\frac{\partial u_i}{\partial t} - \nabla \cdot (\boldsymbol{D}_i \nabla u_i) = f_i(\boldsymbol{u}). \tag{35}$$

The concentration vector $\boldsymbol{u}$ contains one entry u_i per chemical species i, and the diffusion tensors $\boldsymbol{D}_i$ are allowed to vary among species and in space. All chemical reactions are described by the source terms f_i.

8.1. *Previous Approaches and Applications in Biology*

Coupled reaction-diffusion systems exhibit nontrivial stability properties that can give rise to the formation of stable concentration patterns called Turing patterns,[1] or traveling waves called Fisher waves.[126]

Twenty years after the seminal work of Turing,[1] Gierer and Meinhardt used reaction-diffusion systems to formulate their theory of pattern formation in biology.[127] They introduced the Gierer–Meinhardt model, which has become one of the most widely used pattern formation models in biology, with later applications also in computer graphics.[128]

The first biological applications of reaction-diffusion models considered morphogenesis,[1] following the idea that reaction-diffusion patterns of growth factors could explain the geometries and shapes found in nature. Computer simulations linking pattern formation to growth and morphogenesis were done using, e.g. hybrid cellular automata–PDE simulations to explain stalk formation and cell differentiation in slime mold.[2]

Simulations of reaction-diffusion patterns on surfaces first considered spherical objects such as globular tumors.[18] Morphogenesis of more complex surfaces was simulated using an FE method to solve the reaction-diffusion equation on triangulated surfaces.[3] This method allowed treating shapes as complex as branched unicellular algae. Moving-mesh FE techniques were later used to directly couple the motion of the boundary to reaction-diffusion patterns on continuously deforming 2D domains.[129] A different approach uses the solution of an interior Poisson problem to evolve the surface shape.[20]

Besides morphogenesis, reaction-diffusion models also have important applications in cell motility,[130] cell modeling,[131] and cell culture pattern formation.[132] Emerging applications also include spatiotemporal simulations of cell signaling pathways.[133] Since the first ODE model of the chemotaxis pathway in *Escherichia coli* was published by Bray *et al.* in 1993,[134] computer simulations have become increasingly more sophisticated in resolving spatial phenomena. A recent model explicitly includes diffusion of the key signal transduction molecule in the cytoplasmic space.[12] Other examples of reaction-diffusion signaling models include a sporulation control network model[11] and plant shoot meristem simulations.[135]

8.2. Reaction-Diffusion in Particle Methods

Extending the simulation scheme outlined in Sec. 7 to reaction-diffusion problems as governed by Eq. (35), all components of the concentration vector $\boldsymbol{u}$ are represented on the same set of computational particles, supporting property vectors $\boldsymbol{\omega}_p^h$.

Evaluating the reaction terms f_i amounts to a purely local exchange of strength among species in the same particle. Reactions are thus evaluated independently for each particle. The rate of exchange between different u_i is directly given by the reaction kinetics that are evaluated using either a deterministic method based on kinetic ODEs or a stochastic method such as Gillespie's SSA algorithm.[61,62] The latter is possible because individual particles constitute homogeneous reaction spaces with no spatial gradients present inside a single particle. The deterministic solver uses the same time integrator as the diffusion part, and is thus restricted by the time step stability limit; while the stochastic solver directly operates on (scaled) molecule numbers, and is used outside of the time integrator's right-hand side.

8.2.1. *An example with moving reaction fronts*

As an illustrative example, we consider the reaction $a + b \rightarrow 2a$ with rate constant k. Both species a and b diffuse with the same isotropic diffusion constant D. We use a combination of PSE[103] and SSA[61,62] to simulate the example system on the surface of a sphere with diameter ℓ. The initial condition is such that one half of the sphere contains only a, and the other half only b. Since a and b are initially unmixed, diffusion is required in order to bring them together and allow the reaction to start. As b is "eaten up" by a, the reaction front separating the two species moves into the region where b initially was and thus forms a traveling Fisher wave, which propagates in the direction orthogonal to the reaction front.

The wave stops as soon as the sphere contains only a and all of b has been consumed. If we identify the concentration of a with $[a] = u$ and normalize the total concentration to 1 everywhere, mass action kinetics gives the reaction term $f(u) = ku(1 - u)$. Inserting this into Eq. (35) yields a nonlinear PDE. Such systems can exhibit bifurcations (cf. Sec. 2), which is the case in the present example as Fisher waves only exist for speeds $s \geq s^* = 2\sqrt{f'(0)} = 2\sqrt{k}$.[136,137] For $s < s^*$, no moving front exists.

For the stochastic simulations, we denote by X_p the total number of molecules contained in particle p, and define an analog to Avogadro's number, viz. M, the number of molecules per unit mass. Gillespie introduced the product $\mathfrak{h}c$ as the expected number of reactions per unit time. For the binary reaction here, it is $\mathfrak{h} = X_p^a X_p^b$. The rate constant k and the reaction parameter c are related as $k = MVc$.[61] We interpret c as the probability that two molecules of species a and b react, provided they meet in space and time. This probability is independent of the particle volume.[c]

Figure 10(a) shows the total mass, integrated over the surface of the sphere, of a and b as they evolve in time. The front position is also shown, defined as the location where $[a] = [b] = 0.5$. As long as reactions occur, $[a]$ and $[b]$ change and the wave travels at a more or less constant speed s, given by the slope of the dashed curve in Fig. 10(a). If the dimensionless front speed is plotted against the dimensionless reaction parameter, the curves for different diffusion constants D collapse as shown in Fig. 10(b). We also observe that the theoretical scaling[136,137] is well approximated, particularly if the reaction is significantly faster than the diffusion.

9. Conclusions

Spatiotemporal models and numerical computer simulations are of rapidly growing importance in almost all areas of biology. Besides computational data analysis, including image processing, gene and protein sequence analysis, clustering, and machine learning, modeling and simulation are

---

[c] The probability of an encounter to occur, on the other hand, does depend on the particle volume. This is, however, already accounted for in $\mathfrak{h}$, as the number of molecules X_p per particle is an extensive property.

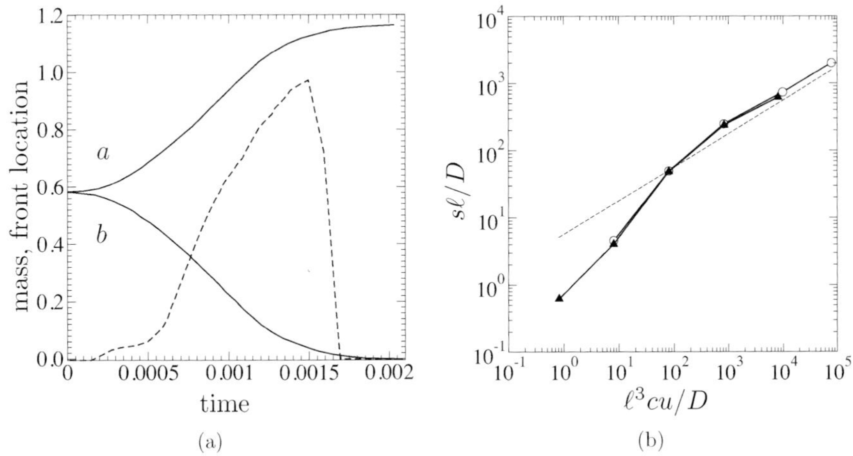

Fig. 10. (a) Evolution of the total mass of *a* and *b* (solid lines) for $M = 10$ molecules per unit mass. See text for problem description. The location of the wave front is shown by the dashed curve. The reaction front moves into the region of *b* until all of *b* is consumed. (b) Dependence of the front speed *s* on the reaction parameter *c*. In dimensionless representation, the two curves for $D = 0.1$ (open circles) and $D = 1.0$ (filled triangles) collapse to one. The theoretical scaling[136,137] is indicated by the dashed line.

important tools in modern biology. "Virtual experiments" *in silico* enable control over all variables and influence factors, allowing us to dissect biological systems, formulate physical models for their constituents, and disentangle coupled processes that are not separable in classical experiments. Moreover, spatiotemporal modeling makes accessible time and length scales unreachable by experiments. This allows studying systems as small as the atoms in an individual protein or as large as complete ecosystems.

The characteristic properties of biological systems frequently complicate the formulation and simulation of spatiotemporal models. Regulation mechanisms, geometric shape complexity, nonlinearity, and couplings across scales increase model complexity, and require powerful and flexible numerical methods as well as efficient high-performance software implementations on supercomputers.[26,32]

We have surveyed some of the most common modeling techniques and categorized them along the dimensions of phenomenological vs.

physical, discrete vs. continuous, and deterministic vs. stochastic. We summarized for each of these model classes some of the available numerical simulation methods with pointers to specialized literature. We then motivated the use of point-based particle methods that do not require any connectivity-based discretization and allow simulating both discrete and continuous systems. While their application to discrete systems is standard, the use of continuum particle methods is less widespread. We have reviewed continuum particle methods in more detail and demonstrated their application to diffusion and reaction-diffusion problems. This was intended as an easy-to-follow introduction. Applications to other phenomena such as flows and waves are well documented in the literature.

Notwithstanding the inherent complexity of biological systems, a wealth of methods and tools are available that enable highly accurate and predictive spatiotemporal simulations. At the same time, biology can serve as an important technology driver to stimulate the development of new methods and further advances in numerical analysis, computational science, and high-performance computing. Numerical methods that efficiently handle multi-scale systems[27–31] and topological changes in complex geometries are at the forefront of research in computational science. In parallel, computer algorithms have to be efficient enough to deal with the vast number of degrees of freedom, and software platforms must be available to effectively and robustly implement these algorithms on multi-processor supercomputers.[32] Just as fluid dynamics — in particular, the nonlinear, multi-scale problem of turbulence — has driven the development of numerical methods in the past, the major scientific goal of modeling an entire cell[138] might continue to do so in the future.

References

1. Turing AM. (1952) The chemical basis of morphogenesis. *Phil Trans R Soc Lond B* **237**: 37–72.
2. Marée AFM. (2000) *From Pattern Formation to Morphogenesis.* PhD thesis. University of Utrecht, Utrecht, The Netherlands.
3. Harrison LG, Wehner S, Holloway DM. (2001) Complex morphogenesis of surfaces: theory and experiment on coupling of reaction-diffusion patterning to growth. *Faraday Discuss* **120**: 277–94.

4. Kaandorp JA, Sloot PMA, Merks RMH, *et al.* (2005) Morphogenesis of the branching reef coral *Madracis mirabilis*. *Proc R Soc Lond B Biol Sci* **272**: 127–33.
5. Grant MR, Mostov KE, Tlsty TD, Hunt CA. (2006) Simulating properties of *in vitro* epithelial cell morphogenesis. *PLoS Comput Biol* **2**: e129.
6. Smith AE, Helenius A. (2004) How viruses enter animal cells. *Science* **304**: 237–42.
7. Dimitrov DS. (2004) Virus entry: molecular mechanisms and biomedical applications. *Nat Rev Microbiol* **2**: 109–22.
8. Dinh AT, Theofanous T, Mitragotri S. (2005) A model for intracellular trafficking of adenoviral vectors. *Biophys J* **89**: 1574–88.
9. Greber UF, Way M. (2006) A superhighway to virus infection. *Cell* **124**: 741–54.
10. Marsh M, Helenius A. (2006) Virus entry: open sesame. *Cell* **124**: 729–40.
11. Marwan W. (2003) Theory of time-resolved complementation and its use to explore the sporulation control network in *Physarum polycephalum*. *Genetics* **164**: 105–15.
12. Lipkow K, Andrews SS, Bray D. (2005) Simulated diffusion of phosphorylated CheY through the cytoplasm of *Escherichia coli*. *J Bacteriol* **187**: 45–53.
13. Dayel MJ, Hom EF, Verkman AS. (1999) Diffusion of green fluorescent protein in the aqueous-phase lumen of the endoplasmic reticulum. *Biophys J* **76**: 2843–51.
14. Lippincott-Schwartz J, Snapp E, Kenworthy A. (2001) Studying protein dynamics in living cells. *Nat Rev Mol Cell Biol* **2**: 444–56.
15. Verkman AS. (2002) Solute and macromolecule diffusion in cellular aqueous compartments. *Trends Biochem Sci* **27**: 27–33.
16. Alberts B, Bray D, Johnson A, *et al.* (1997) *Essential Cell Biology*. New York, NY: Garland Publication, Inc.
17. Rauch EM, Millonas MM. (2004) The role of trans-membrane signal transduction in Turing-type cellular pattern formation. *J Theor Biol* **226**: 401–7.
18. Chaplain MAJ, Ganesh M, Graham IG. (2001) Spatio-temporal pattern formation on spherical surfaces: numerical simulation and application to solid tumour growth. *J Math Biol* **42**: 387–423.
19. Jiang Y, Pjesivac-Grbovic J, Cantrell C, Freyer JP. (2005) A multiscale model for avascular tumor growth. *Biophys J* **89**: 3884–94.
20. Macklin P, Lowengrub J. (2005) Evolving interfaces via gradients of geometry-dependent interior Poisson problems: application to tumor growth. *J Comput Phys* **203**: 191–220.
21. Hense BA, Kuttler C, Müller J, *et al.* (2007) Does efficiency sensing unify diffusion and quorum sensing? *Nat Rev Microbiol* **5**: 230–9.
22. Müller J, Kuttler C, Hense BA, *et al.* (2006) Cell-cell communication by quorum sensing and dimension-reduction. *J Math Biol* **53**: 672–702.

23. Slepchenko BM, Schaff JC, Carson JH, Loew LM. (2002) Computational cell biology: spatiotemporal simulation of cellular events. *Annu Rev Biophys Biomol Struct* **31**: 423–41.
24. Fall CP, Marland ES, Wagner JM, Tyson JJ. (2002) *Computational Cell Biology.* Interdisciplinary Applied Mathematics, Vol. 20. New York, NY: Springer.
25. Strogatz SH. (2000) *Nonlinear Dynamics and Chaos.* Boulder, CO: Westview Press.
26. Takahashi K, Yugi K, Hashimoto K, *et al.* (2002) Computational challenges in cell simulation: a software engineering approach. *IEEE Intell Syst* **17**: 64–71.
27. Koumoutsakos P. (2005) Multiscale flow simulations using particles. *Annu Rev Fluid Mech* **37**: 457–87.
28. Mielke A. (2006) *Analysis, Modeling and Simulation of Multiscale Problems.* Heidelberg, Germany: Springer.
29. Alt W, Chaplain M, Griebel M, Lenz J. (2008) *Polymer and Cell Dynamics: Multiscale Modeling and Numerical Simulations.* Basel, Switerland: Birkhäuser.
30. Kovalenko A, Gusarov S, Stepanova M. (2009) *Multiscale Modeling: Concepts, Theories, and Applications.* Boca Raton, Florida: CRC Press.
31. Chauviere A, Preziosi L, Verdier C. (2009) *Cell Mechanics: From Single Scale-based Models to Multiscale Modeling.* Boca Raton, Florida: Chapman & Hall.
32. Sbalzarini IF, Walther JH, Bergdorf M, *et al.* (2006) PPM — a highly efficient parallel particle-mesh library for the simulation of continuum systems. *J Comput Phys* **215**: 566–88.
33. Reynwar BJ, Gregoria I, Harmandaris VA, *et al.* (2007) Aggregation and vesiculation of membrane proteins by curvaturemediated interactions. *Nature* **447**: 461–4.
34. Periole X, Huber T, Marrink SJ, Sakmar TP. (2007) G protein-coupled receptors self-assemble in dynamics simulations of model bilayers. *J Am Chem Soc* **129**: 10126–32.
35. Chu JW, Voth GA. (2006) Coarse-grained modeling of the actin filament derived from atomistic-scale simulations. *Biophys J* **90**: 1572–82.
36. Kashiwagi A, Urabe I, Kaneko K, Yomo T. (2006) Adaptive response of a gene network to environmental changes by fitness-induced attractor selection. *PLoS ONE* **1**: e49.
37. Burlando B. (1993) The fractal geometry of evolution. *J Theor Biol* **163**: 161–72.
38. Newman TJ, Antonovics J, Wilbur HM. (2002) Population dynamics with a refuge: fractal basins and the suppression of chaos. *Theor Popul Biol* **62**: 121–8.
39. Grasman J, Brascamp JW, Van Leeuwen JL, Van Putten B. (2003) The multifractal structure of arterial trees. *J Theor Biol* **220**: 75–82.
40. Smith TG Jr, Marks WB, Lang GD, *et al.* (1989) A fractal analysis of cell images. *J Neurosci Methods* **27**: 173–80.

41. Aon M, Cortassa S. (1994) On the fractal nature of cytoplasm. *FEBS Lett* **344**: 1–4.
42. Lahiri T, Chakrabarti A, Dasgupta AK. (1998) Multilamellar vesicular clusters of phosphatidylcholine and their sensitivity to spectrin: a study by fractal analysis. *J Struct Biol* **123**: 179–86.
43. Liebovitch LS, Scheurle D, Rusek M, Zochowski M. (2001) Fractal methods to analyze ion channel kinetics. *Methods* **24**: 359–75.
44. Li H, Li Y, Zhao H. (1990) Fractal analysis of protein chain conformation. *Int J Biol Macromol* **12**: 6–8.
45. Li H, Chen S, Zhao H. (1991) Fat fractal and multifractals for protein and enzyme surfaces. *Int J Biol Macromol* **13**: 210–6.
46. Saxton MJ. (1994) Anomalous diffusion due to obstacles: a Monte Carlo study. *Biophys J* **66**: 394–401.
47. Saxton MJ. (2001) Anomalous subdiffusion in fluorescence photobleaching recovery: a Monte Carlo study. *Biophys J* **81**: 2226–40.
48. Saxton MJ. (2007) A biological interpretation of transient anomalous subdiffusion. I. Qualitative model. *Biophys J* **92**: 1178–91.
49. Sbalzarini IF, Mezzacasa A, Helenius A, Koumoutsakos P. (2005) Effects of organelle shape on fluorescence recovery after photobleaching. *Biophys J* **89**: 1482–92.
50. Frenkel D, Smit B. (2002) *Understanding Molecular Simulation: From Algorithms to Applications.* San Diego, CA: Academic Press.
51. Schrödinger E. (1948) *What Is Life? The Physical Aspect of the Living Cell.* Cambridge, UK: Cambridge University Press.
52. Marco E, Wedlich-Soldner R, Li R, *et al.* (2007) Endocytosis optimizes the dynamic localization of membrane proteins that regulate cortical polarity. *Cell* **129**: 411–22.
53. González-Segredo N. (2007) Amphiphilic fluids: mesoscopic modelling and computer simulations. In: Dillon KT (ed.), *Soft Condensed Matter: New Research.* New York, NY: Nova Science, pp. 125–56.
54. Farlow SJ. (1993) *Partial Differential Equations for Scientists and Engineers.* New York, NY: Dover Publications.
55. Strauss WA. (2007) *Partial Differential Equations: An Introduction*, 2nd ed. New York, NY: Wiley.
56. Øksendal BK. (2003) *Stochastic Differential Equations*, 6th ed. Berlin, Germany: Springer.
57. Manninen T, Linne ML, Ruohonen K. (2006) Developing Itô stochastic differential equation models for neuronal signal transduction pathways. *Comput Biol Chem* **30**: 280–91.
58. Saarinen A, Linne ML, Yli-Harja O. (2008) Stochastic differential equation model for cerebellar granule cell excitability. *PLoS Comput Biol* **4**: e1000004.

59. Alfonsi A, Cancès E, Turinici G, *et al.* (2005) Adaptive simulation of hybrid stochastic and deterministic models for biochemical systems. In: Cancès E, Gerbeau JF (eds.). *ESAIM: Proceedings*, Vol. 14, pp. 1–13.
60. Salis H, Sotiropoulos V, Kaznessis YN. (2006) Multiscale Hy3S: hybrid stochastic simulation for supercomputers. *BMC Bioinformatics* 7: 93–112.
61. Gillespie DT. (1976) A general method for numerically simulating the stochastic time evolution of coupled chemical reactions. *J Comput Phys* **22**: 403–34.
62. Gillespie DT. (1977) Exact stochastic simulation of coupled chemical reactions. *J Phys Chem* **81**: 2340–61.
63. Gibson MA, Bruck J. (2000) Efficient exact stochastic simulation of chemical systems with many species and many channels. *J Phys Chem A* **104**: 1876–89.
64. Stundzia A, Lumsden C. (1996) Stochastic simulation of coupled reaction-diffusion processes. *J Comput Phys* **127**: 196–207.
65. Isaacson SA, Peskin CS. (2006) Incorporating diffusion in complex geometries into stochastic chemical kinetics simulations. *SIAM J Sci Comput* **28**: 47–74.
66. Liu JS. (2001) *Monte Carlo Strategies in Scientific Computing*. Springer Series in Statistics. New York, NY: Springer.
67. Rubinstein RY, Kroese DP. (2007) *Simulation and the Monte Carlo Method*, 2nd ed. New York, NY: Wiley.
68. Milizzano F, Saffman PG. (1977) The calculation of large Reynolds number two-dimensional flow using discrete vortices with random walk. *J Comput Phys* **23**: 380–92.
69. Schweitzer F. (2003) *Brownian Agents and Active Particles. On the Emergence of Complex Behavior in the Natural and Social Sciences.* Springer Series in Synergetics. Berlin, Germany, Springer.
70. Gaylord RJ, Nishidate K. (1996) *Modeling Nature: Cellular Automata Simulations with Mathematica*. New York, NY: Springer.
71. Ilachinski A. (2001) *Cellular Automata*. Singapore: World Scientific Publishing.
72. Schiff JL. (2007) *Cellular Automata: A Discrete View of the World*. Wiley Series in Discrete Mathematics and Optimization. Hoboken, NJ: Wiley Interscience.
73. Terano T, Kita H, Kaneda T, *et al.* (2005) *Agent-based Simulation: From Modeling Methodologies to Real-World Applications*. Tokyo, Japan: Springer.
74. Mosler HJ, Schwarz K, Ammann F, Gutscher H. (2001) Computer simulation as a method of further developing a theory: simulating the elaboration likelihood model (ELM). *Pers Soc Psychol Rev* **5**: 201–15.
75. Kloeden PE, Platen E. (1992) *Numerical Solution of Stochastic Differential Equations*. Stochastic Modeling and Applied Probability. Berlin, Germany: Springer.
76. Higham DJ. (2001) An algorithmic introduction to numerical simulation of stochastic differential equations. *SIAM Rev* **43**: 525–46.

77. Evans G, Blackledge J, Yardley P. (1999) *Numerical Methods for Partial Differential Equations.* Berlin, Germany: Springer.
78. Grossmann C, Roos HG, Stynes M. (2007) *Numerical Treatment of Partial Differential Equations.* Berlin, Germany: Springer.
79. Smith GD. (1985) *Numerical Solution of Partial Differential Equations: Finite Difference Methods*, 3rd ed. Oxford, UK: Clarendon Press.
80. Reddy JN. (1993) *An Introduction to the Finite Element Method*, 2nd ed. New York, NY: McGraw-Hill.
81. Zienkiewicz OC, Taylor RL. (2005) *The Finite Element Method*, 6th ed. Oxford, UK: Butterworth-Heinemann.
82. Benkhaldoun F, Vilsmeier R. (1999) *Finite Volumes for Complex Applications*, Vols. 1 and 2. Paris, France: Hermes Science Publications.
83. LeVeque RJ. (2002) *Finite Volume Methods for Hyperbolic Problems.* Cambridge, UK: Cambridge University Press.
84. Kincaid DR, Cheney EW. (2001) *Numerical Analysis: Mathematics of Scientific Computing*, 3rd ed. Pacific Grove, CA: Brooks/Cole.
85. Cottet GH, Koumoutsakos P. (2000) *Vortex Methods — Theory and Practice.* New York, NY: Cambridge University Press.
86. Li S, Liu WK. (2004) *Meshfree Particle Methods.* Berlin, Germany: Springer.
87. Liu GR, Gu YT. (2005) *An Introduction to Meshfree Methods and Their Programming.* Berlin, Germany: Springer.
88. de Berg M, van Krefeld M, Overmars M, Schwarzkopf O. (2000) *Computational Geometry: Algorithms and Applications*, 2nd ed. Heidelberg, Germany: Springer.
89. Schumaker LL. (1987) Triangulation methods. In: Chui CK, Schumaker LL, Utreras FI (eds.). *Topics in Multivariate Approximation.* New York, NY: Academic Press, pp. 219–32.
90. Bänsch E, Morin P, Nochetto RH. (2005) A finite element method for surface diffusion: the parametric case. *J Comput Phys* **203**: 321–43.
91. Novak IL, Gao F, Choi YS, *et al.* (2007) Diffusion on a curved surface coupled to diffusion in the volume: application to cell biology. *J Comput Phys* **226**: 1271–90.
92. Haber J, Zeilfelder F, Davydov O, Seidel HP. (2001) Smooth approximation and rendering of large scattered data sets. In: *Proc IEEE Visualization*, pp. 341–7.
93. Sethian JA. (1999) *Level Set Methods and Fast Marching Methods.* Cambridge, UK: Cambridge University Press.
94. Enright D, Fedkiw R, Ferziger J, Mitchell I. (2002) A hybrid particle level set method for improved interface capturing. *J Comput Phys* **183**: 83–116.
95. Hieber SE, Koumoutsakos P. (2005) A Lagrangian particle level set method. *J Comput Phys* **210**: 342–67.
96. Sbalzarini IF, Hayer A, Helenius A, Koumoutsakos P. (2006) Simulations of (an)isotropic diffusion on curved biological surfaces. *Biophys J* **90**: 878–85.

97. Ahusborde E, Gruber R, Azaiez M, Sawley ML. (2007) Physics-conforming constraints-oriented numerical method. *Phys Rev E Stat Nonlin Soft Matter Phys* **75**: 056704.
98. Courant R, Friedrichs K, Lewy H. (1928) Über die partiellen Differenzengleichungen der mathematischen Physik. *Math Ann* **100**: 32–74.
99. Courant R, Friedrichs K, Lewy H. (1967) On the partial difference equations of mathematical physics. *IBM Res Dev* **11**: 215–34.
100. Bergdorf M, Koumoutsakos P. (2006) A Lagrangian particle-wavelet method. *Multiscale Model Simul* **5**: 980–95.
101. Bergdorf M, Cottet GH, Koumoutsakos P. (2005) Multilevel adaptive particle methods for convection-diffusion equations. *Multiscale Model Simul* 4: 328–57.
102. Haber S. (1967) Midpoint quadrature formulas. *Math Comput* **21**: 719–21.
103. Degond P, Mas-Gallic S. (1989) The weighted particle method for convection-diffusion equations. Part 1: the case of an isotropic viscosity. *Math Comput* **53**: 485–507.
104. Degond P, Mas-Gallic S. (1989) The weighted particle method for convection-diffusion equations. Part 2: the anisotropic case. *Math Comput* **53**: 509–25.
105. Eldredge JD, Leonard A, Colonius T. (2002) A general deterministic treatment of derivatives in particle methods. *J Comput Phys* **180**: 686–709.
106. Harlow FH. (1964) Particle-in-cell computing method for fluid dynamics. *Methods Comput Phys* **3**: 319–43.
107. Hockney RW, Eastwood JW. (1988) *Computer Simulation Using Particles.* Bristol, UK: Institute of Physics Publishing.
108. Trottenberg U, Oosterlee C, Schueller A. (2001) *Multigrid.* San Diego, CA: Academic Press.
109. Monaghan JJ. (1985) Extrapolating B splines for interpolation. *J Comput Phys* **60**: 253–62.
110. Greengard L, Rokhlin V. (1988) The rapid evaluation of potential fields in three dimensions. In: Anderson C, Greengard C (eds.). *Vortex Methods.* Lecture Notes in Mathematics, Vol. 136. Berlin, Germany: Springer-Verlag, pp. 121–41.
111. Verlet L. (1967) Computer experiments on classical fluids. I. Thermodynamical properties of Lennard-Jones molecules. *Phys Rev* **159**: 98–103.
112. Sutmann G, Stegailov V. (2006) Optimization of neighbor list techniques in liquid matter simulations. *J Mol Liq* **125**: 197–203.
113. Barnes J, Hut P. (1986) A hierarchical O($N\log N$) force-calculation algorithm. *Nature* **324**: 446–9.
114. Greengard L, Rokhlin V. (1987) A fast algorithm for particle simulations. *J Comput Phys* **73**: 325–48.
115. Cheng H, Greengard L, Rokhlin V. (1999) A fast adaptive multipole algorithm in three dimensions. *J Comput Phys* **155**: 468–98.

116. Berg H. (1993) *Random Walks in Biology*. Princeton, NJ: Princeton University Press.
117. Brown R. (1828) A brief account of microscopical observations made in the months of June, July, and August, 1827, on the particles contained in the pollen of plants; and on the general existence of active molecules in organic and inorganic bodies. *Philos Mag* **4**: 161–73.
118. Einstein A. (1905) Über die von der molekularkinetischen Theorie der Wärme geforderte Bewegung von in ruhenden Flüssigkeiten suspendierten Teilchen. *Ann Phys* **17**: 549–60.
119. Nadler SB Jr. (1992) *Continuum Theory*. Basel, Switzerland: M. Dekker.
120. Chorin AJ. (1973) Numerical study of slightly viscous flow. *J Fluid Mech* **57**: 785–96.
121. Chandrasekhar S. (1943) Stochastic problems in physics and astronomy. *Rev Mod Phys* **15**: 1–89.
122. Van Kampen NG. (2001) *Stochastic Processes in Physics and Chemistry*, 2nd ed. Amsterdam, The Netherlands: North Holland.
123. Koumoutsakos P, Leonard A, Pépin F. (1994) Boundary conditions for viscous vortex methods. *J Comput Phys* **113**: 52–61.
124. Fisher RA. (1937) The advance of advantageous genes. *Ann Eugen* 7: 335–69.
125. Kolmogorov AN, Petrovsky IG, Piskunov NS. (1937) Étude de l'équation de la diffusion avec croissance de la quantité de matière et son application à un problème biologique. *Bull Univ d'Etat Moscou* (*Bjul Moskowskogo Gos Univ*), Sér Int A **1**: 1–26.
126. Riordan J, Doering CR, ben Avraham D. (1995) Fluctuations and stability of Fisher waves. *Phys Rev Lett* **75**: 565–8.
127. Gierer A, Meinhardt H. (1972) A theory of biological pattern formation. *Kybernetik* **12**: 30–9.
128. Turk G. (1991) Generating textures on arbitrary surfaces using reaction-diffusion. *Comput Graph* **25**: 289–98.
129. Madzvamuse A, Wathen AJ, Maini PK. (2003) A moving grid finite element method applied to a model biological pattern generator. *J Comput Phys* **190**: 478–500.
130. Grimm HP, Verkhovsky AB, Mogilner A, Meister JJ. (2003) Analysis of actin dynamics at the leading edge of crawling cells: implications for the shape of keratocyte lamellipodia. *Eur Biophys J* **32**: 563–77.
131. Plimpton SJ, Slepoy A. (2005) Microbial cell modeling via reacting diffusive particles. *J Phys Conf Ser* **16**: 305–9.
132. Miura T, Maini PK. (2004) Speed of pattern appearance in reaction-diffusion models: implications in the pattern formation of limb bud mesenchyme cells. *Bull Math Biol* **66**: 627–49.
133. Bhalla US. (2004) Models of cell signaling pathways. *Curr Opin Genet Dev* **14**: 375–81.

134. Bray D, Bourret RB, Simon MI. (1993) Computer simulation of the phosphorylation cascade controlling bacterial chemotaxis. *Mol Biol Cell* **4**: 469–82.
135. Jönsson H, Heisler M, Reddy GV, *et al.* (2005) Modeling the organization of the WUSCHEL expression domain in the shoot apical meristem. *Bioinformatics* **21**: i232–40.
136. Benguria RD, Depassier MC. (1996) Speed of fronts of the reaction-diffusion equation. *Phys Rev Lett* 77: 1171–3.
137. Berestycki H, Hamel F, Nadirashvili N. (2005) The speed of propagation for KPP type problems. I: periodic framework. *J Eur Math Soc* 7: 173–213.
138. The 2020 Science Group. (2005) *Towards 2020 Science.* Cambridge, UK: Microsoft Research Cambridge.

Index